A BACKGROUND TO GEOMETRY

A BACKGROUND

(NATURAL, SYNTHETIC AND ALGEBRAIC)

TO GEOMETRY

BY

T. G. ROOM, F.R.S.

Professor of Mathematics in the University of Sydney
Sometime Fellow of St John's College, Cambridge

CAMBRIDGE
AT THE UNIVERSITY PRESS
1967

CAMBRIDGE UNIVERSITY PRESS
Cambridge, New York, Melbourne, Madrid, Cape Town, Singapore, São Paulo, Delhi

Cambridge University Press
The Edinburgh Building, Cambridge CB2 8RU, UK

Published in the United States of America by Cambridge University Press, New York

www.cambridge.org
Information on this title: www.cambridge.org/9780521069502

First published 1967
This digitally printed version 2008

A catalogue record for this publication is available from the British Library

Library of Congress Catalogue Card Number: 67–11529

ISBN 978-0-521-06950-2 hardback
ISBN 978-0-521-09063-6 paperback

PREFACE

The classical format of a school text-book of Euclid's geometry, in which symbolic language is arranged in a fixed pattern of statements, has much to recommend it; even the final triumphant Q.E.D. is not pedagogically supererogatory. The use of symbols and pattern makes it possible to state compactly and precisely every step in the construction of a figure and the proof of a theorem. In this book the format has been adapted to a development of geometry in which 'collinearity' is the fundamental relation, while 'congruence' is an end-product, rather than, as in 'Euclid', an almost self-evident property.

Through the book there run three interwoven strands of development. The central strand might be described as 'making congruence respectable', and this was in fact the aim of the courses of lectures, first given twenty years ago, out of which the book has grown. Over the years the other two strands have been woven in. Along one of them is evolved a sequence of geometric constructions and theorems, based on combinatorial axioms for collinear subsets of a given set of points, which correspond to the operations of an algebraic field and their properties. The other strand follows some of the consequences of prescribing that the geometry shall consist of only finitely-many points, an area of mathematics still open for exploration at a relatively elementary level. Also, as a prelude to the geometry proper, there has been added some discussion in non-arithmetical terms of the ideas of 'order' and 'sense'.

In writing a book which grows, as this has done, out of courses of lectures given over many years, the author receives suggestions from many of his colleagues and from students. It is impossible to acknowledge these in detail, but the author wishes to express his deep sense of gratitude to all those colleagues and students in the University of Sydney who have contributed ideas to this book. There is one member of the Department of Mathematics to whom the author is especially indebted: Philip Kirkpatrick. His careful reading and criticism of both manuscript and proofs has resulted in the correction of many inaccuracies of statement as well as errors of typing and printing.

Because of the use made in the book of an unfamiliar symbolism and format, the printing has presented many difficulties; the sub-editorial staff of the Cambridge University Press has spared no pains in editing the manuscript so that the printers can print what the author intends. To them and to all members of the staff of the Press who have given help and guidance during the process of publication, the author wishes to offer his sincere thanks.

T.G.R.

Sydney, Australia
April 1967

CONTENTS

INTRODUCTION

The chart on p. 2 sets out a pattern for the growth of any body of knowledge and understanding from its beginnings as a mass of uncoordinated observations ('facts') to a fully developed scientific theory. The observations first pass through the mill of classification and selection, and, because of some common characteristics, large sets of observations are grouped together as belonging to the same classes. In the next phase an interpretation of the observed relations among some of these classes is sought in terms of the mathematical relations among symbols in some mathematical system; the body of knowledge constituted by these classes of observations is represented by a system of mathematical axioms, the mathematical consequences of which should correspond to the results of other observations of phenomena within the selected field. From this stage onwards there is continual interplay among observation and experiment, selection and classification of results, and the adaptation of mathematical systems to provide mathematical counterparts of the relations among these results.

It must be appreciated that the chart refers only to what has become known today as 'fundamental research'. We may assume that it is that part that lies at a certain basic level of a three-dimensional chart representing the whole development of science and technology. From every box on this fundamental research chart there should be arrows pointing out of the plane in which it has been drawn to boxes labelled 'practical applications', these boxes being not only themselves interconnected, but also connected by arrows pointing back into various fundamental research boxes. Technology provides new tools whose influence is felt in every phase of fundamental research.

In this book we are concerned only with the very beginnings of Science, namely, the physical relationships of stationary objects on a small part of the Earth's surface, the basic Science usually called 'Geometry'. This is in no sense meant to be a historical account of the growth of the subject, rather it is an allegory. It is the story of how geometry might have developed, adapted specifically to illustrate the pattern exhibited by the chart.

The reader is to imagine himself as some primitive Farmer-Geometer; in the course of his agricultural activities he makes observations, classifies them, and then formulates rules to which all or nearly all of his observations conform. These rules are to be stated in as precise a way as possible, because they form the basic components of the 'mathematical model'; they are the 'axioms' of the 'geometry' which prescribe the relations that are to obtain among the elements and sets of elements in the geometry.

Once an axiom is formulated the reader has to change this role to that of 'Logical Machine' in which role he might be supposed to produce, indefinitely, logical consequences of the axioms, that is, the

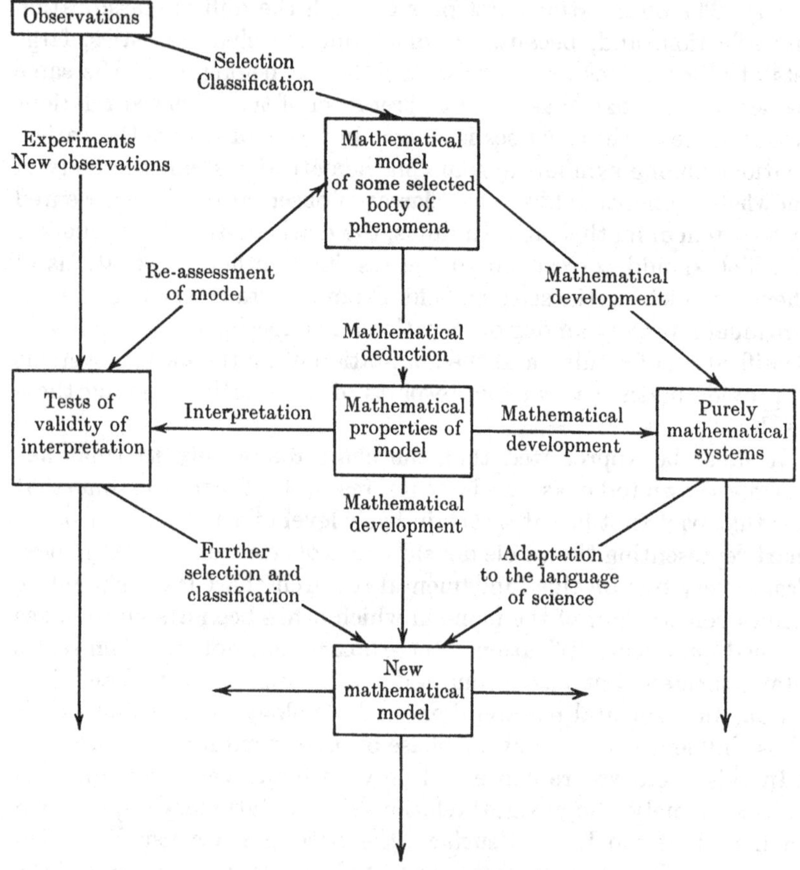

'theorems' of the geometry. From a very few axioms quite an elaborate mathematical structure may be erected; almost any late nineteenth century text-book of geometry with its innumerable problems and riders provides an example of the less worthwhile possibilities of elaboration. On the other hand, the history of science abounds in illustrations of mathematical systems which, when devised, have seemingly no relation to physical reality but which later are found to provide the ideal mathematical model for which the physicist or chemist or biologist has been seeking.

But to return to the Farmer-Geometer; while the Logical Machine has been at work compiling lists of theorems, the Farmer-Geometer has been continuing to make observations, and against these he tests any theorem which seems capable of physical interpretation. In the light of the results of these tests he makes new classifications, and adds to or modifies the set of axioms.

At the beginning of each part of this book the formulation of the rules is discussed, and some of the more immediate and practically applicable deductions are made from them. Each step in the development of the geometry is preceded by a discussion of the physical relations from which the rules are to be derived, written in terms of the observations of the Farmer-Geometer. For simplicity of statement the scope of the main text is limited to two-dimensional geometry; the Farmer-Geometer is supposed to make his observations and conduct his experiments in a large open flat paddock or meadow surrounded by a belt of scattered trees, which thickens into impenetrable forest as the distance from the paddock increases. This forest confines his observations to the paddock and the nearest trees.

The connection between the physical observations and the mathematical model has two links. The observations of the Farmer-Geometer are necessarily gross, and his classifications are broad. All 'objects', whether trees, sticks stuck in the ground, animals grazing in the paddock, or himself as 'Observer' are treated alike as being required to be described only by their relations to each other as they appear to him, viewed along his 'lines of sight', at the instants of observation. The first link between the objects and the model is the representation on a piece of paper of the objects by 'dots' and the lines of sight by 'ink-rulings'. Thus at the first stage the physical relations among objects and lines of sight are translated into visible relations among the dots and ink-rulings of our ordinary text-book-of-Euclid diagram.

The second link is that between this diagram and the mathematical model. Without ascription to them of any inherent properties we give the name 'points' to the entities by which the objects or dots are to be represented. Among these points we prescribe relations which are meant to represent the visual relations among the dots and the ink-rulings, and, alternatively, among the objects and the lines of sight. The subject of our synthesis and investigation is the system of relations among sets of these points which is deducible from these axiomatic relations.

We wish to preserve always the distinction between the abstract 'points' and the 'objects' they represent; to emphasize this distinction, and at the same time to avoid constant repetition of phrases like 'represented by', we make the following convention: if some element of the mathematical model is denoted by say a or P, or some relation is called 'collinearity', then the physical interpretations of these will be denoted by ⁑a, ⁑P, and ⁑collinearity.

It is pertinent here to discuss at some length the nature of geometrical 'theorems', and their relation to 'diagrams' ('figures'). Essentially theorems in geometry are statements about sets of points made in the following form:

Certain relations are prescribed among points of a given set, namely, (at least in the earlier theorems) such and such sets of points are collinear. From these sets of points, possibly after the introduction of some auxiliary elements, other sets of points are 'constructed' by *naming* lines which join pairs of points, and *naming* points which are the intersections of pairs of lines. (These are the equivalents of the physical processes of 'drawing' certain lines and 'finding' their intersections.) Finally the theorem is a statement of an 'incidence relation', a relation, that is, expressible in the form 'such and such a point belongs to the collinear set determined by such and such a pair of points', which is deducible as a consequence of the given incidence relations, and the relations among the points constructed from them. The Desargues configuration (Chapter 1.3) provides the perfect example of such incidence relations.

We might therefore express a geometric theorem in the symbolic form

$$G[\{\mathrm{P}\}] \quad \text{implies} \quad T[\{\mathrm{P}\}, \{C(\mathrm{P})\}]:$$

G:	the given incidence relations among the set of points $\{\mathrm{P}\}$,
$\{C(\mathrm{P})\}$:	a set of points constructed from $\{\mathrm{P}\}$,
T:	the deduced incidence relation among the points $\{\mathrm{P}\}$ and $\{C(\mathrm{P})\}$.

Geometry, viewed in this way, is a branch of 'Symbolic Logic'. But before the subject has developed very far it becomes clear that the format of symbolic logic is too complicated to enable proofs of even simple geometric theorems to be understood without prohibitive effort, while the invention of significant new theorems becomes practically impossible.

The geometric method consists of describing in a mixture of verbal and symbolic language certain relations among the sets of elements of a system, the description being referable to a diagram made up of dots and ink-rulings which gives a visual interpretation of these relations. Then, by logical operations, but with visual help from the diagram, further relations among the sets of elements of the system have to be deduced. These new relations are the 'theorems'.

The theorems could be stated and proved in terms which do not involve reference to any diagram: in fact the steps that are usually written out in a geometric proof do not form the complete set of steps that would be required for a full statement of the complete logical proof. But the intermediate steps, which are omitted in the geometric proof, can be supplied by the reader because of the diagram of the relations among the sets that he can see in front of him.

This makes apparent the possible weakness of the geometric method: the full logical statement must make allowance for all contingencies as each step is taken. In setting out the geometric proof we have to be careful that no unstated assumption has been made as a consequence of some apparent visual relation on the diagram, and further that some supposedly trivial specialization of the relations among the elements, for which a diagram would not make any allowance, does not in fact impose conditions that invalidate the proof of the theorem.

In the main text very few diagrams are provided, and theoretically at least the whole content could be appreciated as a rather difficult exercise in mathematical logic. But every effort has been made to set out as clearly and concisely as possible the relations among the 'given' elements and the construction of other elements from them, so that the reader may draw his own diagrams. He may check his diagrams against those at the end of the book. Moreover, but not for this reason, a definition of 'parallel lines' is introduced very early in the work, so that something very like ordinary Euclidean diagrams can be drawn. But it must be emphasized again that the mathematical model is a logical system, and diagrams are no more than a concession to human weakness; but it is a concession that the reader is strongly advised to make to himself.

The text is divided into four 'Books' marking stages in the development of the geometry. To each of these books there is added an 'Excursus' devoted in the main to the exploration of *finite* geometrical systems which illustrate the relations discussed in the text. In a finite system every point can be named and the members of every collinear set can be listed, so that the verification of a statement of geometrical relations can be reduced to the selection of appropriate combinatorial relations among subsets of a finite set.

From the point of view of the algebraically inclined mathematician the theory of these finite systems may be regarded as complete and well-known, but there is still ample scope for the amateur in the investigation of geometrical relations within particular systems. Indeed problems of current research interest lie not far beyond some of the exercises proposed on the finite geometries.

Note on the printing convention

One of the more difficult phases of the analysis of the basic concepts of mathematics is that of attaching significance to and defining 'order'. Written language consists of recognizable shapes 'ordered in space', spoken language of recognizable sounds 'ordered in time', and constructive thought would appear in effect to be a succession of mental images also 'ordered in time'. It is hard therefore to describe the concept of 'order' without relying on a natural understanding of what constitutes 'order', and, more particularly, without relying on the deeply ingrained acceptance of the printing convention of reading from left to right and down the page.

We can clarify the problem of presentation a little by supposing that the Farmer-Geometer, the writing of whose account of geometry we have taken upon ourselves, has a spoken language but no written language. We assume that he has 'written' the geometry by making marks that he invented himself, but that he makes these marks haphazardly over his writing surface. Ideally, no more than these marks would be necessary as a form of communication of his thought to others. But unless these others were equipped with high cryptographic skill they would not be likely to make much progress in understanding. We assume therefore that the Farmer-Geometer makes use of spoken language to explain the significance of the marks.

In terms of the actual writing of this book, then, we shall assume:

(i) That we can read the ordinary text which explains the steps that are being taken in the development of the geometry.

(ii) That in the formal presentation of the geometry we can recognise 'marks' or 'symbols', that is, when two symbols are drawn, we know whether they are the same or different. We shall in fact use as marks the letters of the alphabet and other symbols common in mathematical writing, but we use them at first only to the extent of recognizing that, for example, A and ∀ are the same, but A and B are different.

(iii) That in a set of marks we can recognize subsets; for example we can assume that we can recognise the pairs AB, CD in the set ABCD on the grounds that the original marks could have been drawn as

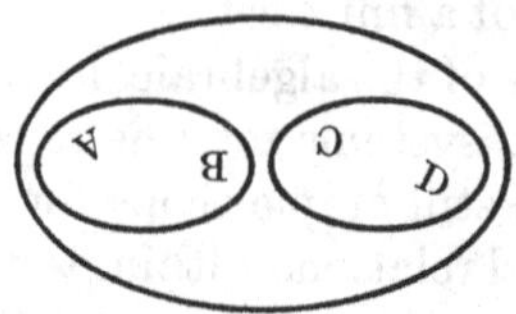

We shall in fact, after this discussion, usually print the symbols the

'right way up' and 'along a line', but this is only a matter of convenience and is not significant in the geometry. *We shall not assume anywhere in the text that AB is different from BA* unless some marks are added in accordance with a prescribed notation. We shall, for example, use '$\mathscr{B}|$ABC' with the clear understanding that this symbol is to convey a meaning different from that conveyed by '$\mathscr{B}|$BAC'. As a consequence, unless there is an explicit indication to the contrary, no compound symbol such as ABCD... is to be regarded as 'ordered'. It is a convenient transcription of the Farmer-Geometer's marks

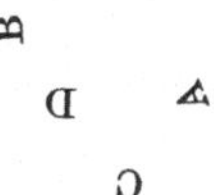

In planning the book the author has had to make the choice between establishing 'order' at once on the basis of axioms which depend as little as possible on the conventions of printing, or of assuming, under the three conditions prescribed above, that printing is intelligible, and on this basis carrying the geometry forward to the stage where 'order' would most naturally be the next phase in the development. It is this latter course that has been chosen, because the concern of the book is primarily with the geometry; the argument above is resumed and 'order' is introduced formally in Chapter 1.6.

BOOK 1

GEOMETRY WITHOUT NUMBERS

CHAPTER 1.1

THE SET OF POINTS: COLLINEAR SUBSETS

We are to imagine ourselves in a state of primitive simplicity, and think of the Farmer-Geometer (F-G) ploughing in his paddock, endeavouring as he does so to synthesize a mathematical system which satisfactorily represents the relations he observes among the objects in the paddock, and, more particularly, among objects which follow the pattern of furrows made by his ploughing.

The first observation that the F-G makes is the practical one that if he wishes to plough a straight furrow he must guide his plough as he walks behind it so that two sticks that he has set up shall always appear to him to be the one directly behind, and more or less hidden by, the other. From this observation he derives his first classification among sets of his objects.

He names his markers ⁑A and ⁑B in such a way that, viewed from some position on the furrow, ⁑A (at the end of the furrow) lies directly behind ⁑B. He then ploughs the furrow so that from every position on it ⁑A appears to be behind ⁑B. On the way to ⁑B he marks two positions ⁑P and ⁑Q on the furrow. Looking back from ⁑B he sees that ⁑Q (say) lies directly behind ⁑P. He now goes beyond ⁑B and gets in a position where ⁑P is obscured by ⁑B. In this way (only he would have to look backwards to do it) he could extend his furrow up to ⁑A, and then beyond ⁑A by walking so that ⁑A covers ⁑B.

He describes this set of objects and markers along the furrow as ⁑collinear; he notes that this ⁑collinear set is certainly not the whole set of objects in the paddock, and further that, at least so far as the properties he is investigating at present are concerned, ⁑A and ⁑B do not play different parts from the other markers.

Let us now devise an abstract formulation which may be interpreted by objects having these properties. The basic undefined entities are 'points' which are taken to represent the 'objects' in the paddock. The objects may be made to assume different relations among themselves (i.e. they may be 'moved about' in the paddock) and any one of them may be replaced by the observer. The set of points corresponds to the set of objects in the various (spatial) relations that they assume.

Points will be denoted by roman capitals A, B, ..., P, ...; sometimes the same point may be designated by two different symbols, so that whenever points are 'named' (i.e. whenever they have symbols assigned to them) we have to specify, if we require each name to refer to a different point, that

'A is not the same as B'

or 'A, B, C, ... are distinct'

and we shall introduce a notation to convey this meaning.

There is at this stage only one relation among points; any given set of points either does or does not belong to a '*collinear set*'; either the given points are 'collinear' or they are not. While the relations we ascribe to points in collinear sets are suggested by the relations of objects along furrows, they are much less specific and can be interpreted in widely different physical diagrams, some of which are described in Excursus 1.

The properties we ascribe to collinear sets are contained in two sets of statements. For the present we formulate these statements as complete English sentences, as a preliminary to formulating them in Chapter 1.2 in fully symbolic form as the 'incidence' axioms $A\mathscr{I}$. The designation 'incidence axioms' is used since the axioms state the ways in which points and collinear sets are incident, i.e. the ways in which points belong to a collinear set and collinear sets contain common points.

The first set of statements is:

1. (i) Among the set of point there are subsets called collinear sets. Any two distinct points determine a collinear set to which each of the points belongs.

The meaning of this is that if we are 'given' three distinct points A, B, C we must be given either 'C belongs to the collinear set determined by A or B' or 'C does not belong to the collinear set determined by A and B'. The parts played by A and B in determining the set are indistinguishable. It is possible that there are no points other than A and B in the set.

1. (ii) If C belongs to the collinear set determined by the distinct points A and B, and B and C are distinct, then A belongs to the collinear set determined by B and C.

It follows of course that (provided A and C are distinct) B belongs to the collinear set determined by A and C, but it does not follow that the three collinear sets determined by pairs of the points A, B, C coincide, since, in diagrammatic form, we might have

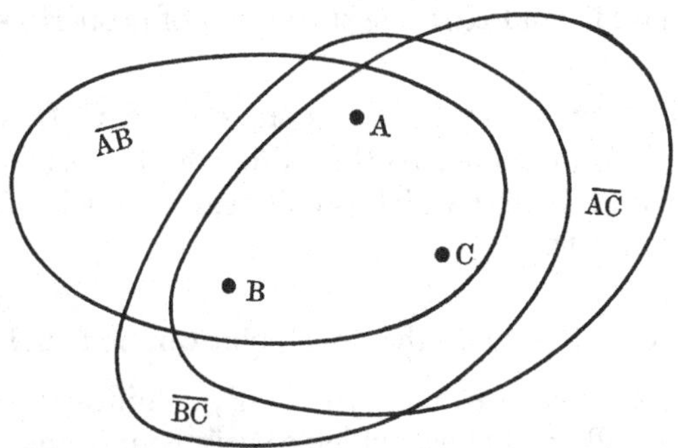

The last axiom of this set is adopted to ensure that a collinear set is determined by any two distinct points of itself, i.e. that the three collinear sets $\overline{AB}$, $\overline{BC}$, $\overline{CA}$ above coincide.

1. (iii) If A, B, C, D are distinct and C belongs to the collinear set determined by A and B, and D belongs to the collinear set determined by B and C, then D belongs to the collinear set determined by A and B.

The F-G's first picture then is of the objects in his paddock and the scattered trees round it as being represented by points of a set among which he has defined subsets by the property that any three points of the subset are collinear. The objects corresponding to a set of three collinear points are such that, if a suitable selection from the three is made, two of them viewed from the third appear to be one directly behind the other.

His next observation is this: if he selects two pairs of pairs of objects, not all four ⁑collinear, and ploughs furrows through and beyond each of these pairs, then usually he finds one place from which both furrows lie along lines of sight; further, he never finds more than one such place. There are certainly some pairs of pairs for which there is no such place, namely, pairs related in such a way that, moving so as only to occupy places ⁑collinear with the objects of one pair, he finds that he has to penetrate more and more deeply into the belt of trees in his efforts to reach a place at which he is ⁑collinear with the objects of the other pairs. But for the present he decides to ignore these exceptions, and formulate a rule which states that for any pair of pairs of objects (not all ⁑collinear) exactly one such place can be found. The exceptional cases will be accounted for later.

This relation we formalize in the statement of properties of collinear sets:

2. If we are given four distinct points, A, B, C, D, such that no three of them belong to the same collinear set, then we can always find a point which is collinear both with A and B and with C and D.

Exercise Prove that no more than one point can be found.

We now turn to what is probably the only significant property that can be stated for all geometries which satisfy conditions 1 and 2: this is usually referred to as the 'principle of duality'.

Suppose that we have any system 'p' whose elements are p-points and p-collinear sets satisfying the axioms 1 and 2. From this construct a system 'd' in which the d-points are taken to be the p-collinear sets. Two distinct d-points correspond to two distinct p-collinear sets and these have a common p-point which is a member of other p-collinear sets. Take the set of d-points d-collinear with two given d-points to be the set of p-collinear sets through the intersection of the two p-collinear sets corresponding to the d-points. Then the elements listed side-by-side in the two columns are identical.

d-point	p-line
d-line	p-point
a set of d-collinear d-points	a set of p-collinear sets with a common member
a set of d-collinear sets with a common member.	a set of p-collinear p-points.

If (X) is a set of points in any geometry satisfying the axioms, a theorem may be written as $G[\{X\}] \Rightarrow T[\{X\},\{C(X)\}]$. $\{X\}$ and $\{C(X)\}$ may be first interpreted as d-points and constructions from them, and then by means of the correspondence as p-collinear sets and constructions from them. Thus, associated with a system 'p' satisfying axioms 1 and 2, there is another system 'd' in which p-points and p-lines are replaced by d-points (which are p-lines) and d-lines (which are concurrent sets of p-lines). Each system is the *dual* of the other. With the exception of Ω^{9*} (Excursus 3.3), every system discussed in this book is self-dual, that is, the statement and proof of any theorem can be interpreted as the statement and proof of the dual theorem within the same system.

The F-G is well aware that the rules he has laid down are not completely in accord with his observations, and that it will be necessary to verify that there are indeed mathematical systems which satisfy the four conditions 1 (i), (ii) and (iii) and 2. If there is any such system, then the axioms are self-consistent, that is, it is not possible, on applying logical processes to the axioms as premises, to deduce both a statement of relations among points and the negative of that statement. Some systems which satisfy these axioms are described in Excursuses 1.1.1 to 1.1.6 and Excursus 1.2, but it must be recognized that all of them have properties that are not deducible from these axioms alone.

The following simple systems satisfy the axioms trivially, some of the axioms being vacuous in relation to them:

(i) The system consisting of only three points (either collinear or non-collinear).

(ii) The system consisting of a single collinear set.

(iii) The system consisting of:

a single collinear set v = {A, B, C, ...} which need not be finite, and

a single point V not belonging to v.

The collinear sets in this system are v, and the pairs {V, A}, {V, B}, {V, C}, ...

We do not wish to have to make provision for these systems among the geometries to be considered and shall use the word geometry to denote a 'non-trivial plane geometry', that is, a non-trivial system satisfying the four conditions 1 and 2. For such geometries we can prove the following structural theorems.

1.1*T*1. (Chapter 1.1 Theorem 1). If a non-trivial system satisfying the four conditions 1 (i), (ii), (iii) and 2 has only a finite number of points, then there is the same number of points in every collinear set.

Proof. Select a collinear set v = {A, B, ..., K} and a point V not belonging to it. Use a, b, ..., k to denote the collinear sets {V, A, ...}, {V, B, ...}, ..., {V, K, ...}. Let s be any collinear set other than v, a, b, ..., k. s and v have exactly one common point; let it be A, and suppose the common points of s and b, s and c, ..., s and k are B′, C′, ..., K′. Since b and c have common the point V, they have no other common point, so that B′ and C′ are different. In the same way no two of the points A, B′, C′, ..., K′ are the same. Now suppose that X is any point of the collinear set s. Then the collinear set {V, X, ...} contains a point of v, that is, a point of the set {A, B, C, ..., K}, so that {V, X, ...} is one of the sets a, b, c, ..., k. It follows that there is a one-to-one correspondence between the points of any two collinear sets, so that the numbers of points in the two sets are the same.

1.1*T*2. In a non-trivial finite geometry the number of collinear sets which contain any point is the same as the number of points in a collinear set.

Proof. The proof is very similar to that of *T* 1, and may be treated as an exercise.

1.1*T*3. A non-trivial finite plane geometry in which there are in a collinear set exactly m points contains altogether $m^2 - m + 1$ points and $m^2 - m + 1$ collinear sets.

Proof. Select any collinear set $v = \{A, B, C, \ldots, K\}$ and any point V not belonging to v. Then every collinear set which contains V contains also one point of v and m—2 other points. Also every point of the geometry belongs to some collinear set which contains V, so that the total number of points in the geometry is

$$1 + m + m(m-2) = m^2 - m + 1.$$

Theorems 1, 2, 3 describe properties of a non-trivial plane geometry, if such a system exists. We have not yet proved that any such system exists. In effect the whole of Books 1 and 2 constitute a proof of existence, but we can construct lists of elements of finite sets in which the elements are named in such a way as to exhibit exactly the relations required to be satisfied by the collinear subsets. Descriptions of some of these systems will be found in Excursus 1.1.

Before we proceed with the development of the geometry by the addition of new axioms we set up the various notations that are to be used to express all relations in a compact symbolism, which, as far as possible, enables theorems and their proofs to be written independently of formal written language. While of necessity a book on mathematics has to be written with descriptions in a particular language, ideally, the statements and proofs of theorems would be unaltered in any translation of the book into another language.

CHAPTER 1.2

VARIOUS NOTATIONS: THE AXIOMS OF INCIDENCE

(i) *Sets*

$\{A, B, C\}$: the (unordered) set of objects A, B, C.

$\{^{\mathcal{C}}A, B, C\}$: the ordered set of objects A, B, C; ('ordered' is 'conformably ordered' as defined in 1.8*D*1).

Operations

$\cap$: intersection or meet.

$\cup$: union.

Relations

$=$: is the same as; $\neq$: is not the same as (but see also the relation '$\#$' below).

$\in$: is a member of; $\notin$: is not a member of.

$\subset$: is contained in ($P \subset Q$ is to include where relevant the possibility $P = Q$); $\not\subset$: is not contained in.

$\supset$: contains; $\not\supset$: does not contain.

$\#|$: the elements listed are all different. Thus $\#|ABC$: the points A, B, C are distinct.

$\Rightarrow$: 'implies', used in the form $[X] \Rightarrow [Y]$; 'the geometrical relations in the statement [X] are sufficient to ensure the validity of the statement [Y]'.

$\Leftrightarrow$: if and only if, equivalent to. $[X] \Leftrightarrow [Y]$ may in various contexts be interpreted as 'both $[X] \Rightarrow [Y]$ and $[Y] \Rightarrow [X]$', that is, '[X] is necessary and sufficient for [Y]', or 'the statements [X] and [Y] are equivalent'.

Connective

&: 'and', used to connect two parts of a statement of relations.

Ranges of indices

$i \in \{1, 2, 3\}$: the range of i is 1, 2, 3.

$\{i, j\} \subset \{1, 2, 3\}$: i and j are different and take the values 1, 2 or 3.

$(i, j) \in \{1, 2, 3\}$: each of i and j is a member of the set $\{1, 2, 3\}$, and they are not necessarily different.

(ii) *Points and collinear sets*

Roman capitals, A, B, P, X, ...: Names for *points*. A Roman capital letter in a statement is always to be understood to be the name of a point.

Join: the collinear set of points determined by two points.

AB: the join of A and B.

$\langle A, B\rangle$: the join of A and B when A and B are composite symbols.

$AB \cap CD$: the intersection or meet of, that is, the common member of the collinear sets AB and CD.

Because of the simplification of statement resulting from using it, the word '*line*' is introduced as, effectively, a synonym for 'join', and hence for 'collinear set', but it is to be understood that this is to convey no sense of a 'whole family' and still less of a 'continuous set' of points collinear with two points. While there may be objects all over the paddock, we are, in any particular investigation, concerned only with certain selected objects. Likewise in the geometry we are concerned only with 'named' points and sets. When we are 'given' A and B, we may 'name' the collinear set AB, but can make use only of those points of the set which we can name as the intersections of AB with other collinear sets. The naming of four points, A, B, C, D not all belonging to the same collinear set is sufficient for the naming of the meet $AB \cap CD$ without the intervention of any physical picture of 'drawing lines' or 'finding points of intersection'.

Roman lower case letters, a, b, p, x, ... will be used for the names of lines.

In a geometrical context the symbols for set relations listed earlier may be interpreted in the following ways:

$A = B$: the same point is denoted by A or B.

$a \cap b = P$: P is the meet of a and b.

$p = AB$: p is the join of A and B.

$\langle a \cap b, C\rangle = p$: p is the join of C to the meet of a and b.

$A \in p$: A lies on p.

$p \supset A$: p passes through A.

$p \cap \{A, B, C\} = \emptyset$: none of the points A, B, C belongs to the collinear set p.

(iii) *Relations*

Phrases such as 'A, B, C are collinear' and 'a, b, c are concurrent' form a considerable part of any statement of geometrical relations, and while we could replace them by '$C \in AB$' and '$c \subset a \cap b$' respectively, these forms mask the symmetry of the relations. To remedy this we introduce what might be described as the 'symbols of dependence', namely:

$\emptyset$: the null set.

$\mathscr{P}$: the set containing a single point.

$\mathscr{L}$: a collinear set.

$\not\mathscr{L}$: a set of points not all collinear.

$\not\mathscr{L}_3$: a set of points, no three of which are collinear.

The symbols are to be used in the following ways:

$\mathscr{L}|$ABCD: the points named after the symbol '|' belong to a collinear set, that is: A, B, C, D are collinear. The printed forms $\mathscr{L}|$ABCD, $\mathscr{L}|$DBAC, etc. are all equivalent.

$\not\mathscr{L}_3|$ABCD: no three of the points A, B, C, D are collinear (and therefore, in particular, no two coincide).

$\#|$xyz...: the elements x, y, z, ..., whatever they are, are all different (distinct).

$\mathscr{L}\#|$ABC: A, B, C are collinear and distinct.

$\mathscr{P}|a \cap b \cap c$: the set common to a, b, c is a single point, i.e. a, b, c are concurrent, that is, provided $\#|$abc,

$$a \cap b \in c \quad \text{or} \quad a \supset b \cap c.$$

(There are many equivalent forms.)

$\emptyset|a \cap b \cap c$: there is no point common to a, b, c, that is, a, b, c are not concurrent, $a \cap b \notin c$, etc.

This notation could be allowed to supersede the terminology of 'meets' and 'joins', 'passing through' and 'lying on', but while the statements

$$C \in AB \quad \text{and} \quad \mathscr{L}|ABC$$

or

$$P = AB \cap CD \quad \text{and} \quad \mathscr{L}|PAB \,\&\, \mathscr{L}|PCD$$

are exactly equivalent in content, the first in each case draws attention to a particular aspect of the relation among the points, and in many contexts makes the significance of the statement easier to grasp.

This multiplicity of notations to express the same relation is one of the strengths of geometry: the method of expressing the relation draws attention to the aspect of the relation which is of immediate relevance. When the 'ordering' relations are being introduced, we shall initially use no fewer than four different ways for expressing the same relation.

The other principal relations to be introduced are:

$\mathscr{I}$: the incidence relations (used only for quoting axioms).

$\mathscr{S}$: separation (Chapter 1.6).

$\mathscr{B}$: betweenness: (Chapter 1.6).

$\mathscr{O}$: (conformable) order (Chapter 1.8).

In general the symbols used to represent geometric relations will be script capitals.

(iv) *Operations and transformers*

There are two ways in which one may regard the statement '**C** is the union of **A** with **B**'. The usual way corresponds to the expression in terms of the symbols for the two sets and an 'operator' symbol '$\cup$', which operates on the pair **A** and **B** to determine a new subset $\mathbf{A} \cup \mathbf{B}$. In a geometrical context there is something to be gained by assuming that there is associated with each subset such as **A** a function or 'transformer', '$\cup_{\mathbf{A}}$', which transforms any subset **B** into the subset $\cup_{\mathbf{A}} \mathbf{B}$. For example, in Chapter 1.4 we associate with every pair of points a transformer $\mathscr{H}_{AB}$, which is a geometric construction built on A, B, and any point C, $C \in AB$; the construction determines a point D, the 'harmonic transform of C by A, B'. The same geometrical property is expressed by

$D = \mathscr{H}_{AB} C$: D is the harmonic transform of C by A, B and by

CD $\mathscr{H}$ AB: the pairs {C, D} and {A, B} are harmonically related.

Because 'transformations' and 'relations' are two equivalent expressions of the same property we use script capitals also for transformers. The transformers to be defined are:

$\mathscr{H}$: the harmonic transformer (Chapter 1.4),
$\mathscr{D}$: the displacement transformer (Chapter 2.1),
$\mathscr{J}$: the involutory transformer (Chapter 2.5),
$\mathscr{Q}$: the polar transformer (Excursus 2.4),
$\mathscr{M}$: the multiplicative transformer (Chapter 3.3),
$\mathscr{R}$: the reflection transformer (Chapter 4.2).

Gothic capitals will be used to designate the specific geometric system under consideration:

$\mathfrak{N}$: a geometric net (Chapter 2.2),
$\mathfrak{F}$: the corresponding algebraic field (Chapter 2.2),
$\mathfrak{f}, \mathfrak{g}, \ldots$: units in an extension of a primary field (Chapter 3.4),
$\mathfrak{C}$: co-ordinate net (Chapter 2.4),
$\mathfrak{R}$: region (Chapter 1.7). Also, with a different meaning, as a subscript in the symbol
$\mathfrak{F}_{\mathfrak{R}}$: the real field (Chapter 3.4).

(v) *The format of theorems*

Italic capital letters will be used in the numbering of axioms, theorems and formal statements generally.

A: axiom,
C: construct (construction),
D: definition,
G: given (statement of data),
N: notation (for a new notation being introduced),

P: proof,
S: select ('take an arbitrary element of the designated set'),
T: theorem,
X: exercise.

Axioms are numbered independently of Chapter numbers, for example, $A\mathscr{S}$ 3: the third axiom of the set which establishes the relation of separation.

Theorems, notations, definitions and *exercises* are numbered within the chapters in which they occur, for example, as 1.2*D*1, 3.3*T*5. Some of the theorems are enunciated without proof, the proof being intended as an exercise. These are designated by numbers such as 3.2*T*4*X*.

Similar formal patterns will be used for the statement of axioms and definitions, for the description of new notations and constructions, and for the statement and proof of theorems. The sequence to be adopted for the statement and proof of a theorem, taking extracts from 1.3*T*2 as an example, is the following:

1.3*T*2 (Verbal and possibly imprecise statement of the result to be proved.) The general figure as axiom $\Rightarrow$ existence of simply-special figure.

G (data)
$\mathscr{L}_3|$KABC

(given four points K, A, B, C, no three collinear)

M: $\mathrm{M} \in \mathrm{AC}$ & $\mathrm{M} \notin \mathrm{KB}$ & $\# |\mathrm{MAC}$

(given also M lying on AC and not on KB and that M is not A or C)

C (construct)
N: $\mathrm{N} = \mathrm{AB} \cap \mathrm{KM}$

(this construction is in effect only the naming of the point N)

T $\mathscr{L}|$KLMN

(symbolic statement of the theorem)

P *s* (further selection of elements to be made in order to complete the proof)

C″: $\mathrm{C}'' \in \mathrm{KC}$ & $\# |\mathrm{KCC'C''}$.

c (auxiliary construction)
L′: $\mathrm{L}' = \mathrm{BC}'' \cap \mathrm{B'C'}$

t $\mathscr{L}|$LMN (lemma)

p (proof of lemma)

T $\mathscr{L}|$LMN

(final statement of the theorem in terms of the data).

In terms of these notations we now formalise the axioms of incidence which were discussed in Chapter 1.1.

AXIOMS OF INCIDENCE, 1 AND 2

A$\mathscr{I}$1 Properties prescribed for a collinear set.

A (i)	Every collinear set contains at least three points.
*G*1	#|AB
A (ii)	A and B determine a unique subset of the set of points, namely the 'collinear set' denoted by AB.
A (iii)	A ∈ AB, B ∈ AB.
*G*2	C : C ∈ AB & #|CB
A (iv)	A ∈ BC.
*S*1	D : D ∈ BC
A (v)	D ∈ AB.

Since D is arbitrarily selected on AB, (iv) implies:

1.2*T*1. C ∈ AB & #|BC ⇒ AB = BC.

1.2*T*2*X*. A collinear set is determined by any two points of itself.

*G*1	#|AB, #|CD
*G*2	C ∈ AB, D ∈ AB
T	CD = AB.

A$\mathscr{I}$2 Any two collinear sets have at least one common point.

G	$\mathscr{L}_3$|ABCD
A	$\mathscr{P}$|AB ∩ CD.

1.2*T*3*X*. If two collinear sets have two distinct common points, then they coincide.

*G*1	#|ABCD
*G*2	P: P ∈ AB & P ∈ CD
	Q: Q ∈ AB & Q ∈ CD
	#|PQ
T	AB = CD.

We need to be somewhat careful in the use of the description 'collinear set'. If A, B determine a collinear set AB, and C ∈ AB, then {A, B, C} is a set of points which are collinear; it is preferable however not to describe them as 'a collinear set'; they are a subset of the collinear set AB, and we have already prescribed several ways of denoting this relation among the three points.

Without other information we do not know any specific points in the collinear set AB other than A and B. The set may be regarded as containing other points which are latent, waiting to be identified. For example, to be given $\mathscr{L}_3|$ABCD is to have specified at least one other point in the collinear set AB, namely, AB $\cap$ CD. This is a process we have referred to before, namely, that of 'naming', or 'constructing', a point, and one of the main themes of this geometry is the devising of operations which will enable us to name increasingly comprehensive sets of points.

CHAPTER 1.3

THE DESARGUES FIGURE

As an introduction to the Desargues figure, we consider a 'practical' problem with which the F-G may be supposed to have been faced. Five objects ⁑A, ⁑A′, ⁑B, ⁑B′, ⁑L are supposed to be such that the place, from which furrows could be driven one through ⁑A, ⁑B, and one through ⁑A′, ⁑B′, is inaccessible, but he wishes to drive a furrow that shall pass through ⁑L and this place.

After considerable experimenting he devises a solution for the corresponding problem when all the points are accessible. Expressed in geometrical language this is

$$
\begin{array}{ll}
G & \mathscr{L}_3 \# | \mathrm{ABA'B'L} \\
S & \mathrm{C:C \in LB} \ \& \ \# | \mathrm{CLB} \\
C1 & \mathrm{K:K = AA' \cap BB'} \\
C2 & \mathrm{C':C' = KC \cap B'L} \\
C3 & \mathrm{M:M = CA \cap C'A'} \\
C4 & \mathrm{N:N = AB \cap A'B'}
\end{array}
$$

Then, as a matter of experience, he finds that $\mathrm{LM \supset N}, = \mathrm{AB \cap A'B'}$, but finds it outside the range of his capability to prove this result on the basis of $A\mathscr{I}$ 1, 2. We may imagine, in fact, circumstances in which he is able to make an observation which casts doubt on the universal validity of the relation $\mathrm{LM \supset N}$ as a consequence of the prescribed construction. Let us suppose that he has arranged nine objects in the paddock corresponding to the points K, A, B, C, A′, B′, C′, L and M, and has observed that, a long way off in the forest, there is a conspicuous tree, ⁑N, which satisfies the conditions of lying along the lines of sight ⁑AB, ⁑A′B′ and ⁑LM. One evening the tribe occupying the part of the forest between the object ⁑M and the tree ⁑N stages a large barbecue and in consequence a broad stream of heated air rises across the line of sight ⁑LM, but not across ⁑AB or ⁑A′B′. The effect of this is that, while the tree ⁑N is still visible along the lines of the sight ⁑AB and ⁑A′B′, it wavers to and fro in and out of the line of sight ⁑LM.

The F-G of course knows nothing of convection currents or refraction, but he does make the quite reasonable assumption that 'space' outside his field of direct observation may be different in some ways from the 'space' he can observe. That, in fact, while its properties may be represented by the axioms $A\mathscr{I}$ 1, 2, the concurrence of the three collinear sets AB, A′B′, LM in his construction is not a logical consequence of those axioms.

The method of postulating a region in the plane in which lines are refracted was precisely that adopted by Hilbert when he first showed that the final incidence in the Desargues construction is not a consequence of axioms such as $A\mathscr{I}$ 1, 2 alone. Later (Excursus 2.4) when the geometry has been sufficiently developed we can give, in modified form, Hilbert's proof. A finite non-Desarguesian geometry appropriate to the present stage of development is described in Excursus 1.3.2.

There is another facet of this problem, which is discussed in Excursus 1.4, namely, that if the geometry is embedded in a system which has points outside the system satisfying $A\mathscr{I}$ 1, 2 (i.e. in a three dimensional geometry) with, of course, suitable additional incidence axioms, then '$N \in LM$' is a consequence of the other given incidence relations.

We now state the existence of the Desargues figure formally as an axiom.

$A\mathscr{I}3$ The Desargues axiom.

$G1$ $\mathscr{L}_3|KABC$
$G2$ $A' : A' \in KA \;\&\; \#|A'AK$
$B' : B' \in KB \;\&\; \#|B'BK$
$C' : C' \in KC \;\&\; \#|C'CK$
C $L : L = BC \cap B'C'$
$M : M = CA \cap C'A'$
$N : N = AB \cap A'B'$
A $\mathscr{L}|LMN.$

We may suppose $\mathscr{L}|A'B'C'$ for otherwise the statement is trivial, since $\{L, M, N\} \subset A'B'$ if $\mathscr{L}|A'B'C'$. We also suppose in general that there are no coincidences or additional collinearities among the points. It may happen that the nature of the geometry itself forces these additional conditions on the figure; if it does so then either the validity of the axiom for such a system (if it is not inherent in the definition of the system) has to be stated, or the conclusion $\mathscr{L}|LMN$ has to be stated as a theorem.

For example, the configurations $\Gamma^{(2)}$ and $\Gamma^{(3)}$ involve many additional incidences, but the properties of the Desargues figure are immediate consequences of the definition of the geometries. Thus in $\Gamma^{(2)}$ (Excursus 1.1.1) take

$$\{{}^{\mathscr{O}}K, A, B, C\} = \{{}^{\mathscr{O}}6, 2, 4, 5\}.$$

The only remaining points are the collinear set {0, 1, 3}, which, if taken to be {A,′ B′, C′}, is also {L, M, N} when the construction is completed!

The corresponding properties of $\Gamma^{(3)}$ (Excursus 1.1.2) may be exhibited by arranging the ten points of the Desargues figure as part of the cyclic pattern of the geometry.

For example:

A	B	C	C′	N	L	.	.	K	.	M	A′	B′
B	C	C′	N	L	.	.	K	.	M	A′	B′	A
N	L	.	.	K	.	M	A′	B′	A	B	C	C′
.	.	K	.	M	A′	B′	A	B	C	C′	N	L

The ten required collinearities are clear, and in addition we have $\mathscr{L}|$NA′B′C, $\mathscr{L}|$MA′C′B, $\mathscr{L}|$LB′C′A, $\mathscr{L}|$KLMN.

But we shall assume that the axiom has been stated for a figure in which neither the peculiarities of the geometry, nor the selection of the seven points K, A, B, C, A′, B′, C′ is such that any two of the ten points coincide, and, at present, that no set of four is collinear. At the end of the chapter we discuss special cases in which one or more sets of four points are collinear.

The statements made in defining the Desargues figure are essentially statements about sets of three points; we are not concerned with 'lines' as such but only with sets of three points related by collinearity. In the form of statement above there is a step-by-step 'construction' from the seven given points, i.e. an underlying implication that we are making dots and strokes on paper. Let us write out the statement again using only assertions of collinearity:

G A, B, C; A′, B′, C′; L, M, N; K are ten points with the following relations of collinearity:

$$\mathscr{L}|\begin{cases} \text{KAA}', \text{KBB}', \text{KCC}' \\ \text{LBC}, \text{LB}'\text{C}' \\ \text{MCA}, \text{MC}'\text{A}' \\ \text{NAB}, \text{NA}'\text{B}' \end{cases}$$

A $\mathscr{L}|$LMN.

Alternatively the figure could be organised in such a way as to begin with any three concurrent lines in it. For example, from lines through A we could arrange the collinearities corresponding to those above in the following ways:

G

$$\mathscr{L}|\begin{cases} \text{AKA}', \text{ACM}, \text{ABN} \\ \text{BCL}, \text{MNL}, \\ \text{KBB}', \text{A}'\text{NB}' \\ \text{KCC}', \text{A}'\text{MC}' \end{cases}$$

A $\mathscr{L}|$LB′C′

Associated with the Desargues figure is the figure (which in fact turns out to be the same figure again) obtained by assuming the collinearity of L, M, N and leading back to the concurrence of AA′, BB′, CC′. This we can prove if we assume $A\mathscr{I}$ 3. The theorem is:

1.3*T*1. *Converse (and dual) of Desargues*

*G*1	$\mathscr{L}\vert$LMN
*G*2	#|AA′ & {A, A′} ⊄ LM
*G*3	B : B ∈ NA & #|ABN
*G*4	B′ : B′ ∈ NA′ & #|A′B′N
C	C : C = LB ∩ MA
	C′ : C′ = LB′ ∩ MA′
T	$\mathscr{P}\vert$AA′ ∩ BB′ ∩ CC′.

We can set the data out as nine statements of collinearity, thus

$$\mathscr{L}\big|\begin{cases}\text{NAB, NA}'\text{B}'\text{, NLM}\\ \text{LCB, MCA}\\ \text{LC}'\text{B}'\text{, MC}'\text{A}'\\ \text{KAA}'\text{, KBB}'\end{cases}$$

These are precisely the same statements as in the Desargues axiom, with the points renamed according to the table:

	K;	A, B, C;	A′, B′, C′;	L, M, N
renamed	N;	A, A′, M;	B, B′, L;	C′, C, K.

It follows from $A\mathscr{I}$ 3 that

$$\mathscr{L}|\text{KCC}',$$

i.e.

$$\text{CC}' \supset \text{AA}' \cap \text{BB}'.$$

Alternatively we can exhibit the data of the converse theorem as the dual of the statement of the Desargues axiom.

The Desargues figure is completely determined by K and three distinct pairs of points with each of which K is collinear. This accounts for seven of the ten points; the remaining three are the set {L, M, N} which the Desargues axiom requires shall be collinear. That is, the construction leads from a point K to a collinear set {L, M, N} which we may designate k. Writing the original constructions and the dual side by side we shall have:

K	k	k ⊃ {L, M, N}
$\mathscr{L}\vert$KAA′	$\mathscr{P}\vert$k ∩ a ∩ a′	a ⊃ {L, B, C}
$\mathscr{L}\vert$KBB′	$\mathscr{P}\vert$k ∩ b ∩ b′	b ⊃ {M, C, A}

$\mathscr{L}\|KCC'$	$\mathscr{P}\|k \cap c \cap c'$	$c \supset \{N, A, B\}$
$L = BC \cap B'C'$	$l = \langle b \cap c, b' \cap c' \rangle$	$l \supset \{A, A', K\}$
$M = CA \cap C'A'$	$m = \langle c \cap a, c' \cap a' \rangle$	$m \supset \{B, B', K\}$
$N = AB \cap A'B'$	$n = \langle a \cap b, a' \cap b' \rangle$	$n \supset \{C, C', K\}$
$\mathscr{L}\|LMN$	$\mathscr{P}\|l \cap m \cap n$	$K = l \cap m \cap n$

The Desargues construction will be used frequently in the course of proofs of theorems, and it is convenient therefore to have a compact method of conveying the data and the deduction. We shall adopt the following:

1.3*N*1. (i) The construction set out in $A\mathscr{I}$ 3:

$$K\begin{Bmatrix} AA' \\ BB' \\ CC' \end{Bmatrix} \Rightarrow \mathscr{L}|LMN$$

(ii) The construction of the converse, 1.3*T*1

$$\begin{Bmatrix} L \\ M \\ N \end{Bmatrix} \begin{Bmatrix} AA' \\ BB' \\ CC' \end{Bmatrix} \Rightarrow \mathscr{P}|AA' \cap BB' \cap CC' \quad (=K).$$

It is of course essential in using this notation that every letter should be in its correct relative position. (We are making full use of the printing convention!)

The Desargues figure consists of a set of ten points arranged in ten sets of three which are collinear, three of these sets containing any point of the set. It is symmetrical in the ten points, and we expect therefore to be able to devise notations which would display this symmetry. The number ten conveys the suggestion half of 5·4, and we try therefore the set of symbols $\{P_{ij}\}$, where $\{i, j\} \subset \{1, 2, 3, 4, 5\}$, as the names of the points, with the obvious guess $\{P_{ij}, P_{jk}, P_{ki}\}$ for the collinear sets. We find, in fact, that the points may be named thus:

K	A	B	C	A′	B′	C′	L	M	N
P_{45}	P_{14}	P_{24}	P_{34}	P_{15}	P_{25}	P_{35}	P_{23}	P_{31}	P_{12}

The statement in 1.3*N*1 (i) then appears as

$$P_{45}\begin{Bmatrix} P_{14} & P_{15} \\ P_{24} & P_{25} \\ P_{34} & P_{35} \end{Bmatrix} \Rightarrow \mathscr{L}|P_{23}P_{31}P_{12},$$

a statement which is valid for any permutation of the indices.

We conclude this Chapter with some discussion of the Desargues figures in which some sets of four points are collinear. Let us initially restrict $A\mathscr{I}3$ to the 'general' case, namely,

$$\begin{array}{lll} \mathrm{A}\mathscr{I}\,3 & G & \#|\mathrm{KABCA'B'C'LMN} \\ & & \&\ \mathscr{L}|\mathrm{KAA'}\ \&\ \mathscr{L}|\mathrm{KBB'}\ \&\ \mathscr{L}|\mathrm{KCC'} \\ & & \&\ \mathscr{L}|\mathrm{LBC}\ \&\ \mathscr{L}|\mathrm{MCA}\ \&\ \mathscr{L}|\mathrm{NAB} \\ & & \&\ \mathscr{L}|\mathrm{LB'C'}\ \&\ \mathscr{L}|\mathrm{NC'A'}\ \&\ \mathscr{L}|\mathrm{NA'B'} \\ & & \&\ \text{no set of four of the ten points is collinear.} \\ & A & \mathscr{L}|\mathrm{LMN} \end{array}$$

We are to show that we can deduce the existence of various 'special' configurations in which one or more sets of four points are collinear. Consider first the 'simply-special' figure in which there is exactly one set of four collinear points; let us assume that K belongs to one of the collinear sets which does not, in the general figure, contain K. If $\mathrm{K} \in \mathrm{BC}$, then the figure collapses completely, since $\mathscr{L}|\mathrm{KBCB'C'}$ and L is indeterminate (except as $\mathrm{L} = \mathrm{MN} \cap \mathrm{BC}$, and then the axiom is nugatory). The only collinear set to which K can be made to belong is therefore {L, M, N}, that is, k. Thus the possible additional collinearities are $\mathscr{L}|\mathrm{LKMN}$, $\mathscr{L}|\mathrm{A'BCL}$, $\mathscr{L}|\mathrm{AB'C'L}$, $\mathscr{L}|\mathrm{LKAA'}$, etc. Because of the symmetry of the general figure, all these are equivalent. If we require more than one set of four points to be collinear our choice is limited by the requirement that the figure shall not collapse.

All relevant forms of Desargues data include an unrestricted set of four points, say, $\mathscr{L}_3|\mathrm{KABC}$, the various specializations being dependent on the freedom of choice of the other points. The specific choices may be taken to be:

general: three points chosen one from each of three assigned collinear sets.

$$\mathrm{A'} \in \mathrm{KA}, \quad \mathrm{B'} \in \mathrm{KB}, \quad \mathrm{C'} \in \mathrm{KC}.$$

simply-special: two points chosen one from each of two assigned collinear sets.

$$\mathrm{B'} \in \mathrm{KB}, \quad \mathrm{C'} \in \mathrm{KC}, \quad \text{but } \mathrm{A'} = \mathrm{KA} \cap \mathrm{BC}.$$

doubly-special: one point chosen from an assigned collinear set.

$$\mathrm{C'} \in \mathrm{KC}, \quad \text{but } \mathrm{A'} = \mathrm{KA} \cap \mathrm{BC}, \quad \mathrm{B'} = \mathrm{KB} \cap \mathrm{CA}.$$

triply-special: completely determined by K, A, B, C.

$$\mathrm{A'} = \mathrm{KA} \cap \mathrm{BC}, \quad \mathrm{B'} = \mathrm{KB} \cap \mathrm{CA}, \quad \mathrm{C'} = \mathrm{KC} \cap \mathrm{AB}.$$

To make the configuration triply-special is as far as we may go without specializing our initial choice of points. For suppose we take $\mathscr{L}_3|$KABC and construct A′, B′, C′, L, M, N. The only additional set of points that can be made collinear without inducing incidences which make any Desargues-type statement nugatory is {K, L, M} (or {K, M, N} or {K, N, L}). That is, after L and M have been constructed, we require that our initial set of four points {K, A, B, C} should have been such that K $\in$ LM. We have seen that such a figure exists in $\Gamma^{(3)}$, and shall see later (Excursus 2.1), that any geometry which satisfies a Desargues axiom, and contains a quadruply-special Desargues figure includes a set of 13 points forming $\Gamma^{(3)}$ in which the ten points of this figure are embedded.

We are going to take the point of view that if the axiom $A\mathscr{I}$ 3 is expressed in terms of data for the general configuration, then it is necessary to prove that the relation still holds when the data have been specialized by imposing additional conditions of collinearity. That is, if the data are the nine relations of collinearity and no others, and the axiom is that there is in consequence a tenth condition of collinearity, then we have to prove that, when the data include additional relations of collinearity, the final relation of collinearity is a consequence. We shall prove the following sequence of theorems:

general axiom $\Rightarrow$ simply-special theorem,
general axiom $\Rightarrow$ doubly-special theorem,
simply-special axiom $\Rightarrow$ triply-special theorem.

Later we shall see that an axiom based on the simply-special figure (instead of the general figure) implies the existence of the doubly-special and triply-special figures. None of the three statements can be reversed, but we give only a plausible reason for this, namely, that an axiom stated for a set of points with a certain freedom of choice is, in general, unlikely to be strong enough to prove a theorem which allows additional freedom of choice in the elements providing the data. Thus, when we are given $\mathscr{L}_3|$KABC, then, in the simply-special case (with the specialisation $\mathscr{L}|$A′BCL), B′ and C′ may be freely chosen from the collinear sets KB and KC, but A′ is fixed, while in the general case B′, C′ and A′ all may be chosen freely from assigned collinear sets.

The data for the simply-special figure may be presented in two different forms (i) the data are eight statements of collinearity of three points and one statement of collinearity of four points ($\mathscr{L}|$A′BCL) and the conclusion is that there are three other points collinear ($\mathscr{L}|$LMN), or (ii) the data are ten statements of collinearity for three points (the additional relation being $\mathscr{L}|$LKM) and the conclusion is that a fourth point is collinear with one of the sets of three ($\mathscr{L}|$KLMN).

1.3 *T*2 $A\mathscr{I}\ 3 \Rightarrow$ existence of the simply-special figure (first form).

G $\mathscr{L}_3|\mathrm{KABC}$
$\mathrm{B'}:\mathrm{B'}\in\mathrm{KB}\ \&\ \#|\mathrm{KBB'}$
$\mathrm{C'}:\mathrm{C'}\in\mathrm{KC}\ \&\ \#|\mathrm{KCC'}$
$C1$ $\mathrm{A'}:\mathrm{A'}=\mathrm{KA}\cap\mathrm{BC}$
$C2$ $\mathrm{M}:\mathrm{M}=\mathrm{CA}\cap\mathrm{C'A'}$
$C3$ $\mathrm{N}:\mathrm{N}=\mathrm{AB}\cap\mathrm{A'B'}$
$C4$ $\mathrm{L}:\mathrm{L}=\mathrm{BC}\cap\mathrm{B'C'}$
T $\mathscr{L}|\mathrm{LMN}$
P s $\mathrm{C''}:\mathrm{C''}\in\mathrm{KC}\ \&\ \#|\mathrm{KCC'C''}$
c $\mathrm{L'}:\mathrm{L'}=\mathrm{BC''}\cap\mathrm{B'C'}$
$\mathrm{M'}:\mathrm{M'}=\mathrm{C''A}\cap\mathrm{C'A'}$
$t1$ $\mathscr{L}|\mathrm{L'M'N}$

$$p\quad \mathrm{K}\begin{Bmatrix}\mathrm{AA'}\\ \mathrm{BB'}\\ \mathrm{C''C'}\end{Bmatrix} \Rightarrow \mathscr{L}|\mathrm{L'M'N}\quad (A\mathscr{I}\ 3)$$

$t2$ $\mathscr{L}|\mathrm{LMN}$

$$p\quad \begin{Bmatrix}\mathrm{C}\\ \mathrm{C'}\\ \mathrm{C''}\end{Bmatrix}\begin{Bmatrix}\mathrm{L'M'}\\ \mathrm{LM}\\ \mathrm{BA}\end{Bmatrix} \Rightarrow \mathscr{P}|\mathrm{L'M'}\cap\mathrm{LM}\cap\mathrm{AB}=\mathrm{N}\qquad 1.3T1$$

T $\mathscr{L}|\mathrm{LMN}$

No four of the ten points in the Desargues figure used either in $t1$ or in $t2$ are collinear, that is, we have used the general form of the figure to establish the special form.

We may now organize the data so as to be able to re-interpret this proof as a proof for the second form of the simply-special conditions. Take the data and construction as in $1.3T2$ up to and including construction $C3$, then replace $C4$ by $C4'$

$$\mathrm{L}:\mathrm{L}=\mathrm{B'C'}\cap\mathrm{MN}.$$

The theorem, which expresses the same relations of collinearity as in $T2$, and to which the same proof applies, now becomes,

$$T'\quad \mathscr{L}|\mathrm{LBC}.$$

That is, we have proved

$$\mathrm{A'}\begin{Bmatrix}\mathrm{KA}\\ \mathrm{C'M}\\ \mathrm{B'N}\end{Bmatrix} \Rightarrow \mathscr{L}|\mathrm{LBC}$$

on a basis of $\mathscr{L}|\mathrm{A'LBC}$ and eight sets of three collinear points. So that, since the data for the two theorems are the same, and the conclusion in each case is $\mathscr{P}|\mathrm{MN}\cap\mathrm{BC}\cap\mathrm{B'C'}$,

1.3 *T*3 The two forms of the simply-special Desargues theorem are equivalent.

The configuration dual to the first form of the simply-special configuration is one in which $\mathscr{P}|a \cap b' \cap c' \cap l$, that is,

$$\mathscr{P}|BC \cap A'C' \cap A'B' \cap KA',$$

a condition which is equivalent to $A' = BC \cap KA$. Thus the simply-special figure and its dual are identical in structure.

The proof of *T* 2 is in no way affected if there is the further condition $\mathscr{L}|B'ACM$, and the Desargues figures used in the proof are not restricted by any overt additional conditions of collinearity, so that the doubly-special Desargues figure in which the data include two sets of four collinear points, $\mathscr{L}|A'BCL$ and $\mathscr{L}|B'CAM$, exists also as a consequence of $A\mathscr{I}$ 3. There is, however, less freedom of choice of the ten points in both *t* 1 and *t* 2 than there is in the general figure: after $\mathscr{L}_3|KABC$ the only freedom of choice to be exercised is in the selection of the two points C′ and C″. There is in fact a hidden collinearity condition in the data of the lemma *t* 1 and of the lemma *t* 2, namely, in *t* 1 in the figure

$$K\begin{Bmatrix} AA' \\ BB' \\ C''C' \end{Bmatrix} \Rightarrow \mathscr{L}|L'M'N$$

we have $\mathscr{L}|\langle AB' \cap A'B, C', C''\rangle$

and in *t* 2 in the figure

$$\begin{Bmatrix} C \\ C' \\ C'' \end{Bmatrix}\begin{Bmatrix} L'M' \\ LM \\ BA \end{Bmatrix} \Rightarrow N.$$

we have $\mathscr{L}|\langle LL' \cap AM, MM' \cap BL, N\rangle$. These do not contravene the explicitly stated conditions of $A\mathscr{I}$ 3.

1.3*X*1 general axiom ⇒ existence of figure with a certain hidden collinearity.

G $\mathscr{L}_3|KABH$
$\{C, C'\} \subset KH \;\&\; \#|KHCC'$
C $A' : A' = BH \cap KA$
$B' : B' = AH \cap KB$
$L : L = BC \cap B'C'$
$M : M = CA \cap C'A'$
$N : N = AB \cap A'B'$
T $\mathscr{L}|LMN$
P(*X*) *s* $D : D \in BC \;\&\; \#|DBCL$
c $D' : D' = KD \cap B'C'$

The triply-special Desargues figure depends on the choice of only four points, no three collinear. The proof that its existence is a consequence of the general axiom follows the pattern of $T\,2$:

1.3 *T4X* Simply-special axiom ⇒ triply-special.

G $\mathscr{L}_3|$KABC
C A′:A′ = BC ∩ KA
B′:B′ = CA ∩ KB
C′:C′ = AB ∩ KC
L:L = BC ∩ B′C′
M:M = CA ∩ C′A′
N:N = AB ∩ A′B′
T $\mathscr{L}|$LMN
$P(X)$ s C″:C″ ∈ KC & #|KCC′C″.

In Excursus 1.3.2 a system of 91 points is described which satisfies the combinatorial axioms $A\mathscr{I}$ 1, 2 but contains sets of ten points which satisfy all but one of the incidence conditions in any one of the foregoing forms of the Desargues configuration, and yet which do not satisfy the final condition. Later, when coordinates have been introduced, another non-Desarguesian geometry is described (Excursus 2.2). Some form of Desargues axiom is therefore necessary to the development of the geometry, and in all that follows we shall assume, unless some contrary statement is made,

A$\mathscr{I}$ 1 Two points determine a line and any line is determined by any two of its points.

A$\mathscr{I}$ 2 Any two distinct lines have a single common point, and either

A$\mathscr{I}$ 3 Desargues in the general form, or

A$\mathscr{I}$ 3* Desargues in the simply-special form.

We shall find that the geometry can be developed to considerable depth (at least to the extent of providing a complete model for the F-G's observations) using only $A\mathscr{I}$ 3*, but at one critical place we do not seem to be able to proceed without assuming $A\mathscr{I}$ 3.

CHAPTER 1.4

THE HARMONIC TRANSFORMER, $\mathscr{H}$

At two stages at least in the development of the geometry the F-G has to be supposed to devise mathematical properties which are not suggested to him by the physical relations of the objects in the paddock, but which are purely mathematical consequences of his axioms. In each case, as it appears in due course, the abstract piece of pure mathematics is precisely what is required as a basis for a vital step forward in the development of the mathematical model.

One such piece of pure mathematics is a consequence of the Desargues axiom. It is this: there exists a simple construction by which, from three given points of a collinear set, to be considered as a pair and a single point, using certain arbitrarily selected auxiliary points not in the collinear set, a fourth point in the collinear set can be determined which depends only on the given three points and not at all on the selection of the auxiliary points. The construction itself is fairly simple; the difficulties arise in the proof that the point constructed is independent of the auxiliary points.

1.4*C*1 Construction of the fourth harmonic point.

G $\mathscr{L}\#|$ABC
S L:L$\notin$AB
M:M$\in$LC, $\#|$MLC
C N:N = AM $\cap$ BL
K:K = BM $\cap$ AL
D:D = AB $\cap$ KN
= AB $\cap$ $\langle$AM $\cap$ BL, AL $\cap$ BM$\rangle$.

D is determined by the given points A, B, C and the arbitrarily selected points L, M; we have to prove that a different choice of L, M leads to the same point D. We may organise the selection of these points rather differently by selecting arbitrarily three lines, one each through A, B and C, say then:

S l: l $\supset$ A, m: m $\supset$ B, n: n $\supset$ C
$\#|$l, m, n, AB & $\emptyset|$l $\cap$ m $\cap$ n

i.e. instead of choosing L, we choose the lines BL = m and CL = n, and instead of choosing M on CL we choose l through A and take M = l $\cap$ CL.

We have to prove now that D is independent of the choice of l, m, n, and we shall have done this when we have shown that (i) for a given choice of l and m two different choices of n lead to the same point D and (ii) since A and B are symmetrically related in the figure, that, for a given choice of m and n, two different choices of l lead to the same point D.

1.4*T*1 (i) D is unaffected by the choice of n.

G $\mathscr{L}\#|\mathrm{ABC}$
$S1$ $\mathrm{L}:\mathrm{L}\notin\mathrm{AB}$
$\mathrm{L}':\mathrm{L}'\in\mathrm{BL},\ \#|\mathrm{BLL}'$
$S2$ $\mathrm{M}:\mathrm{M}\in\mathrm{CL},\ \#|\mathrm{MCL}\ \&\ \mathscr{L}|\mathrm{AML}'$
$C1$ $\mathrm{M}':\mathrm{M}' = \mathrm{CL}'\cap\mathrm{AM}$
$C2$ $\mathrm{N}:\mathrm{N} = \mathrm{AM}\cap\mathrm{BL}$
$C3$ $\mathrm{K}:\mathrm{K} = \mathrm{BM}\cap\mathrm{AL}$
$\mathrm{K}':\mathrm{K}' = \mathrm{BM}'\cap\mathrm{AL}'$
T $\mathscr{L}|\mathrm{NKK}'.$
P $\begin{Bmatrix}\mathrm{B}\\ \mathrm{A}\\ \mathrm{C}\end{Bmatrix}\ \begin{Bmatrix}\mathrm{LL}'\\ \mathrm{MM}'\\ \mathrm{KK}'\end{Bmatrix} \Rightarrow \mathscr{P}|\mathrm{LL}'\cap\mathrm{MM}'\cap\mathrm{KK}' = \mathrm{N}$ (1.3*T*1)
T $\mathrm{NK}\cap\mathrm{AB} = \mathrm{NK}'\cap\mathrm{AB} = \mathrm{D}$

1.4 *T*1 (ii) *X* D is unaffected by the choice of m.

The construction 1.4*C*1 involves the points A and B symmetrically, since the selection and construction could be rephrased as:

S $\{\mathrm{L},\mathrm{M}\}:\mathscr{L}\#|\mathrm{CLM},\ \#|\mathrm{CL},\ \mathrm{AB}$
C $\{\mathrm{N},\mathrm{K}\}:\mathrm{N} = \mathrm{AM}\cap\mathrm{BL},\ \mathrm{K} = \mathrm{AL}\cap\mathrm{BM}$
$\mathrm{D}:\mathrm{D} = \mathrm{AB}\cap\mathrm{NK}.$

The two compound statements are unaltered, except in interchanging the names of the points N and K, when the points A and B are interchanged.

Under the construction 1.4*C*1, any pair {A, B} of distinct points defines from any point C, collinear with A and B, a unique point D collinear with A and B. That is, any pair of points {A, B} defines a geometrical *function*, or *operator*, or *transformer*, say $\mathscr{H}_{\mathrm{AB}}$, which transforms any point C collinear with A and B into a point D. We could express this as a functional relation, in the form $\mathscr{H}_{\mathrm{AB}}(\mathrm{C}) = \mathrm{D}$, or, since the parenthesis contribute nothing to meaning or intelligibility, as a transformation, in the form $\mathscr{H}_{\mathrm{AB}}\mathrm{C} = \mathrm{D}$.

1.4*N*1 $\mathscr{H}_{AB}C = D$: D is obtained from $\mathscr{L}\#|ABC$ by the construction 1.4*C*1.

$\mathscr{H}_{AB}$: *the harmonic transformer.*

We have already proved:

1.4*T*2 $\mathscr{H}_{AB} = \mathscr{H}_{BA}$

(with the meaning: for all C, $C \in AB$, $\mathscr{H}_{AB}C = \mathscr{H}_{BA}C$) but the system of four points has in fact a much higher degree of symmetry than this; we are to prove that the construction establishes a relation between the members of an (unordered) pair of (unordered) pairs of points, all collinear.

1.4*T*3 The harmonic relation is symmetric in the two pairs of points.

G $\mathscr{H}_{AB}C = D \;\&\; \#|CD$
T $\mathscr{H}_{CD}A = B$
*C*1 {L, M, N, K, D} as in *T*1
*C*2 H:H = CL ∩ DK
P:P = DL ∩ BH
*t*1 $\mathscr{L}|\langle H, B, CK \cap DL\rangle$

p $A\begin{Bmatrix} MN \\ CD \\ KL \end{Bmatrix} \Rightarrow \mathscr{L}|\langle H, B, CK \cap DL\rangle$ (triply-special Desargues)

*t*2 $\mathscr{H}_{CD}A = B$
p In 1.4*C*1

replace	A	B	C	L	M	N	K	D
by	C	D	A	K	L	H	P	B

$B = CD \cap \langle CK \cap DL, CL \cap DK\rangle$.

We might therefore regard the construction 1.4*C*1 both as defining a transformer based on two points and as stating a symmetrical relation among four collinear points arranged as a pair of pairs. When we wish to emphasize the second aspect, we shall use the following terminology and notation:

1.4*N*2 Two pairs of points {A, B}, {C, D} related by the construction 1.4*C*1 are two 'harmonically conjugate pairs' or the points of either pair are 'harmonically conjugate' with regard to the other. The relation will be expressed as $AB\mathscr{H}CD$.

$AB\mathscr{H}CD$ is one of eight similar forms in which the relation can be printed: $AB\mathscr{H}CD \Leftrightarrow DC\mathscr{H}AB \Leftrightarrow \mathscr{H}_{AB}C = D \Leftrightarrow \mathscr{H}_{DC}A = B$, etc.

In general we are concerned only with distinct points, but if the construction 1.4*C*1 is effected for a figure with B = C, we find

1.4*T*4 $\mathscr{H}_{AB}B = B$.

P $K = AL \cap BM = L$, $N = AM \cap BL = M$, $KN \cap AB = B$.

1.4*T*5*X* *G* {A, B, C, L, M, N, K, D, H} as in 1.4*T*3.

T $CH\mathscr{H}LM$.

1.4*T*6*X* If two sets of four collinear points are 'in perspective' and one of them forms two harmonic pairs, then so does the other.

*G*1 $\mathscr{L}|ABC$ & $D = \mathscr{H}_{AB}C$
*G*2 $\mathscr{L}|A'B'C'D'$ & $\#|AB, A'B$ & $\mathscr{P}|AA' \cap BB' \cap CC' \cap DD'$
T $D' = \mathscr{H}_{A'B'}C'$.

1.4*X*1 In geometry $\Gamma^{(3)}$ show, for some selected set $\mathscr{L}\#|ABCD$, that

$$D = \mathscr{H}_{AB}C = \mathscr{H}_{AC}B = \mathscr{H}_{BC}A.$$

1.4*X*2 Show that for any selected three collinear points A, B, C in $\Gamma^{(4)}$

$$\mathscr{H}_{AB}C = C.$$

1.4*X*3 *C*1 Construction as in 1.4*C*1.

*C*2 $X : X = MD \cap NB$
$Y : Y = MB \cap NC$
T $\mathscr{L}|AXY$.

1.4*X*4 *G*1 $\mathscr{L}|PQR$

*G*2 $l : \{P, Q, R\} \cap l = \emptyset$
*C*1 $\{U, V, W\} : \{^{c}U, V, W\} = l \cap \{^{c}QR, RP, PQ\}$
*C*2 $\{X, Y, Z\} : X = \mathscr{H}_{QR}U$, $Y = \mathscr{H}_{RP}V$, $Z = \mathscr{H}_{PQ}W$
T (i) $\mathscr{L}|UYZ$
(ii) $\mathscr{P}|PX \cap QY \cap RZ$.

1.4*X*5 *G* {A, B, C, L, M, N, K, D, L′, N′, K′} as in 1.4*T*1 (ii).

T $\mathscr{L}|\langle D, AL \cap BL', AL' \cap BL\rangle$.

1.4*X*6 *G* The triply-special Desargues figure, 1.3*T*4

T $LA'\mathscr{H}BC$.

1.4*X*7 If in *X* 6 we have also $\mathscr{L}|KLMN$, then

$$K = \mathscr{H}_{MN}L = \mathscr{H}_{NL}M = \mathscr{H}_{LM}N.$$

CHAPTER 1.5

PARALLELS

The F-G now turns his attention to the pairs of pairs of objects which determine ⁑collinear sets which do not appear to have any common member. He is able to recognise two types of such pairs. In the first case he finds that by setting up objects in suitable positions ⁑collinear with one or other of the pairs, he obtains a physical appearance of a relation among pairs of these objects which suggests that, if he could penetrate deeply enough into the surrounding forest, he would be able to find a position which could be occupied by an object belonging to both his ⁑collinear sets. For the present then he is prepared to make the assumption that his geometry is such that the pairs of lines corresponding to these do have a common point even though the position corresponding to the point is inaccessible.

The second type of pairs of ⁑collinear sets presents itself very clearly as he ploughs two consecutive furrows. His practice is to set his first furrow 'straight' by sighting along two objects, but to plough the adjacent furrow by visual reference to the first. He makes no attempt at present to formalize this process of 'visual reference' although he realises that he will later have to evolve some explanation of it. However that may be, he is aware (i) that his second furrow does correspond to a line, and (ii) that the physical process that he used in determining the second furrow from the first seems to imply that the two lines by which they are representable do not have any common point, or at least any common point in the sense that he was prepared to accept the postulation of an 'inaccessible' point in the previous case.

He calls a pair of lines related in this way 'parallels'. We shall write the relation as:

$$\mathrm{a} \parallel \mathrm{b}\text{: `a is parallel to b'.}$$

Leaving for later reference whatever property parallels may have which corresponds to the process of visual reference by which the furrows were constructed, the properties which parallels appear to have which can be fitted into the framework of the geometry as so far developed are these:

(i) The relation of parallelism is *symmetric*, i.e., $\mathrm{a} \parallel \mathrm{b}$ and $\mathrm{b} \parallel \mathrm{a}$, are equivalent statements.

(ii) The relation is *transitive*, i.e. $\mathrm{a} \parallel \mathrm{b}$ & $\mathrm{b} \parallel \mathrm{c} \Rightarrow \mathrm{a} \parallel \mathrm{c}$.

(iii) Through any point a single line can be constructed ('named') parallel to a given line.

This last raises two questions which have to be settled. First, suppose the point lies on the line itself; what is the parallel line through it? It appears to be the line itself, and the F-G therefore adds (tentatively, since this is a matter of definition, the consistence of which will have to be tested later):

(iv) The relation is *reflexive*, i.e. always $a \parallel a$.

(i), (ii) and (iv) are the properties required in order that parallelism shall be an 'equivalence relation' or 'equable'.

The second question raised by (iii) is this, is it true that through *any* point a single line can be drawn parallel to a given line? That is, is this property to hold for the inaccessible points also? The F-G assumes that the answer is yes, and proceeds accordingly.

The next task is to secure these properties as consequences of assigned incidence properties. As a first attempt we see that this can be done by postulating:

(i) On every line there is a single 'ideal' point U*.

(ii) If U* is the ideal point on a, then a line b is parallel to a if and only if it passes through U*.

(iii) U* is also the ideal point on every line parallel to a.

So far then each family of parallel lines is identified by a single ideal point U*. Suppose now that U* and V* are distinct ideal points (belonging to different parallel families), and P is any point not collinear with U*, V*. Then both PU* and PV* satisfy the condition of being parallel to U*V*, so that U*V* cannot be among the lines intended to be included in the original statement (iii).

We can circumvent this difficulty by making the postulate contained in the following definition:

1.5 *D*1 (Parallel postulate).

(i) In any system of geometry determined by $A\mathscr{I}$ 1, 2 some line is selected to be the *ideal line*.

(ii) All points of this line are *ideal points*.

(iii) Two distinct lines are *parallel* if and only if their common point is an ideal point.

(iv) Any line is *parallel* to itself.

(v) The relation of parallelism does not apply to the ideal line itself (i.e., the ideal line is neither parallel to nor not parallel to any other line).

The notation to be used has already been introduced, namely:

1.5 *N*1 $a \parallel b$: a is parallel to b (a and b are parallel), the relation of parallelism having the 'equivalence' properties:

(i) $a \parallel b \Rightarrow b \parallel a$.

(ii) a ∥ a.

(iii) a ∥ b & b ∥ c ⇒ a ∥ c.

and the further property that the parallel to a given line through a given point is unique, i.e.

(iv) $\left.\begin{array}{r}b \supset P \ \&\ b \parallel a \\ \text{and } c \supset P \ \&\ c \parallel a\end{array}\right\} \Rightarrow b = c.$

The relation is determined by an ideal line u, thus:

1.5*D2* $\mathscr{P}|a \cap b \cap u \ \&\ \#|au \ \&\ \#|bu \overset{\text{def}}{\Leftrightarrow} a \| b$ and the line through B parallel to a is constructed as ⟨B, a ∩ u⟩.

Points and lines which are *not ideal* will be called *proper* and we shall use the notation.

1.5*N2* ℘|AB...: the points A, B, ... are *proper*

℘|ab...: the lines a, b, ... are *proper*

i.e. A, B, ... do not lie on the ideal line, and none of a, b, ... is the ideal line.

Thus frequently the statement of the data of a theorem will be of the form

G $\mathscr{L}$#℘|ABCD.

'Given A, B, C, D non-collinear distinct points, none of them ideal.'

The construction above for the line through a point B and parallel to a, as ⟨B, a ∩ u⟩, is in no sense a 'practical' solution to the corresponding problem for the F-G, since he has no way of naming the point a ∩ u.

The ideal line, like every other line, is determined by any two of its points, so that in the F-G's paddock two pairs of ⁑parallel lines (not all parallel), since they correspond to two pairs of lines whose meets are distinct ideal points, must be sufficient to enable him to determine lines parallel to any given line. The next theorems describe how he may do this, when he has fixed a basic set of four objects at the intersections of two pairs of parallel furrows (in ordinary Euclidean language, at the vertices of a parallelogram).

We assume that there is in the paddock a set of four objects corresponding to four points K, L, M, N which are such that KL ∥ MN, KM ∥ LN, and we wish to devise a 'practical' means of finding a second accessible point on the line through a point Z which is parallel to a given line XY.

We have to proceed by two stages, first naming a second point on the line through Z parallel to one of our base lines, and then adapting this construction to the more general problem. In each case a solution can be found which depends on a single application of the Desargues theorem.

1.5*T*1 Construction of a line parallel to one of the base lines.

*G*1 $\mathrm{KL} \parallel \mathrm{MN}$, $\mathrm{KM} \parallel \mathrm{LN}$, $\mathrm{KL} \neq \mathrm{MN}$, $\mathrm{KM} \neq \mathrm{LN}$
*G*2 Z, $\mathrm{Z} \notin \mathrm{KL}, \mathrm{LN}, \mathrm{NM}, \mathrm{MK}$
C $\mathrm{R} = \mathrm{KN} \cap \mathrm{LM}$
$\mathrm{T} = \mathrm{ZN} \cap \mathrm{KM}$
$\mathrm{S} = \mathrm{TR} \cap \mathrm{KZ}$
$\mathrm{P} = \mathrm{SL} \cap \mathrm{KM}$
T $\mathrm{ZP} \parallel \mathrm{KL}$
P *t* $\mathscr{P}|\mathrm{KL} \cap \mathrm{MN} \cap \mathrm{ZP}$ (i.e. $\mathrm{KL} \parallel \mathrm{MN} \parallel \mathrm{ZP}$)

p $\begin{Bmatrix}\mathrm{T}\\\mathrm{R}\\\mathrm{S}\end{Bmatrix}\begin{Bmatrix}\mathrm{KL}\\\mathrm{ZP}\\\mathrm{NM}\end{Bmatrix} \Rightarrow \mathscr{P}|\mathrm{KL} \cap \mathrm{MN} \cap \mathrm{PZ}.$

(Notice that we have used the only $A\mathscr{I}3^*$ since $\mathscr{L}|\mathrm{MPTK}$.)

1.5*X*1 Devise a construction for the case when $\mathrm{Z} \in \mathrm{KM}$.

1.5*T*2 Construction of parallel to a general line.

*G*1, 2 as in *T*1
*G*3 $\mathrm{XY}: \mathrm{XY} \not\ni \mathrm{L}$
C $\mathrm{P}: \mathrm{P} \in \mathrm{KM}$ & $\mathrm{ZP} \parallel \mathrm{KL}$ (as in *T*1)
$\mathrm{F}: \mathrm{F} = \mathrm{XY} \cap \mathrm{KL}$
$\mathrm{G}: \mathrm{G} = \mathrm{XY} \cap \mathrm{LN}$
$\mathrm{E}: \mathrm{E} = \mathrm{FZ} \cap \mathrm{LP}$
$\mathrm{Q}: \mathrm{Q} = \mathrm{KM} \cap \mathrm{EG}$
T $\mathrm{ZQ} \parallel \mathrm{XY}$
P *t*1 $\mathrm{E}\begin{Bmatrix}\mathrm{FZ}\\\mathrm{LP}\\\mathrm{GQ}\end{Bmatrix} \Rightarrow \mathscr{L}|\langle \mathrm{LG} \cap \mathrm{PQ}, \mathrm{FL} \cap \mathrm{ZP}, \mathrm{FG} \cap \mathrm{ZQ}\rangle$

*t*2 $\mathrm{ZQ} \cap \mathrm{FG} = \mathrm{ZQ} \cap \mathrm{XY}$ (*C*)
*t*3 $\mathrm{ZQ} \cap \mathrm{XY} \in \langle \mathrm{LG} \cap \mathrm{PQ}, \mathrm{FL} \cap \mathrm{ZP}\rangle$ (*t*1)
$= \langle \mathrm{LN} \cap \mathrm{KM}, \mathrm{KL} \cap \mathrm{ZP}\rangle$ (*C*)
= ideal line. (*C*, *T*1)
T $\mathrm{ZQ} \parallel \mathrm{XY}$.

1.5*X*2 Devise a construction in the cases (i) $\mathrm{Z} \in \mathrm{KM}$, (ii) $\mathrm{L} \in \mathrm{XY}$.

CHAPTER 1.6

SEPARATION, BETWEENNESS, ORDER

In this Chapter we resume the discussion begun in the Note on the Printing Convention.

The F-G considers again the observations that he made of the markers along a furrow. So far he has made use only of the property that he interpreted as collinearity, and noticed, but postponed for later investigation, the further property which could be described in these terms:

'As he walks northwards along the furrow he reaches marker ⁑A before he reaches marker ⁑B, ⁑B before ⁑C, and so on. Whereas, as he walks southwards along the furrow he reaches ⁑C before ⁑B...' There are two phenomena here for which he has no equivalents in the geometry as so far devised, one, that covered by the descriptions northwards and southwards, and the other, that of a relation between two objects which corresponds to the adverb 'before'. He sees that he can simplify the description if he frees it from the distinction between northwards and southwards, and regards the relation he is seeking as involving three points rather than two. Thus he recasts his description of his observations in the form:

(i) If an observer at ⁑A looking towards ⁑B sees that ⁑B obscures ⁑C, then looking from ⁑C towards ⁑B, he sees that ⁑B obscures ⁑A.

(ii) If from ⁑A the object ⁑B obscures ⁑C, then from ⁑B the object ⁑C neither obscures nor is obscured by ⁑A.

(iii) From any three ⁑collinear objects one can be selected such that, viewed from either of the others, this object obscures the third (with (ii) this means 'one and only one' can be selected).

(iv) If from ⁑A, the object ⁑B obscures ⁑C, and from ⁑B the object ⁑C obscures ⁑D, then (when ⁑B has been removed) from ⁑A the object ⁑C obscures ⁑D.

That is, he has a three term relation among his markers which he calls '⁑betweenness' and, when the markers ⁑A, ⁑B and ⁑C are related as in (i), (ii) and (iii), draws as

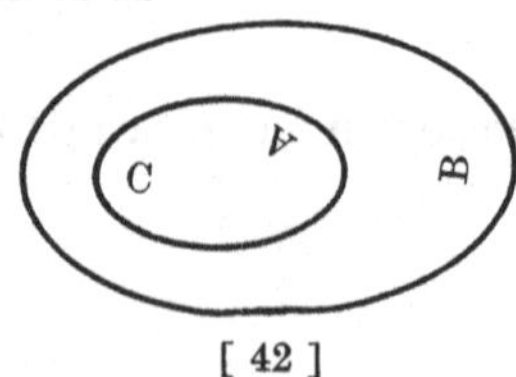

This relation we shall print as

$$\mathrm{B} \quad \mathscr{B} \quad \mathrm{AC}$$

and read as 'B is between A and C'. Its properties are

(i) $\mathscr{B}$ is a relation among three distinct proper collinear points.

(ii) $G \quad \mathscr{L}\#\wp|\mathrm{ABC}$, then A, B, C satisfy one and only one of the relations

$$\mathrm{A} \quad \mathscr{B} \quad \mathrm{BC},$$
$$\mathrm{B} \quad \mathscr{B} \quad \mathrm{AC},$$
$$\mathrm{C} \quad \mathscr{B} \quad \mathrm{AB}.$$

(iii) $G \quad \mathscr{L}\#\wp|\mathrm{ABCD}$, then

$$\mathrm{B}\,\mathscr{B}\,\mathrm{AC} \quad \& \quad \mathrm{C}\,\mathscr{B}\,\mathrm{BD} \Rightarrow \mathrm{B}\,\mathscr{B}\,\mathrm{AD} \quad \text{and} \quad \mathrm{C}\,\mathscr{B}\,\mathrm{AD}.$$

This set of statements for the property of 'betweenness' (which can be regarded as combinatorial statements about any set of elements, but are to be interpreted here as statements about points of a collinear set) could be adopted as axioms, but we need a rather stronger relation if we are to connect sets of points on different lines and if we are to apply the relation to ideal points. It is preferable therefore to introduce this stronger relation by axioms, then to specify its connection with $\mathscr{B}$ and deduce the required properties of $\mathscr{B}$.

The new relation to be introduced is 'separation', and to clarify the conception we think first of four objects ⁑A, ⁑B, ⁑C, ⁑D, at the corners of a ⁑square, and in proceeding round the ⁑square we pass through the objects in the ⁑order in which their names have been printed. The objects then split into two pairs of equivalent pairs ⁑A, ⁑C and ⁑B, ⁑D. Starting from either member of one pair and proceeding round the square in either ⁑direction we have to pass through exactly one member of the other pair before we arrive at the second member of the first pair.

For the next step we have to look more closely at the consequence of introducing ideal elements. There is a single ideal point in a collinear set but in the physical picture we have to think of this single point as lying ⁑beyond the accessible objects in either ⁑direction. This is in fact implicit in the assumption that we may designate any point of the collinear set as the ideal point. Replace then the four objects at the corners of the ⁑square by four objects of a ⁑collinear set, which, as we proceed along the ⁑line in one ⁑direction are encountered in the printed ⁑order above. As we go from ⁑A to ⁑C we pass through ⁑B, from ⁑B to ⁑D through ⁑C, from ⁑C onwards through ⁑D and 'through' the ideal point to ⁑A, and finally from ⁑D 'through' the ideal point and through ⁑A to ⁑B.

The essence of the physical relation among the four objects is this:

the set of marks, A, B, C, D, can be split in the following ways into complementary subsets whose members can be put into one-to-one correspondence:†

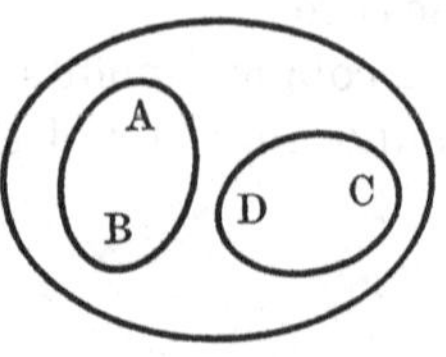

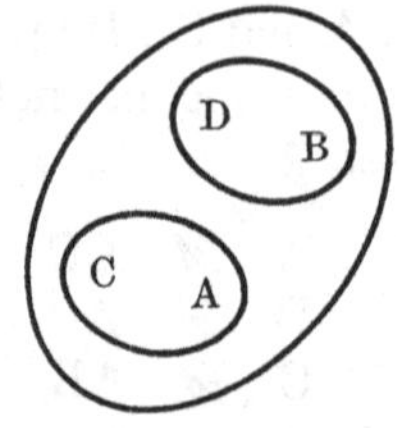

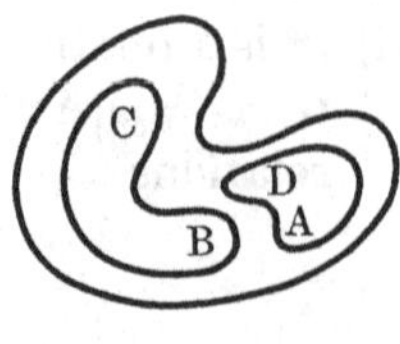

What we are going to assume is that, when we are 'given' a set of four collinear points, we are at the same time given an explicit division of the points into two subsets, that is, we are told which of the (three) possible splittings is to apply to them.

We shall print the specified splitting (with an appropriate choice of letters) in the form

1.6*N*1(i) AC $\mathscr{S}$ BD,

to be read as 'A and C separate B and D'. Thus given four distinct points A, B, C, D, there is among them one and only one of the three possible relations

AB $\mathscr{S}$ CD
AC $\mathscr{S}$ BD
AD $\mathscr{S}$ BC.

To find the connection among the separation relations for five points we think first of points A, B, C, D, E representing ⁑corners of a ⁑pentagon, from which we see that in the physical sense four of the relations are

(i) AC $\mathscr{S}$ BD.
(ii) AD $\mathscr{S}$ CE.
(iii) AC $\mathscr{S}$ BE.
(iv) AD $\mathscr{S}$ BE.

Write (i) and (ii) in the pattern

(i) A ¦ C $\mathscr{S}$ D ¦ B,

(ii) A ¦ D $\mathscr{S}$ C ¦ E,

then it is clear that C and D occur symmetrically. Thus any axiom

† In future we shall write 'The *four* marks can be split...into *two* pairs...' We have used the circumlocution above because we have not yet defined numbers, but the statement we now make, that the two sentences are equivalent, is tantamount to a definition of 'four' and 'two'. We shall use certain other numbers in the same informal way.

which makes (iii) a consequence of (i) and (ii) makes also (iv) a consequence.

Consider next the possible ⁑diagrams for the arrangements of A, B, C, D, E, satisfying (i) and (ii). Since neither ⁑mirror-images nor ⁑cyclic-permutations have to be distinguished, there is exactly one way in which we can represent (i), namely,

A
D B
C

From relation (ii) above we next have to place E and again there is a single possibility

A
E B
DC

Thus, in the formulation of axioms, (i) and (ii) must determine the separation relation among every set of four of the five points. Suppose we prescribe that (i) and (ii) imply (iii). We have already seen that they then imply also (iv). The only remaining set of four points is B, C, D,E, and for them we have

(iv)	D	A	$\mathscr{S}$	B	E
(i)	D	B	$\mathscr{S}$	A	C

and as a consequence, the required relation

(v) DB $\mathscr{S}$ CE.

The statement, (i) and (ii) $\Rightarrow$ (iii),

is therefore exactly sufficient to ensure that the mathematical model of the separation relation meets all the requirements of the physical picture. Formally then we take as the *Axioms of Separation*:

A$\mathscr{S}$**1** Among any four distinct collinear points A, B, C, D there is a separation relation $\mathscr{S}$ such that the points satisfy one and only one of the three relations

AB $\mathscr{S}$ CD,
AC $\mathscr{S}$ BD,
AD $\mathscr{S}$ BC.

If any two points of the four coincide, then there is no separation relation among the four.

A𝒮2 *G*1 AC 𝒮 DB,
*G*2 AD 𝒮 CE,
A AC 𝒮 BE.

It should be noted that *G*1, *G*2 imply that #|ABCD, #|ACDE, since the separation relations exist, and that #|BE, since otherwise *G*1 and *G*2 are inconsistent.

Since on the basis of these axioms we are going to establish the printing convention, it is important to be certain that we have not already assumed too much.

The relation AC 𝒮 BD can be assumed to be a conventional printed form of

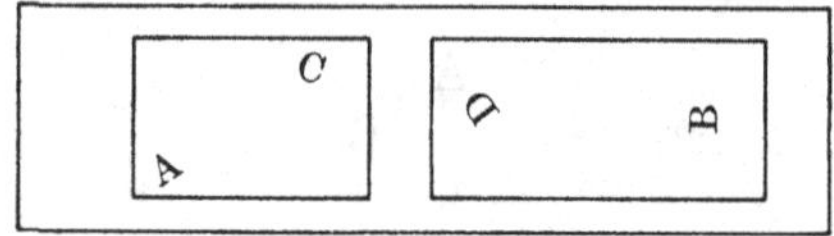

and to emphasize this, it might be reprinted as $\overline{\overline{\text{AC}}\ \overline{\text{BD}}}$ where the bars represent the boxes. We do not now have to specify that A, B, C, D are distinct since the marks are recognisably different.

In these terms, how can we frame *A*𝒮1? First we need a new symbol which conveys the meaning 'not' and, if we use the accepted symbol ~, then the statement can be drawn as

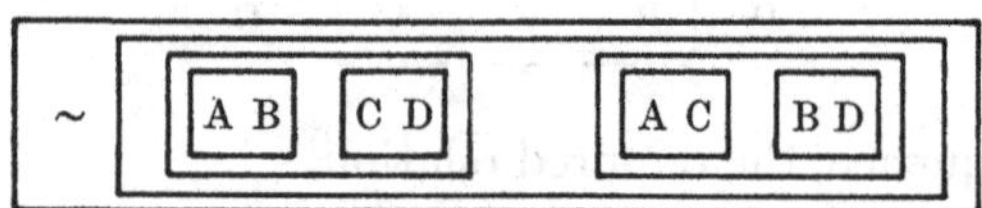

and printed as

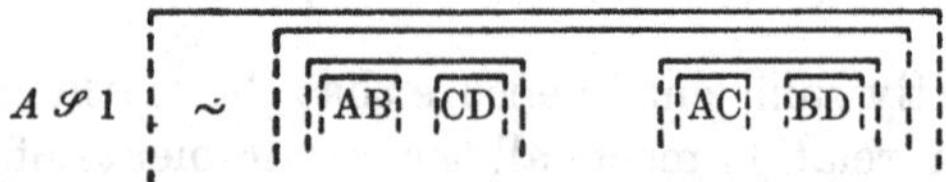

We need not make any further statement, since the effect of the statement above is to assert that the elements can be paired off in only one way to give a valid separation relation.

For *A*𝒮2 we need another new symbol to express that some relation is a consequence of others and again we may use the accepted symbol ⇒ and print *A*𝒮2 as

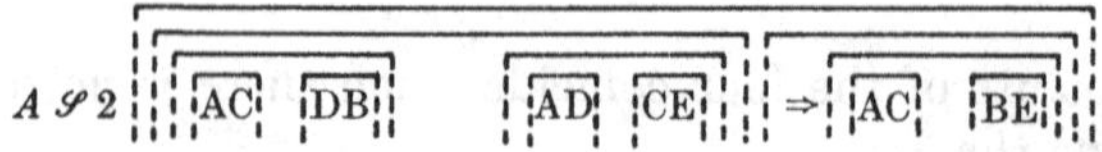

Now that we know that the marks can be drawn anywhere over the writing surface and still convey the proper meaning, provided the appropriate boxes are drawn (and the reader has sufficient cryptographic insight to deduce, from a reasonable mass of material, the significance of $\sim$ and $\Rightarrow$), we may resume the use of the much more convenient $\mathscr{S}$ notation and conventional printing.

As a consequence of $A\mathscr{S}1$ and $A\mathscr{S}2$ we can deduce:

1.6*T*1 The complete set of separation relations among five collinear points is determined by those prescribed in $A\mathscr{S}2$.

$G1$	AC	$\mathscr{S}$	DB,
$G2$	AD	$\mathscr{S}$	CE,
T (i)	AD	$\mathscr{S}$	BE,
T (ii)	BD	$\mathscr{S}$	CE.

These have been proved in the course of the earlier discussion; but we need to verify that the proofs can be expressed in terms of a set of marks made at random over a page and boxed in as subsets. Using for convenience rulings instead of the boxes and brackets instead of some of the rulings, we have for T (i) and T (ii) respectively:

$$\{(\overline{\overline{\overline{\mathrm{AC}}}\;\overline{\overline{\mathrm{DB}}}}\quad \overline{\overline{\overline{\mathrm{AD}}}\;\overline{\overline{\mathrm{CE}}}})\quad (\Rightarrow \overline{\overline{\overline{\mathrm{AC}}}\;\overline{\overline{\mathrm{BE}}}})\}$$

$$\text{(i)}\qquad [\Rightarrow \{(\overline{\overline{\overline{\mathrm{AD}}}\;\overline{\overline{\mathrm{CE}}}}\quad \overline{\overline{\overline{\mathrm{AC}}}\;\overline{\overline{\mathrm{DB}}}})\quad (\Rightarrow \overline{\overline{\overline{\mathrm{AD}}}\;\overline{\overline{\mathrm{BE}}}})\}],$$

$$\text{(ii)}\qquad [\Rightarrow \{(\overline{\overline{\overline{\mathrm{DA}}}\;\overline{\overline{\mathrm{BE}}}}\quad \overline{\overline{\overline{\mathrm{DB}}}\;\overline{\overline{\mathrm{AC}}}})\quad (\Rightarrow \overline{\overline{\overline{\mathrm{DB}}}\;\overline{\overline{\mathrm{CE}}}})\}].$$

For further convenience we introduce a notation which indicates that either of two separation relations may hold among a set of four points (that is, that a particular one of the relations does not hold):

1.6*N*1(ii) AC $\not\mathscr{S}$ BD $\Leftrightarrow$ AB $\mathscr{S}$ CD or AD $\mathscr{S}$ BC.

Our next task is to establish a definition of 'order' in terms which involve no more than recognisable marks and recognisable subsets. We assume that we are given a set of objects, each of which is represented by a distinguishable mark on the writing-surface, that the separation relation among every set of four is represented on the writing-surface by the appropriate pattern such as $\boxed{\boxed{\mathrm{WY}}\qquad\boxed{\mathrm{XZ}}}$ and that the separation relations among every set of five are consistent with $A\mathscr{S}2$. We suppose also that we can identify at once the separation relation among any required four, even though in practice the operation would require our looking at the sets of four one by one until we found the one we were seeking.

Select one of the objects arbitrarily and rename it 'A' (that is, in future for this object we make the mark 'A' to represent it on the writing surface). Let P and H be (the marks corresponding to) any other two objects. Our aim is to show that there is one and only one object, which we shall rename 'B', that satisfies both conditions (i) AP $\mathscr{S}$ BH (unless P itself is the object) and (ii) there is no object Z such that HZ $\mathscr{S}$ AB, a condition which we shall express at present as

$$\{Z : HZ \quad \mathscr{S} \quad AB\} = \emptyset.$$

A, P, H split the set of marks into two mutually exclusive subsets

$$\{X\} = \{X : AP \mathscr{S} \ HX\} \quad \text{and} \quad \{Y\} = \{Y : AP \ \not\mathscr{S} \ HY\}.$$

There are two cases to consider:

(i) $\{X\} = \emptyset$. In this case rename P as B.

(ii) $\{X\} \neq \emptyset$. Take $Q \in \{X\}$, and form the sets

$$\{X'\} = \{X' : AQ \mathscr{S} HX'\} \quad \text{and} \quad \{Y'\} = \{Y' : AQ \not\mathscr{S} HY'\}.$$

Then, as a consequence of T (ii), $\{X'\}$ is a proper subset of $\{X\}$, since

$$[AP \quad \mathscr{S} \quad HQ \ \& \ AQ \quad \mathscr{S} \quad HX'] \Rightarrow AP \quad \mathscr{S} \quad HX',$$

that is, $X' \in \{X\}$ while $Q \in \{X\}$ and $Q \notin \{X'\}$.

We now select $R \in \{X'\}$ and continue repeating the operation of splitting the set as above: since we are given a set of *objects* (i.e. the set is finite), after a number of repetitions we shall obtain an object T such that $\{X^* : AT \mathscr{S} HX^*\} = \emptyset$. We then rename T as B.

With A and B fixed we start again: select (new) objects P, $\# | PAB$, and $H \in \{Z : BZ \mathscr{S} AP\}$, and repeat the process. The result is to isolate an object, which we rename C, for which

$$\{X : BC \quad \mathscr{S} \quad PX\} = \emptyset,$$

while as a consequence of the relations assumed, every object Y of the set, for which $\# | YABC$, has the separation property BY $\mathscr{S}$ AC.

We may proceed in this way until the objects have been renamed A, B, C, D ... L, M, N with the relations

$$\{X : AX \quad \mathscr{S} \quad BC\} = \emptyset,$$
$$\{X : BX \quad \mathscr{S} \quad CD\} = \emptyset,$$
$$\text{-----------------}$$
$$\{X : LX \quad \mathscr{S} \quad MN\} = \emptyset,$$
$$\{X : MX \quad \mathscr{S} \quad NA\} = \emptyset,$$
$$\{X : NX \quad \mathscr{S} \quad AB\} = \emptyset.$$

We now look back over the operation to see how it could have been represented only by marking off subsets. Having selected A, the first round of operations isolated, according to our choice of P and H, one of the two objects we ultimately renamed B and N (but whichever it is we rename it B), which have the property that for any K, $\#|ABK$, $\{X : AB\,\mathscr{S}\,KX\} = \emptyset$. We could suppose that this property of A and B was expressed by making the marks '$\overline{\overline{AB}}\ \overline{\overline{K}}$.'. That is: 'there is no point X such that AB $\mathscr{S}$ KX.' In this notation, but using brackets instead of bars, the succeeding operations have the effect of locating the objects that can be named C, D ... with the properties [(BC) (A.)], [(CD) (B.)] ... [(LM) (K.)], [(MN) (L.)], [(NA) (M.)], [(AB) (N.)].

The situation is in fact this: (i) among the given total set of consistent separation relations, the separation relation among any four objects can be found, (ii) there are pairs of objects such as AB for which every separation relation in which they appear is of the form AX $\mathscr{S}$ BY; that is, A and B never appear together as a pair in a box, (iii) each object belongs to exactly two such pairs.

The operations that we performed above, after we had arbitrarily selected an object and named it A, were directed first towards finding one of the two objects paired with A. This object we named B. Next we found the second object paired with B and named it C, and so on. Because the separation relations were prescribed for all sets of four points and were consistent, the operation terminated when we named as N the last of the objects, this being also the object other than B that was paired with A.

The complete operation has had of necessity to be explained as a succession of steps, but we could imagine that it had been performed by some divine coup d'oeil; that the F-G, having drawn the table showing all the separation relations, picked out all the missing pairs at a glance. We could moreover replace the whole table of separation relations by the simple table of 'unseparated' pairs. For six objects, for example, the complete set of separation relations would be determined by a diagram such as

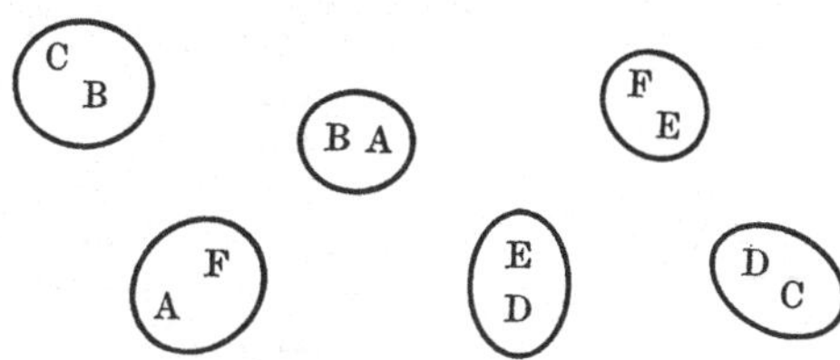

The significance of this table is that for every pair X, Y such that $\#|ABXY$, we have AB $\mathscr{S}$ XY, with corresponding relations for each of the six pairs.

There is only one more obstacle to intelligible printing by the scattering of pairs of symbols over the page, namely, that presented by the repetition of letters. We may overcome this by adding distinguishing marks to a symbol at its successive occurrences and including on the page an auxiliary table of distinguishing marks. Thus in an ordinary text we may have to have E^*, E^0, E^Δ, ... (each regarded as a single symbol, with the significance of the undifferentiated letter E).

A detailed scheme could be worked out that would make reading possible in a text consisting of sporadic pairs, but the principles are sufficiently illustrated by printing in this manner 'PA, PB, PC', on the assumption that this constitutes the whole of the text to be carved on one rock-face. Set up linearly, the text would appear as:

α Symbol to help pick out 'first-and-last'.
$*$ Distinguishing marks.
0
Δ
β End of distinguishing marks.
P^* Text.
A
$,^*$
P^0
B
$,^0$
P^Δ
C
α Symbol to help pick out 'first-and-last'.

The actual arrangement as sporadic pairs could be some such as:

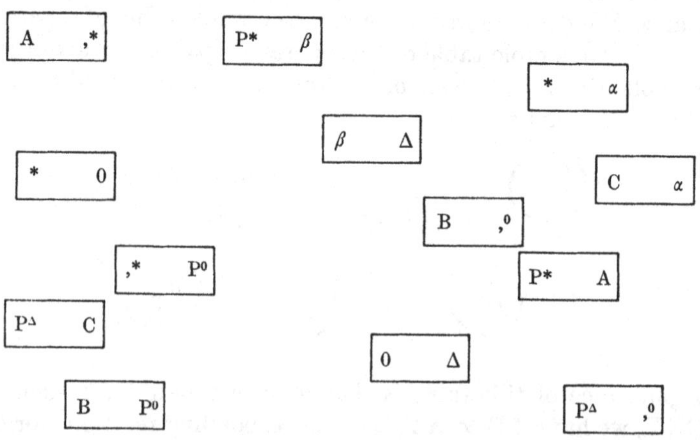

The set of thirteen symbols is now arranged in 'separation order', but the symbol α imposes 'betweenness order' on the rest. The 'sense'

(left-to-right) in which the text is to be read, that is, as 'PA, ...' and not as 'CP, ...', is determined by the placing of the distinguishing marks.

We may now employ, without qualms, the sequence symbol $\{^{\mathscr{O}}\,\mathrm{A}, \mathrm{B}, \mathrm{C}, \ldots\}$: 'the members of the set $\{\mathrm{A}, \mathrm{B}, \mathrm{C}\}$ are to be treated as a sequence in printing order', and with some understanding of what lies behind the idea of 'printing order' in the evolution of a geometry.

From this point onwards we use the conventions: (i) a set of symbols printed as ABCD... represents a set of marks scattered over the writing surface; (ii) if certain symbols to be prescribed later, as for example $\mathscr{S}|$, are adjoined to the set then in the composite symbol $\mathscr{S}|$ABCD... the letters can be read in the way which is ordinarily understood to be 'in order'.

We return now to the formal discussion of the separation relation and its consequences. There is one other substantial theorem:

1.6*T*2 $G1$ AB $\mathscr{S}$ PQ
$G2$ AB $\mathscr{S}$ PR
$G3$ $\#|$QR
T (i) AP $\not\mathscr{S}$ QR
T (ii) AB $\not\mathscr{S}$ QR.

P We consider the consequences of assuming in turn the three possible separation relations among A, P, Q, R.

$g1$ PA $\mathscr{S}$ QR
$t1$ PA $\mathscr{S}$ BR
p $G1$, $g1$.

T (i) P. $t1$ contradicts $G2$ and therefore the assumption $g1$ is false, i.e. PA $\not\mathscr{S}$ QR.

$g2$ QA $\mathscr{S}$ PR
$t2$ QA $\mathscr{S}$ RB
p $G1, g2$
$g3$ RA $\mathscr{S}$ PQ
$t3$ RA $\mathscr{S}$ BQ
p $G2, g3$.

Since $g1$ is false, one of the statements $g2$ and $g3$ must be true, and thus one of the statements $t2$ and $t3$ must be true, and if either of them is true, then AB $\not\mathscr{S}$ QR.

This is the important theorem, but corresponding theorems with different arrangements of the $\mathscr{S}$ relations provide useful exercises.

1.6*T*3(*X*) *G*1 AB $\mathscr{S}$ PQ
*G*2 AB $\mathscr{S}$ PR
*G*3 #|QR
T AB $\mathscr{S}$ QR.

1.6*T*4(*X*) *G*1 AB $\mathscr{S}$ PQ
*G*2 AB $\mathscr{S}$ PR
T AB $\mathscr{S}$ QR.

1.6*T*5(*X*) *G*1 AB $\mathscr{S}$ PQ
*G*2 $U \in AB, U \notin \{A, B, P, Q\}$
T either AU $\mathscr{S}$ PQ and BU $\mathscr{S}$ PQ
or BU $\mathscr{S}$ PQ and AU $\mathscr{S}$ PQ.

We now specify, when a particular point has been assigned as ideal, the connections between $\mathscr{S}$ and $\mathscr{B}$, thus

1.6*D*1 Definition of $\mathscr{B}$ in terms of $\mathscr{S}$ and an assigned ideal point.

#|ABCU & ideal|U

D UB $\mathscr{S}$ AC $\Leftrightarrow$ B $\mathscr{B}$ AC; 'B is between A and C'.

Betweenness is a property related to the selection of ideal point. Where it is desirable to specify the ideal point, U, we shall write B $\mathscr{B}_U$ AC.

From $A\mathscr{S}$ 2 and its consequences, *T*1 and *T*2, we may derive a variety of statements in terms of $\mathscr{B}$ by selecting different points to be ideal. For convenience we restate the $\mathscr{S}$ relations:

$$T1 \quad \left.\begin{array}{r} AX\ \mathscr{S}\ YP \\ \&\ AY\ \mathscr{S}\ XQ \end{array}\right\} \Rightarrow \left\{\begin{array}{l} \quad\ AX\ \mathscr{S}\ PQ \\ \text{and } AY\ \mathscr{S}\ PQ \\ \text{and } PY\ \mathscr{S}\ XQ. \end{array}\right.$$

$$T2 \quad \left.\begin{array}{r} AX\ \mathscr{S}\ PY \\ \&\ AX\ \mathscr{S}\ PZ \\ \&\ \#|YZ \end{array}\right\} \Rightarrow \left\{\begin{array}{l} \quad\ AP\ \mathscr{S}\ YZ \\ \text{and } AX\ \mathscr{S}\ YZ \\ \text{and } XP\ \mathscr{S}\ YZ. \end{array}\right.$$

1.6*T*6 A ideal in *T*1

$$\left.\begin{array}{r} X\ \mathscr{B}\ YP \\ \&\ Y\ \mathscr{B}\ XQ \end{array}\right\} \Rightarrow \left\{\begin{array}{l} \quad\ X\ \mathscr{B}\ PQ \\ \text{and } Y\ \mathscr{B}\ PQ. \end{array}\right.$$

1.6*T*7 X ideal in *T*1

$$\left.\begin{array}{r} A\ \mathscr{B}\ PY \\ \&\ Q\ \mathscr{B}\ AY \end{array}\right\} \Rightarrow \left\{\begin{array}{l} \quad\ A\ \mathscr{B}\ PQ \\ \text{and } Q\ \mathscr{B}\ PY. \end{array}\right.$$

The same patterns are obtained when Y is taken as ideal. To apply $T2$ we need to introduce $\not\mathscr{B}$ with the meaning

1.6*N*2 X $\not\mathscr{B}$ YZ: either Y $\mathscr{B}$ XZ or Z $\mathscr{B}$ XY.

1.6*T*8 A ideal in $T2$

$$\left.\begin{array}{l} \,\mathrm{X}\ \mathscr{B}\ \mathrm{PY} \\ \&\,\mathrm{X}\ \mathscr{B}\ \mathrm{PZ} \\ \&\,\#|\mathrm{YZ} \end{array}\right\} \Rightarrow \left\{\begin{array}{l} \mathrm{P}\ \not\mathscr{B}\ \mathrm{YZ} \\ \mathrm{X}\ \not\mathscr{B}\ \mathrm{YZ}. \end{array}\right.$$

1.6*T*9 P ideal in $T2$

$$\left.\begin{array}{l} \,\mathrm{Y}\ \mathscr{B}\ \mathrm{AX} \\ \&\,\mathrm{Z}\ \mathscr{B}\ \mathrm{AX} \\ \&\,\#|\mathrm{YZ} \end{array}\right\} \Rightarrow \mathrm{A}\ \not\mathscr{B}\ \mathrm{YZ}.$$

Now in $T7$ replace

Y Q A P

by K L M N;

we obtain

1.6*T*10

$$\left.\begin{array}{l} \mathrm{L}\ \mathscr{B}\ \mathrm{KM} \\ \mathrm{M}\ \mathscr{B}\ \mathrm{KN} \end{array}\right\} \Rightarrow \mathrm{L}\ \mathscr{B}\ \mathrm{KN}.$$

Order. We have seen that a consistent set of separation relations among the points of a set leads to the identification of a set of pairs of points, which can be put into one-to-one correspondence with the set of points (meaning there are 'as many' pairs as there are points), with the properties: (i) every separation relation in which one of these pairs, say A, B, appears, is of the form AX $\mathscr{S}$ BY; (ii) each point belongs to exactly two of the pairs; (iii) now that the printing convention has been established, the pairs can be made visible as the *adjacent pairs* in an arrangement such as ABCD ... LMN together with the pair AN.

We shall say that in these circumstances the set is *ordered by separation* and shall use the notation

1.6*N*3 $\mathscr{S}$|ABCD ... LMN; {A, B, C, D, ..., L, M, N} is ordered by separation or, if the condition of collinearity is also to be expressed,

$\mathscr{L}\mathscr{S}$|ABCD ... LMN.

This relation among the points is identical with that expressed by

$\mathscr{S}$|BCD ... LMNA,

$\mathscr{S}$|CD ... LMNAB etc.

and $\mathscr{S}$|NML ... DCBA etc.

Because of this last set of possibilities 'ordering by separation' differs from '‡cyclic ordering'; in the set of possible printed arrangements all the cyclic orders are included and all the 'reversed' cyclic orders. In particular, of course, a set of only three elements cannot be ordered by separation. For four elements

$$\mathscr{S}|\mathrm{ABCD} \Leftrightarrow \mathrm{AC} \;\; \mathscr{S} \;\; \mathrm{BD}.$$

For five elements we have, from 1.6*T* 1,

$$\mathrm{AC} \;\; \mathscr{S} \;\; \mathrm{BD} \;\&\; \mathrm{AD} \;\; \mathscr{S} \;\; \mathrm{CE} \Leftrightarrow \mathscr{S}|\mathrm{ABCDE}.$$

We can reduce ordering by separation to a relation which more closely resembles what is ordinarily understood by (physical) ‡ordering by adjoining a new element U (in the case of collinear points, by adjoining the ideal point) to the set and replacing 'separation' by 'betweenness in relation to U'.

Thus if

$$\mathscr{S}|\mathrm{ABCD} \ldots \mathrm{LMNU}$$

we have from 1.6*T* 10

$$\left.\begin{matrix} \mathrm{B} & \mathscr{B}_{\mathrm{U}} & \mathrm{AC} \\ \mathrm{C} & \mathscr{B}_{\mathrm{U}} & \mathrm{AD} \end{matrix}\right\} \Rightarrow \mathrm{B} \;\; \mathscr{B}_{\mathrm{U}} \;\; \mathrm{AD} \quad \text{etc.}$$

$$\text{- - - - - - - - - - - -}$$

$$\mathrm{M} \;\; \mathscr{B}_{\mathrm{U}} \;\; \mathrm{LN}.$$

We shall say the set is *ordered by betweenness* and write

$$\mathscr{B}_{\mathrm{U}}|\mathrm{ABCD} \ldots \mathrm{LMN}$$

where we now read the printed order as conveying actual betweenness relations, so that if X, Y, Z occur in the symbol in the following way

$$\mathscr{B}_{\mathrm{U}}|\mathrm{A} \ldots \mathrm{X} \ldots \mathrm{Y} \ldots \mathrm{Z} \ldots \mathrm{N}$$

then

$$\mathrm{Y} \;\; \mathscr{B}_{\mathrm{U}} \;\; \mathrm{XZ}.$$

Again, no distinction can be made between

$$\mathscr{B}|\mathrm{ABC} \ldots \mathrm{LMN}$$

and

$$\mathscr{B}|\mathrm{NML} \ldots \mathrm{CBA}.$$

1.6*N*4 $\mathscr{B}|\mathrm{ABC} \ldots \mathrm{LMN}$: the betweenness relations among the elements are those specified in the 'natural order' of reading the symbol.

If collinearity is also to be expressed we write

$$\mathscr{L}\mathscr{B}|\mathrm{ABC} \ldots \mathrm{LMN}.$$

If the ideal point U is to be specified we may write $\mathscr{L}\mathscr{B}_{\mathrm{U}}|\mathrm{ABC} \ldots$. We have $\mathscr{L}\mathscr{B}_{\mathrm{U}}|\mathrm{ABC} \ldots \Leftrightarrow \mathscr{L}\mathscr{S}|\mathrm{UABC} \ldots$

If we return to the primitive picture of the relation of separation, namely, a set of pairs scattered over the writing surface, then ordering by betweenness is obtained by adjoining a new element U to the set {A, ..., N} and pairing it with, say, A and N, so that there are the two new pairs UA, UN, while the pair AN is erased. This device merely serves to pick out A and N from the other elements, a result we could equally well achieve by simply omitting the pair AN.

That is: $\mathscr{S}|$ABC ... LMN describes the set of visibly ‡adjacent pairs together with AN, $\mathscr{B}|$ABC ... LMN describes the set of visibly ‡adjacent pairs. We could designate AN the 'terminal pair' but shall use the more evocative term 'first-and-last'.

1.6*N*5 (i) In relation to $\mathscr{S}|$ABC ... MN, the pairs {A, B}, {B, C}, ... {M, N}, {A, N} will be referred to as *adjacent pairs*.

(ii) In relation to $\mathscr{B}|$ABC ... MN, the pairs {A, B}, {B, C}, ..., {M, N} will be referred to as *adjacent pairs*, and {A, N} will be referred to as the *first-and-last pair*.

At this stage it is convenient to introduce conventional names for the elements of an ordered set, thus:

1.6*N*6 When the elements of a set are denoted by any of the following names or symbols it is to be understood that they satisfy the betweenness relations specified:

$\mathscr{B}|$first	second	third ...	r-th ...	n-th
$\mathscr{B}|$1	2	3 ...	r ...	n
$\mathscr{B}|P_1$	P_2	P_3 ...	P_r ...	P_n

but it must be remembered that, for example,

$$\mathscr{B}|1\ \ 2\ \ 3 \ldots n \quad \text{and} \quad \mathscr{B}|n \ldots 3\ \ 2\ \ 1$$

are statements of the same set of relations.

We can adapt this notation to express the 'construction' of an ordering for a set originally not named in order. Suppose the set is $\{\alpha_i\}$ and the data about the set include all the necessary statements of their betweenness relations. We rename the elements $\{\beta_j\}$ where $\mathscr{B}|\beta_1\beta_2 \ldots \beta_n$, which we can abbreviate as $\mathscr{B}|\{\beta_j\}$. We shall express this construction in the following form.

1.6.*C*1 Construction of an ordering of a set $\{\alpha_i\}$:

$$\{\beta_j\} = \{\alpha_i\} \ \& \ \mathscr{B}|\{\beta_j\},\ (i, j) \in \{1, \ldots, n\}.$$

On the basis of *N*6 we may introduce, at least tentatively, the idea of 'number', in effect by 'counting' the elements in the set. Thus any (finite) set of elements can be arranged in order by selecting a member of the set and associating with it the symbol '1', selecting another and associating with it '2', and so on until the set is exhausted. If the last symbol in the sequence is n, then the *number* of elements in the set is n.

By combining sets with different numbers of elements and sets of replicas of a set we can arrive at definitions of addition, subtraction (of a 'smaller number' from a 'larger') and multiplication. In particular we shall apply these operations to the indices of sets of symbols like $P_1, P_2, \ldots, P_n$ and shall use $P_{n-r}, P_{n-1}, P_n, P_{n+1}, P_{n+s}, P_{kn}$, etc. with their customary meanings.

On this understanding we may use relations among members of the set $\{r\}$ of subscripts of the elements P_r to express betweenness and separation relations. Thus:

1.6*T*11 *G* $\mathscr{B}|P_1P_2\ldots P_n$

T $P_j \;\mathscr{B}\; P_iP_k \Leftrightarrow (i-j)(j-k) > 0.$

1.6*T*12 *G* $\mathscr{S}|P_1P_2\ldots P_n$

T $P_hP_j \;\mathscr{S}\; P_iP_k \Leftrightarrow (h-i)(h-k)(j-i)(j-k) < 0.$

Because this usage is purely notational and not directly concerned with the logical development, it is not important that these operations with number have not been introduced with full formality; we shall introduce them properly in the course of the discussion on 'Displacement' (Chapter 2.1). But we shall shortly need to distinguish formally between sets containing 'even' and 'odd' numbers of elements. These we can define in the following terms:

Even-set: an even-set S of elements can be split into two subsets L and L′ with the properties:

(i) $L \cap L' = \emptyset$;
(ii) $L \cup L' = S$;
(iii) a one-to-one correspondence exists between the elements of L and the elements of L′.

The empty set $\emptyset$ will be treated as an even-set.

Odd-set: an odd-set T can be split into two subsets M and M′ with the properties

(i) $M \cap M' = \emptyset$;
(ii) $M \cup M' = T$;
(iii) M contains exactly one element, and M′ is an even-set.

Up to this stage the discussion of betweenness and separation has been applicable to any sets; we may suppose that the set is built up element by element, the separation relation among the first four being given, and as each new element is named its separation relations with the points already named are prescribed. It would seem then that separation relations can be arbitrarily assigned within the restrictions $A\mathcal{S}2$.

When however the relations are applied explicitly to collinear sets we find that assigned separation relations in one collinear set effectively determine the separation relations in all other collinear sets. We have to suppose that the F-G experiments with various possibilities and discovers that in fact the visible separation relations for all collinear sets marked ABCD on the diagram below are the same.

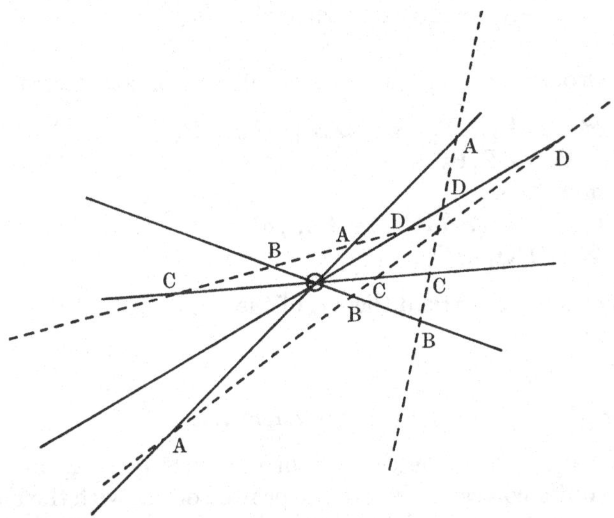

This property is to be the basis of the next axiom, but we can express it more compactly if we introduce first the notation:

1.6*N*7 Sets of points in perspective.

$$\text{ABC}\ldots \overline{\overline{\wedge}}^{\text{V}}\, \text{A}'\text{B}'\text{C}'\ldots,$$

to be read† as 'ABC... are *in perspective* from V with A′B′C′...':

$$\text{AA}' \cap \text{BB}' \cap \text{CC}' \cap \ldots = \text{V}.$$

V is the *centre of perspective*, and the symbol 'V' may be omitted (i.e. $\overline{\overline{\wedge}}$ printed instead of $\overline{\overline{\wedge}}^{\text{V}}$) when the name of the centre is irrelevant. The relation of 'perspective' applies to any sets of points, and not

† We have to read the two sets ABC..., A′B′C′... in 'printed order': this is a convention which is not part of the geometry and is used only for convenience of notation.

only collinear sets, although in the present context we shall be concerned mostly with collinear sets. The axiom is

$A\mathscr{S}$ **3** Preservation of separation under perspective

$G1$ $\mathscr{L}\#|ABCD$ & $\mathscr{L}\#|A'B'C'D'$ & $\#|AB, A'B'$
$G2$ $V: V \notin AB$ & $V \notin A'B'$,
$ABCD \overset{V}{\overline{\wedge}} A'B'C'D'$
$G3$ $AC \; \mathscr{S} \; BD$
A $A'C' \; \mathscr{S} \; B'D'$.

An immediate consequence of this axiom is

1.6*T*13 Ordering by separation of a finite set of concurrent lines.

$G1$ $\mathscr{L}\mathscr{B}|P_1P_2\dots P_n$ or $\mathscr{L}\mathscr{S}|P_1P_2\dots P_n$
$G2$ $\wp|V: V \notin P_1P_2$
$G3$ $m: m \not\ni V$
C $Q_i: Q_i = m \cap VP_i$, $i \in \{1, \dots, n\}$
T $\mathscr{L}\mathscr{S}|Q_1Q_2\dots Q_n$.

We may therefore extend the use of the notation $\mathscr{S}|$ to

1.6*N*8 $\mathscr{S}|\{VP_i\}$, $i \in \{1, \dots, n\}$

or $\mathscr{S}|\{l_i\}$, $i \in \{1, \dots, n\}$ & $\mathscr{P}|l_1 \cap l_2 \cap \dots \cap l_n$.

The lines $VP_1, \dots, VP_n$, or the concurrent lines $l_1, \dots, l_n$, are in the separation order corresponding to the printed order, with the meaning that any line not through V meets the lines VP_i in a set of points in the prescribed separation order.

When we apply $A\mathscr{S}$ 3 in cases when V is ideal we obtain the following results:

1.6*T*14(*X*) G $\mathscr{L}\mathscr{B}|ABC$
$\mathscr{L}|A'B'C'$
$G2$ (i) $AA' \| BB' \| CC'$ or
$G2$ (ii) $A = A'$, $BB' \| CC'$
T $\mathscr{L}\mathscr{B}|A'B'C'$.

While $A\mathscr{S}$ 1, 2 have no restricting effect on the class of geometry (since they merely describe properties of a four-term relation), $A\mathscr{S}$ 3 imposes quite severe limitations. For example, given any set of collinear points, we may name them A, B, ..., M, N in any way, and then prescribe the consistent set of separation relations

$$\mathscr{S}|AB\dots MN.$$

There is nothing in the separation axioms that precludes more than one scheme of consistent separation relations among the points of a given collinear set.

With the assumption of $A\mathscr{S}$ 3 the position is quite different. The class of geometries in which this axiom is valid is much narrower than the class of projective geometries to which the axioms have so far been applied; in particular, as we shall see, after making one further apparently minor assumption, all geometries with only finite numbers of points are excluded. As a first example of this, consider the geometry $\Gamma^{(3)}$. In it we have

$$\begin{aligned} P_1P_0P_4P_6 \overline{\overline{\wedge}}^{P_u} P_1P_2P_5P_7 &\overline{\overline{\wedge}}^{P_9} P_1P_0P_6P_4 \\ &\overline{\overline{\wedge}}^{P_w} P_1P_4P_0P_6 \\ &\overline{\overline{\wedge}}^{P_8} P_1P_6P_4P_0. \end{aligned}$$

I.e. if applied to $\Gamma^{(3)}$, $A\mathscr{S}$ 3 would imply both $P_1P_0 \;\mathscr{S}\; P_4P_6$ and $P_1P_4 \;\mathscr{S}\; P_0P_6$, which are incompatible relations.

Before proceeding to a more general discussion of geometries with finite numbers of points, we establish a connection between the harmonic relation $\mathscr{H}$ (1.4*C* 1) and the separation relation $\mathscr{S}$.

1.6*T*15 If A, B are harmonically conjugate with regard to C, D, then A, B separate C, D.

G 1	$\mathscr{L}\#\|$ABCD
G 2	AB $\mathscr{H}$ CD
T	AB $\mathscr{S}$ CD
C 1	as in 1.4*C* 1
C 2	H: H = KN ∩ LM
P	ABCD $\overline{\overline{\wedge}}^{N}$ MLCH $\overline{\overline{\wedge}}^{K}$ BACD.

Either	(i)	AB $\mathscr{S}$	CD	and	BA	$\mathscr{S}$	CD,
or	(ii)	AC $\mathscr{S}$	BD	and	BC	$\mathscr{S}$	AD,
or	(iii)	AD $\mathscr{S}$	BC	and	BD	$\mathscr{S}$	AC.

Only statements (i) are consistent.

1.6*X*1

G 1	CC′ $\mathscr{H}$ AA′
G 2	CC′ $\mathscr{H}$ BB′ & $\#\|$AA′BB′CC′
T	AA′ $\mathscr{S}$ BB′.

In order to exclude the finite geometries we need the further axiom

***A*$\mathscr{S}$ 4** Distinctness of harmonic conjugates

G 1	$\mathscr{L}\#\|$ABC
C	D : AB $\mathscr{H}$ CD
A	$\#\|$CD.

This excludes immediately $\Gamma^{(2)}$; in the notation of 1.4*C*1 take the points of the geometry

	0	1	2	3	4	5	6
to be	A	B	K	C	M	N	L.

Then D: CD $\mathscr{H}$ AB ⇒ C = D(= 3).

As a prelude to the main theorem we use the harmonic relation and $A\mathscr{S}$ 4 to prove

1.6*T*16 Between any two points there is at least one other point.

G $\mathscr{L}$#|ABU & ideal|U
C M:MU $\mathscr{H}$ AB
T M $\mathscr{B}$ AB (*T* 15 and $A\mathscr{S}$ 4).

1.6*T*17 Beyond any two points there is at least one other point.

G $\mathscr{L}$#|ABU & ideal|U
C R:AR $\mathscr{H}$ BU
T B $\mathscr{B}$ AR.

The significance of this theorem is that, given any finite set of collinear proper points, which we may suppose ordered as

$\mathscr{L}\mathscr{B}$|AB ... MN,

we can construct points R, S such that†

$\mathscr{L}\mathscr{B}$|RAB ... MNS.

As a consequence of these two theorems the F-G is forced into the recognition that while his paddock contains only a finite number of namable objects, the mathematical model that he has devised enables him to name unlimited sequences of distinct points.

1.6*T*18 Under $A\mathscr{S}$ 1–4 a geometry cannot consist of only a finite number of points on a line, i.e. given A, B we can construct iteratively unlimited sequences of points, $M_1, M_2, \ldots, M_r, \ldots$, both between A and B (*T* 16), and beyond A and B (*T* 17), in which, if #|rs then #|M_rM_s.

G $\mathscr{L}$#|ABU & ideal|U
C M_1:M_1U $\mathscr{H}$ AB
M_2:M_2U $\mathscr{H}$ M_1B
. . . .

† That is, we adjoin {R, A} and {S, N} to the set of pairs, and the missing pair, i.e., the first-and-last is now {R, S}.

$M_r : M_r U \quad \mathscr{H} \quad M_{r-1}B$

....

T (i) $\mathscr{L}\mathscr{B}|AM_1M_2 \dots M_r \dots B$

(ii) $\# | rs \Rightarrow \# | M_r M_s$

P *t*1 $M_1 \ \mathscr{B} \ AB, \ M_2 \ \mathscr{B} \ M_1B, \ M_3 \ \mathscr{B} \ M_2B, \dots$ (*T* 16)

*t*2 $\mathscr{L}\mathscr{B}|AM_1M_2 \dots B$ (*T* 10).

1.6*T*19(*X*) Iterative construction of an unlimited sequence of points beyond B.

*G*1 $\mathscr{L}\# | ABU$ & ideal|U

C $R_1 : AR_1 \quad \mathscr{H} \quad UB$

$R_2 : BR_2 \quad \mathscr{H} \quad UR_1$

$R_3 : R_1R_3 \quad \mathscr{H} \quad UR_2$

....

$R_{r+1} : R_{r-1}R_{r+1} \quad \mathscr{H} \quad UR_r$

T (i) $\mathscr{L}\mathscr{B}|ABR_1R_2 \dots R_\gamma R_{\gamma+1} \dots$

(ii) $\# | rs \Rightarrow \# | R_r R_s$.

For most of the finite geometries that we are to discuss the condition $A\mathscr{S}$ 4 is nugatory: the geometry is specified in such a way that there is in it no set of points $\mathscr{L}\# | ABC$ such that $\mathscr{H}_{AB}C = C$. Consequently, if the geometry satisfied conditions $A\mathscr{S}$ 1, 2, 3 it would have the property proved in *T* 19, namely, that a collinear set of points $R_1, R_2, \dots, R_r, \dots, R_s, \dots$ can be constructed in which, for all pairs {r, s},

$$\# | rs \Rightarrow \# | R_r R_s.$$

That is, such a finite geometry cannot satisfy the essential geometric separation condition $A\mathscr{S}$ 3.

CHAPTER 1.7

SEGMENTS AND REGIONS

To the F-G the physical concept is initially quite simple: a region is some part of his paddock enclosed by a fence. But the prescription of a mathematical model of a physical ⁑region turns out to be rather complicated, even when, as we shall assume here, the fence or ⁑boundary of the ⁑region is restricted to consisting of a finite number of straight pieces.

There are two principal problems, the first is that of prescribing a test which will enable us to determine whether a given object is ⁑inside or ⁑outside a given ⁑region. For, even though the fence may consist entirely of straight pieces, a description of only some part of the fence can provide no information about the rest of the fence, and consequently we cannot decide, even for an object 'near' the part of the fence described, whether it lies 'inside' or 'outside' the fence.

The second problem is to show that no matter how complicated the line of fencing may be, we can always find two corners which can be joined by a straight fence entirely within the ⁑region. This added piece of fence has the effect of dividing the given ⁑region into two ⁑regions, and the purpose we have in mind when obtaining this property is this: to erect a finite succession of straight pieces of fence which cut the ⁑region up into ⁑triangular regions. This operation of 'triangulation' we shall need when we have to devise a mathematical equivalent for the ⁑area of a ⁑region.

The properties of regions that are to be discussed in this section are purely descriptive, and depend entirely on the betweenness relation. 1.6T16 plays at one stage a quite important role: since between two points we can construct other points, we can assume that in the mathematical model of the fences we can always construct a fence which passes between two objects so that one is on each ⁑side of this fence.

Throughout this section the ideal line is fixed, and betweenness is defined in relation to it.

As a preliminary to the discussion of regions in the plane, we have to consider 'regions' on a line, which means, in effect, that we devise another way of expressing the relation so far denoted by

$$\mathrm{P} \ \mathscr{B} \ \mathrm{AB} \quad \text{or} \quad \mathscr{L}\mathscr{B}|\mathrm{APB}.$$

We begin by dividing the points of a line, in relation to a given proper point of itself, into two sets, either of which is determined by any of its points. These sets we shall call 'half-lines' (they are often called 'rays', but the analogy between the 'half-line' and the 'half-plane' that is to be defined later makes 'half-line' preferable in this context).

1.7*D*1 *Half-line*: ⌜A(P).

G ℘|l
℘|A : A ∈ l.
S ℘|P : P ∈ l & ≠|PA
D ⌜A(P) : the set of points Q such that A$\mathscr{B}$ PQ, together with P.

1.7*D*2 *Complementary half-line*: ⌜A($\tilde{\text{P}}$).

G and *S* as in *D*1.
D ⌜A($\tilde{\text{P}}$) : the set of points Q′ such that A $\mathscr{B}$ PQ′.

1.7*N*1 *End-point*: A is the end-point of ⌜A(P) and of ⌜A($\tilde{\text{P}}$).

A itself belongs neither to ⌜A(P) nor to ⌜A($\tilde{\text{P}}$), but every point of l other than A and the ideal point, U, belongs either to ⌜A(P) or to ⌜A($\tilde{\text{P}}$). I.e.

$$\text{l} = \text{AP} = {}^\ulcorner\text{A(P)} \cup {}^\ulcorner\text{A}(\tilde{\text{P}}) \cup \text{A} \cup \text{U}.$$

Also

$${}^\ulcorner\text{A(P)} \cap {}^\ulcorner\text{A}(\tilde{\text{P}}) = \emptyset.$$

We next define a 'segment' as the intersection of the sets constituting two half-lines:

1.7*D*3 *Segment*: ⌜AB⌝

$${}^\ulcorner\text{AB}^\urcorner = {}^\ulcorner\text{A(B)} \cap {}^\ulcorner\text{B(A)}.$$

Since ⌜A(B), ⌜B(A) are respectively the sets of points P for which either P$\mathscr{B}$AB or B$\mathscr{B}$AP or P = B and either P$\mathscr{B}$AB or A$\mathscr{B}$BP or P = A, it follows that

1.7*T*1 P ∈ ⌜AB⌝ ⇔ P $\mathscr{B}$ AB.

1.7*N*2 *End-points*: A, B are the *end-points* of ⌜AB⌝.

⌜AB⌝ is not defined when A = B.

For convenience, we now list the various equivalent forms of the betweenness relation:

If U is ideal and AB$\mathscr{S}$ PU, then P$\mathscr{B}$AB (P is between A, B),
$\mathscr{L}\mathscr{B}$|APB (the points are in the betweenness order A, P, B), and
P ∈ ⌜AB⌝ (P lies on the segment AB).

So far the question of whether there are necessarily any points in the sets ⌜AB⌝, ⌜A(B), ⌜A($\tilde{\text{B}}$) has not been discussed, but 1.6*T*16, 17 answer the question: unlimited sequences of points can be constructed which belong to $\ulcorner AB \urcorner$ (T16) (and therefore to $\ulcorner A(B)$), and to $\ulcorner A(\tilde{B})$ (T17).

From the two pairs of half-lines determined by a pair of points A, B we might expect to obtain four sets of points on the line, but in fact we have

$$\text{(i)} \quad \ulcorner A(B) \cap \ulcorner B(A) = \ulcorner AB \urcorner,$$

$$\text{(ii)} \quad \ulcorner A(B) \cap \ulcorner B(\tilde{A}) = \ulcorner B(\tilde{A}),$$

$$\text{(iii)} \quad \ulcorner A(\tilde{B}) \cap \ulcorner B(A) = \ulcorner A(\tilde{B}),$$

while

$$\text{(iv)} \quad \ulcorner A(\tilde{B}) \cap \ulcorner B(\tilde{A}) = \emptyset,$$

since if $K \in \ulcorner A(\tilde{B})$ and $K \in \ulcorner B(\tilde{A})$ we should have both $A \mathscr{B} BK$ and $B \mathscr{B} AK$, which are inconsistent.

Suppose now we are given a finite set of collinear proper points; since we are given the betweenness relations among all sets of three of them, we may arrange them in betweenness order (1.6N4), and can identify the first-and-last, say X and Y. Every other point of the set lies between X and Y, i.e. every point of the set except X and Y lies on $\ulcorner XY \urcorner$. By 1.6T17, we can find points on both $\ulcorner X(\tilde{Y})$ and $\ulcorner Y(\tilde{X})$, i.e.

1.7*T*2 Given any finite set of collinear proper points a segment can be found which contains them all.

G $\Sigma = \{\mathscr{L}\#\wp|AB\ldots K\}$
C $X, Y\colon \{X, Y\} \subset \Sigma \ \&\ (S \in \Sigma \ \&\ \#|SXY) \Rightarrow S \mathscr{B} XY$
S $\wp|PQ\colon P \in \ulcorner X(\tilde{Y}),\ Q \in \ulcorner Y(\tilde{X})$ (1.6T17)
T $\{A, B, \ldots, K\} \subset \ulcorner PQ \urcorner$.

We shall need to make considerable use of this construction for an 'origin' in a collinear set, with all the points of a given (finite) subset 'on one side' of it, and therefore set it out formally as a 'permissible construction'.

1.7*C*1 'Origin' on a half-line in relation to a given set of points.

G $\mathscr{L}\#\wp|AB\ldots K$
S $\wp|O\colon \{B, \ldots, K\} \subset \ulcorner O(A)$.

A, of course, may be replaced by any point of the set $\{A, B, \ldots, K\}$.

The next objective is to formulate definitions, analogues for the plane of 1.7D1, 2 for the line, of a *half-plane* and the *complementary half-plane*. Assume *G* $\wp|p, A \ \&\ A \notin p$, and define in relation to p two sets of points denoted temporarily by Π_A and $\Pi_{\tilde{A}}$, consisting respectively of points 'on the same side' of p as A, and points 'on the opposite side', thus:

$$\Pi_A : X \in \Pi_A \Leftrightarrow AX \cap p \notin \ulcorner AX \urcorner \text{ or } X = A,$$
$$\Pi_{\bar{A}} : Y \in \Pi_{\bar{A}} \Leftrightarrow AY \cap p \in \ulcorner AY \urcorner.$$

For such a definition to be consistent, we must have the property

$$X \in \Pi_A \Rightarrow \Pi_A = \Pi_X$$

and as a first step towards proving this we need the lemma

1.7*T*3*X* $X \in \Pi_A \Rightarrow A \in \Pi_X$
$Y \in \Pi_{\bar{A}} \Rightarrow A \in \Pi_{\bar{Y}}$.

If we are given $B \in \Pi_A$ and $C \in \Pi_A$, then, in order that the definition of Π_A should be consistent, we must have $B \in \Pi_C$. In terms of physical geometry this is equivalent to: if a ‡line p does not meet either of the ‡side-segments AB, AC of the ‡triangle ABC, then it does not meet the ‡side-segment BC.

1.7*T*4 Consistency of the proposed definition of a half-plane.

*G*1 $\mathcal{L}\wp|ABC$
*G*2 $\wp|p: (A, B, C) \notin p$
C $P, Q, R: \{^{\mathcal{O}}P, Q, R\} = p \cap \{^{\mathcal{O}}BC, CA, AB\}$
*G*3 $R \in \ulcorner AB \urcorner$
T Either $P \in \ulcorner BC \urcorner$ or $Q \in \ulcorner CA \urcorner$ but not both.

This theorem states effectively that if a line does not pass through any of the ‡vertices of a triangle it meets either two or none of the ‡side-segments. The theorem is more easily proved if we state it as a theorem for separation rather than betweenness.

1.7*T*5 Fundamental separation relation for the intersections of a line with four given lines.

G $\#|abcd$, no three concurrent†
*C*1 $A = b \cap c$, $B = c \cap a$, $C = a \cap b$, $A' = a \cap d$, $B' = b \cap d$, $C' = c \cap d$
*G*2 $p : p \not\ni (A, B, C, A', B', C')$
*C*2 $P = p \cap a$, $Q = p \cap b$, $R = p \cap c$
*G*3 $C'R \;\; \mathcal{S} \;\; AB$

*G*4 (i) $B'Q \;\; \mathcal{S} \;\; AC$	or	*G*4	(ii) $B'Q \;\; \mathcal{S} \;\; AC$
T (i) $A'P \;\; \mathcal{S} \;\; BC$		*T*	(ii) $A'P \;\; \mathcal{S} \;\; BC$

† *T*4 is derived from *T*5 by selecting d as the ideal line.

P c $\mathrm{K} = \mathrm{RC} \cap \mathrm{B'C'}$
        $\mathrm{S} = \mathrm{RQ} \cap \mathrm{B'C'}$
    $t1$ $\mathrm{B'A'} \;\mathscr{S}\; \mathrm{KC'}$
    p $\mathrm{ABRC'} \doublebarwedge^{\mathrm{C}} \mathrm{B'A'KC'}$
        $\mathrm{AB} \;\mathscr{S}\; \mathrm{RC'}$ $(G3)$
    $t2$ (i) $\mathrm{B'S} \;\mathscr{S}\; \mathrm{KC'}$ or (ii) $\mathrm{B'S} \;\not\mathscr{S}\; \mathrm{KC'}$
    p $\mathrm{ACQB'} \doublebarwedge^{\mathrm{R}} \mathrm{C'KSB'}$
        (i) $\mathrm{AC} \;\mathscr{S}\; \mathrm{QB'}$ or (ii) $\mathrm{AC} \;\not\mathscr{S}\; \mathrm{QB'}$ $(G4)$
    $t3$ (i) $\mathrm{A'S} \;\mathscr{S}\; \mathrm{KC'}$ or (ii) $\mathrm{A'S} \;\mathscr{S}\; \mathrm{KC'}$
    p $t1, t2$ $1.6T2, 4$
T P $\mathrm{A'SKC'} \doublebarwedge^{\mathrm{R}} \mathrm{A'PCB}$ and $(t3)$.

Returning now to the half-planes, we are given

$$\mathrm{B} \in \Pi_{\mathrm{A}}, \quad \text{i.e.} \quad \mathrm{AB} \cap \mathrm{p} \notin \ulcorner \mathrm{AB} \urcorner,$$
$$\mathrm{C} \in \Pi_{\mathrm{A}}, \quad \text{i.e.} \quad \mathrm{AC} \cap \mathrm{p} \notin \ulcorner \mathrm{AC} \urcorner;$$

it follows that $\mathrm{BC} \cap \mathrm{p} \notin \ulcorner \mathrm{BC} \urcorner$, i.e. that $\mathrm{B} \in \Pi_{\mathrm{C}}$, and thus:

1.7*T*6 $\mathrm{B} \in \Pi_{\mathrm{A}} \Rightarrow \Pi_{\mathrm{A}} = \Pi_{\mathrm{B}}$
$\mathrm{B} \in \Pi_{\tilde{\mathrm{A}}} \Rightarrow \Pi_{\tilde{\mathrm{A}}} = \Pi_{\mathrm{B}}$.

In relation to p, any point A, not on p, determines two sets of points Π_{A}, $\Pi_{\tilde{\mathrm{A}}}$, each completely determined by any of its points. If u is the ideal line, then:

$$\Pi_{\mathrm{A}} \cap \Pi_{\tilde{\mathrm{A}}} = \emptyset, \quad \Pi_{\mathrm{A}} \cap \mathrm{p} = \emptyset, \quad \Pi_{\tilde{\mathrm{A}}} \cap \mathrm{p} = \emptyset, \quad \Pi_{\mathrm{A}} \cap \mathrm{u} = \emptyset,$$
$$\Pi_{\tilde{\mathrm{A}}} \cap \mathrm{u} = \emptyset \quad \text{and} \quad \Pi_{\mathrm{A}} \cup \Pi_{\tilde{\mathrm{A}}} \cup \mathrm{p} \cup \mathrm{u} = \text{the whole plane.}$$

We now introduce the formal definition and notations for half-planes:

1.7*D*4 (i) Half-plane: $\ulcorner \mathrm{a(P)}$
(ii) Complementary half-plane: $\ulcorner \mathrm{a}(\tilde{\mathrm{P}})$.

G $\wp|\mathrm{a}$
S $\wp|\mathrm{P}, \mathrm{P} \notin \mathrm{a}$
D (i) $\ulcorner \mathrm{a(P)} = \{\mathrm{Q} : \mathrm{PQ} \cap \mathrm{a} \notin \ulcorner \mathrm{PQ} \urcorner\} \cup \mathrm{P}$
D (ii) $\ulcorner \mathrm{a}(\tilde{\mathrm{P}}) = \{\mathrm{Q'} : \mathrm{PQ'} \cap \mathrm{a} \in \ulcorner \mathrm{PQ'} \urcorner\}$.

1.7*N*3 Base of the half-plane: a is the *base* of $\ulcorner \mathrm{a(P)}$.
If a = LM, we shall write

$$\ulcorner \mathrm{LM(P)}, \quad \ulcorner \mathrm{LM}(\tilde{\mathrm{P}})$$

to denote the half-planes.

1.7*T*7*X* If the end-points of a segment lie in a given half-plane (or if one lies in the half-plane and the other in its base), then every point of the segment lies in the half-plane.

*G*1 $\wp | \mathrm{a}, \mathrm{P};\ \mathrm{P} \notin \mathrm{a}$
*G*2 Q: either $\mathrm{Q} \in \ulcorner \mathrm{a}(\mathrm{P})$ or $\mathrm{Q} \in \mathrm{a}$
*G*3 R: $\mathrm{R} \in \ulcorner \mathrm{PQ} \urcorner$
T $\mathrm{R} \in \ulcorner \mathrm{a}(\mathrm{P})$.

1.7*T*8*X* Any line parallel to the base of a half-plane and passing through a point of the half-plane lies entirely in the half-plane.

*G*1 $\wp | \mathrm{a}, \mathrm{P};\ \mathrm{P} \notin \mathrm{a}$
*G*2 $\wp | \mathrm{b}, \mathrm{b} \supset \mathrm{P}, \mathrm{b} \parallel \mathrm{a}$
S $\wp | \mathrm{Q}, \mathrm{Q} \in \mathrm{b}$
T $\mathrm{Q} \in \ulcorner \mathrm{a}(\mathrm{P})$.

Consider now the regions determined by two parallel lines; there are two half-planes, and a 'strip', defined thus:

1.7*N*4 Regions determined by two parallel lines.

G $\# \wp | \mathrm{a}, \mathrm{b};\ \mathrm{a} \parallel \mathrm{b}$
S $\wp | \mathrm{A}, \mathrm{B};\ \mathrm{A} \in \mathrm{a}, \mathrm{B} \in \mathrm{b}$
N (i) half-plane, $\ulcorner \mathrm{a}(\tilde{b}) = \ulcorner \mathrm{a}(\tilde{\mathrm{B}}) \cap \ulcorner \mathrm{b}(\mathrm{A}) = \ulcorner \mathrm{a}(\tilde{\mathrm{B}})$
(ii) strip, $\ulcorner \mathrm{ab} \urcorner = \ulcorner \mathrm{a}(\mathrm{B}) \cap \ulcorner \mathrm{b}(\mathrm{A})$
(iii) half-plane, $\ulcorner \mathrm{b}(\tilde{\mathrm{a}}) = \ulcorner \mathrm{b}(\tilde{\mathrm{A}})$.

In analogy with theorem *T*2 for a set of points on a line, there is a theorem for any finite set of points in the plane, namely, that we can find a strip, with sides parallel to any given line, that contains all points of the set.

1.7*T*9 A strip containing a given set of points.

*G*1 $\# \wp | \mathrm{A}_1, \mathrm{A}_2, \ldots, \mathrm{A}_r$
*G*2 $\wp | \mathrm{l}$
*S*1 $\wp | \mathrm{m}, \mathrm{m} \nparallel \mathrm{l}$
*C*1 $\{\mathrm{a}_i\}$: $\mathrm{a}_i \supset \mathrm{A}, \mathrm{a}_i \parallel \mathrm{l}, i = 1, \ldots, r$
*C*2 $\{\mathrm{M}_i\}$: $\mathrm{M}_i = \mathrm{a}_i \cap \mathrm{m}$
*S*2 M′, M″: $\{\mathrm{M}', \mathrm{M}''\} \subset \mathrm{m}, \ulcorner \mathrm{M}'\mathrm{M}'' \urcorner \supset \{\mathrm{M}_1, \ldots, \mathrm{M}_r\}$
*C*3 l′, l″: $\mathrm{l}' \supset \mathrm{M}', \mathrm{l}'' \supset \mathrm{M}'', \mathrm{l} \parallel \mathrm{l}' \parallel \mathrm{l}''$.
T $\ulcorner \mathrm{l}'\mathrm{l}'' \urcorner \supset \{\mathrm{A}_1, \ldots, \mathrm{A}_r\}$
P $\ulcorner \mathrm{l}'\mathrm{l}'' \urcorner \supset \{\mathrm{a}_1, \ldots, \mathrm{a}_r\} \supset \{\mathrm{A}_1, \ldots, \mathrm{A}_r\}$.

Since l is arbitrary we can find a corresponding strip parallel to any other line. We define therefore

1.7*D*5 A 'parallelogram region'.

G $\# \wp | \mathrm{l}, \mathrm{l}', \mathrm{m}, \mathrm{m}' : \mathrm{l} \parallel \mathrm{l}', \mathrm{m} \parallel \mathrm{m}', \mathrm{l} \# \mathrm{m}$
D Parallelogram region: $\ulcorner \mathrm{ll}' \urcorner \cap \ulcorner \mathrm{mm}' \urcorner$.

1.7*T*10 Parallelogram region containing a given set of points.

*G*1 $\# \wp | \mathrm{A}_1, \ldots, \mathrm{A}_r$
*G*2 $\# \wp | \mathrm{l}, \mathrm{m}, \mathrm{l} \# \mathrm{m}$
*C*1 $\ulcorner \mathrm{l}'\mathrm{l}'' \urcorner \supset \{\mathrm{A}_1, \ldots, \mathrm{A}_r\}, \mathrm{l}' \parallel \mathrm{l}'' \parallel \mathrm{l}$
$\ulcorner \mathrm{m}'\mathrm{m}'' \urcorner \supset \{\mathrm{A}_1, \ldots, \mathrm{A}_r\}, \mathrm{m}' \parallel \mathrm{m}'' \parallel \mathrm{m}$ (*T*9)
T $\ulcorner \mathrm{l}'\mathrm{l}'' \urcorner \cap \ulcorner \mathrm{m}'\mathrm{m}'' \urcorner \supset \{\mathrm{A}_1, \ldots, \mathrm{A}_r\}$.

The parallelogram region is a more complex entity than the next regions to be considered, namely, the regions determined by any two lines, that is, by the various half-planes determined by the lines, but its relation to the strip justifies its slightly premature introduction.

The regions determined by two non-parallel proper lines we shall call

1.7*D*6 Angular regions.

G $\mathscr{L} \wp | \mathrm{VAB}$
D Angular regions:
$$\angle \mathrm{V(AB)} = \ulcorner \mathrm{VA(B)} \cap \ulcorner \mathrm{VB(A)}$$
$$\angle \mathrm{V(A\tilde{B})} = \ulcorner \mathrm{VA(\tilde{B})} \cap \ulcorner \mathrm{VB(A)}$$
$$\angle \mathrm{V(\tilde{A}B)} = \ulcorner \mathrm{VA(B)} \cap \ulcorner \mathrm{VB(\tilde{A})}$$
$$\angle \mathrm{V(\tilde{A}\tilde{B})} = \ulcorner \mathrm{VA(\tilde{B})} \cap \ulcorner \mathrm{VB(\tilde{A})}.$$

1.7*N*5 *Vertex* of the angular region: V in *D*6.

Boundary of $\angle \mathrm{V(AB)}$: $\ulcorner \mathrm{V(A)} \cup \ulcorner \mathrm{V(B)} \cup \mathrm{V}$.

It is to be noted that as $\ulcorner \mathrm{VA(B)}$ does not include points of the line VA, none of the four angular regions includes any point of VA or VB.

1.7*T*11(*X*) Any segment whose end-points lie one on each of the two half-lines bounding an angular region, lies in the angular region.

*G*1 $\angle \mathrm{V(AB)}$
*G*2 $\mathrm{P} \in \ulcorner \mathrm{V(A)}, \mathrm{Q} \in \ulcorner \mathrm{V(B)}$
T $\ulcorner \mathrm{PQ} \urcorner \subset \angle \mathrm{V(AB)}$.

1.7*T*12(*X*) Condition that an angular region should contain a given half-line with end-point at its vertex.

G $\mathscr{L}_3 \wp | \mathrm{ABCD}$
S $\mathrm{P} \in \ulcorner \mathrm{A(B)}, \mathrm{Q} \in \ulcorner \mathrm{A(C)}$
C $\mathrm{R}: \mathrm{R} = \mathrm{AD} \cap \mathrm{PQ}$
T $\ulcorner \mathrm{A(D)} \subset \angle \mathrm{A(BC)} \Leftrightarrow \mathscr{L}\mathscr{B} | \mathrm{PRQ}$.

This theorem provides an analogue, for half-lines with a common end-point, of the betweenness relation for points on a line, but it should be noticed that there is no analogue of the property that given three collinear proper points PQR, then either P $\mathscr{B}$ QR or Q $\mathscr{B}$ RP or R $\mathscr{B}$ PQ. If $D \in {}^{\ulcorner}AB(\tilde{C}) \cap {}^{\ulcorner}AC(\tilde{B})$ then none of the three angular regions determined by pairs of the half-lines $^{\ulcorner}A(B)$, $^{\ulcorner}A(C)$, $^{\ulcorner}A(D)$ contains the third half-line. But, with the reservations that (i) we obtain a separation order and not a betweenness order, and (ii) that we may have initially to split the half-lines into the sets lying in two complementary half-planes, we can establish an order relation among half-lines with a common end-point.

1.7*T*13 Given a set of half-lines with a common end-point then there can be constructed: either (i) an ordering of the set by separation in which the angular region determined by each pair of half-lines in the ordering contains no other half-lines of the set, or (ii) an ordering by betweenness in which the angular region determined by every pair in the ordering, except the first-and-last, contains no other half-line of the set, while the angular region determined by the first-and-last contains all the other half-lines of the set. (If the first-and-last are complementary half-lines—on the same line—then 'angular region' is replaced by 'one of the half-planes'.)

*G*1 $\# \wp | VA_1 \dots A_n : \# | {}^{\ulcorner}V(A_1)$
*S*1 $v: v \supset V,\ v \neq VA_1, \quad i \in \{1, \dots, n\}$
*S*2 $l: l \parallel v,\ l \subset {}^{\ulcorner}v(A_1)$
*C*1 $\{A_i'\}: A_i' = l \cap VA_1, \quad i = \{1, \dots, n\}$
*S*3 $R, S: \{R, S\} \subset l,\ \{A_i'\} \subset {}^{\ulcorner}RS^{\urcorner}$
*S*4 $m: m \parallel v,\ m \subset {}^{\ulcorner}v(\tilde{A}_1)$
*C*2 $\{B_j\}: B_j = l \cap {}^{\ulcorner}V(A_1), \quad j \in \{j_1, \dots, j_r\}, \quad 1 \in \{j\}$
$\{B_k\}: B_k = m \cap {}^{\ulcorner}V(A_1), \quad k \in \{k_{r+1}, \dots, k_n\}, \quad \{j\} \cap \{k\} = \emptyset.$

($\{B_j\}$, $\{B_k\}$ are complementary subsets of $\{A_i'\}$, consisting of the intersections with l and m of those half-lines of the set $\{{}^{\ulcorner}V(A_i)\}$ which meet respectively l and m.)

*C*3 $R', S': R' = VR \cap m, \quad S' = VS \cap m$
*C*4 $\{C_\rho\}: \{C_\rho\} = \{B_j\}\ \&\ \mathscr{B}|\{C_\rho\}, \quad \rho \in \{1, \dots, r\}$
$\{C_{r+\sigma}\}: \{C_{r+\sigma}\} = \{B_k\}\ \&\ \mathscr{B}|\{C_{r+\sigma}\}, \quad \sigma \in \{1, \dots, n-r\}$
T(i) *G* $\{B_k\} \neq \emptyset$,
$\left.\begin{array}{l} {}^{\ulcorner}RR'(\tilde{A}_1) \cap \{{}^{\ulcorner}V(A_i)\} \neq \emptyset \\ {}^{\ulcorner}SS'(\tilde{A}_1) \cap \{{}^{\ulcorner}V(A_i)\} \neq \emptyset \end{array}\right\} \quad i \in \{1, \dots, n\}$
T $\mathscr{S}|\{{}^{\ulcorner}V(C_i)\}$ and each pair in the ordering defines an angular region containing no half-line of the set.

T (ii) Either $G \quad \{B_k\} \neq \emptyset$
and either $\ulcorner RR'(\tilde{A}_1) \cap \{\ulcorner V(A_1)\} = \emptyset$
or $\ulcorner SS'(\tilde{A}_1) \cap \{\ulcorner V(A_1)\} = \emptyset$.
Name the points so that

$$\ulcorner SS'(\tilde{A}_1) \cap \{\ulcorner V(A_1)\} = \emptyset.$$

Or $G \quad \{B_k\} = \emptyset$ (a lucky selection of v!)
T Each of $\angle V(C_g C_{g+1})$, $g \in \{1, \ldots, n-1\}$ contains no half-line $\ulcorner V(A_1)$, but $\angle V(C_1 C_n)$ contains all $\ulcorner V(A_h)$, $h \in \{2, \ldots, n-1\}$.

The next region to be discussed is the *triangular region*, which is determined by three half-planes.

1.7 *D* 7 *Triangle*: Any three non-collinear proper points together with the segments determined by pairs of them, form a *triangle*.

1.7 *N* 6 If A, B, C are the points, then for the triangle the following terminology is used

A, B, C: *vertices*
BC, CA, AB: *side-lines*
$\ulcorner BC \urcorner$, $\ulcorner CA \urcorner$, $\ulcorner AB \urcorner$: *side-segments*
$\angle A(BC)$, $\angle B(CA)$, $\angle C(AB)$: *angles* (angular regions)
$\triangle ABC = \ulcorner BC \urcorner \cup \ulcorner CA \urcorner \cup \ulcorner AB \urcorner \cup A \cup B \cup C$: the *triangle* or, in explicit contexts, the *boundary* of the *triangular region*
$\mathfrak{R}ABC = \ulcorner A(BC) \cap \ulcorner B(AC) \cap \ulcorner C(AB)$: the *triangular region bounded by* $\triangle ABC$.

The theorem 1.7 *T* 4 shows that any line not through a vertex meets either two or none of the side-segments, i.e. if it does not contain any vertices, then it meets $\triangle$ in two points or no points.

We can use this theorem to classify the regions into which three lines divide the plane. There are apparently eight of them:

$$\mathfrak{R}_1 = \ulcorner BC(A) \cap \ulcorner CA(B) \cap \ulcorner AB(C) = \mathfrak{R}ABC$$

three such as $\mathfrak{R}_{2,A} = \ulcorner BC(\tilde{A}) \cap \ulcorner CA(B) \cap \ulcorner AB(C)$

three such as $\mathfrak{R}_{3,A} = \ulcorner BC(A) \cap \ulcorner CA(\tilde{B}) \cap \ulcorner AB(\tilde{C})$

and $\mathfrak{R}_4 = \ulcorner BC(\tilde{A}) \cap \ulcorner CA(\tilde{B}) \cap \ulcorner AB(\tilde{C})$.

By suitably renaming the points A, B, C we may suppose, for a given line which meets two of the side-segments, that

$$l \cap \ulcorner BC \urcorner = P, \quad l \cap \ulcorner CA \urcorner = Q, \quad l \cap \ulcorner A(\tilde{B}) = R,$$

and that $\mathscr{B}|PQR$. Consider now points K, $K \in l$, variously related to {P, Q, R}:

(i) $\mathscr{B}|KRQP : K \in \mathfrak{R}_{3,A}$
(ii) $\mathscr{B}|RKQP : K \in \mathfrak{R}_{2,B}$
(iii) $\mathscr{B}|RQKP : K \in \mathfrak{R}_{ABC}$
(iv) $\mathscr{B}|RQPK : K \in \mathfrak{R}_{2,A}$.

Since such a line may be drawn through any given point K (namely, the join of K to a point of one of the side-segments), the regions determined by three lines have the following properties:

1.7*T*14 Regions determined by three lines.

(i) $\ulcorner BC(\tilde{A}) \cap \ulcorner CA(\tilde{B}) \cap \ulcorner AB(\tilde{C}) = \emptyset$
(ii) *G*1 $K \in \mathfrak{R}ABC$
S $\wp|P : KP \not\ni (A, B, C)$
T $\ulcorner K(P) \cap \triangle = \{1 \text{ point}\}$
(iii) *G*2 $\wp|K : K \notin \mathfrak{R}ABC \;\&\; K \notin \triangle ABC$
S $\wp|P : KP \cap \{A, B, C\} = \emptyset$
T $\ulcorner K(P) \cap \triangle = \emptyset$ or $\{2 \text{ points}\}$.

For convenience of reference we pick out the main result:

1.7*T*15 Condition that a point lies inside a triangle.

G $\triangle ABC$; $\wp|K : K \notin \triangle$.
S $\wp|P : KP \cap \{A, B, C\} = \emptyset$
T $K \in \mathfrak{R}ABC \Leftrightarrow \ulcorner K(P) \cap \triangle = \{1 \text{ point}\}$.

When we come to consider 'sense' in the plane we shall need

1.7*T*16 The separation relations for points on lines which do not meet the boundary of a triangular region.

*G*1 $\mathscr{L}\wp|ABC$
*C*1 $a, b, c : a = BC, \quad b = CA, \quad c = AB$
*S*1 $\wp|g : \ulcorner g(A) \supset \{B, C\}$
(i.e. $g \cap \triangle ABC = \emptyset$)
*C*2 $P, Q, R : \{^{\mathscr{O}}P, Q, R\} = g \cap \{^{\mathscr{O}}a, b, c\}$
S2 $G : G \in g \;\&\; GQ \;\mathscr{S}\; PR$
*S*3 $g' : g' \supset G \;\&\; \ulcorner g'(A) \supset \{B, C\}$
*C*3 $P', Q', R' : \{^{\mathscr{O}}P', Q', R'\} = g' \cap \{^{\mathscr{O}}a, b, c\}$
T $GQ' \;\mathscr{S}\; P'R'$

P c $\mathrm{P''}:\mathrm{P''} = \mathrm{g'} \cap \mathrm{AP}$

$t1$ $\mathrm{GPQR} \overline{\wedge}^{\mathrm{A}} \mathrm{GP''Q'R'}$ & $\mathrm{GQ} \ \mathscr{S} \ \mathrm{PR}$

$\Rightarrow \mathrm{GQ'} \ \mathscr{S} \ \mathrm{P''R'}$

$t2$ $S1, C2$ & $S3, C3$

$\Rightarrow \mathrm{QQ'} \ \mathscr{S} \ \mathrm{AC}$

$t3$ $\mathrm{ACQQ'} \overline{\wedge}^{\mathrm{P}} \mathrm{P''P'GQ'}$ & $\mathrm{AC} \ \mathscr{S} \ \mathrm{QQ'}$ ($t2$)

$\Rightarrow \mathrm{GQ'} \ \mathscr{S} \ \mathrm{P'P''}$

T $\mathrm{GQ'} \ \mathscr{S} \ \mathrm{P''R'}$ ($t1$) & $\mathrm{GQ'} \ \mathscr{S} \ \mathrm{P'P''}$ ($t3$)

$\Rightarrow \mathrm{GQ'} \ \mathscr{S} \ \mathrm{P'R'}$.

That is, if G is outside the triangle, then, on every line g through G which does not meet the boundary of the triangular region, G stands in the same separation relation to the points in which g meets the side-lines of the triangle.

The last part of this Chapter is to be devoted to the definition and investigation of some of the properties of a 'simple closed polygon'.

1.7 ***D*8** A *simple closed polygon* is a set of points ordered by separation together with the segments whose end-points are the adjacent pairs in the ordering, the points being such that no two of these segments have a common point.

The last clause sets out the condition that the polygon is 'simple'. The definition is framed to preclude the degenerate case in which all the points are collinear, but we have to make allowance for the possibility of three or more vertices being collinear. Such sets of vertices are typified in the following diagrams:

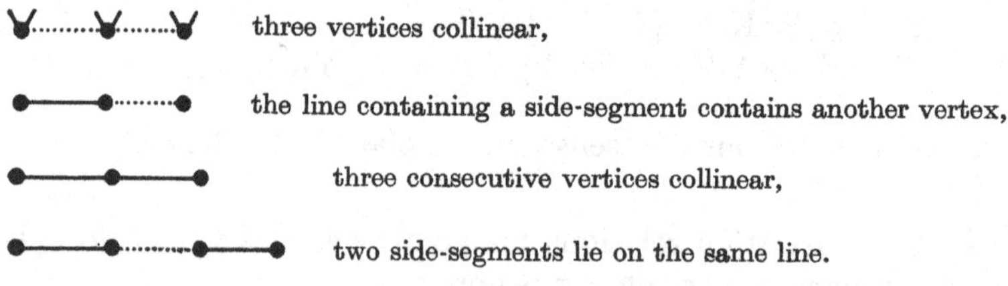

three vertices collinear,
the line containing a side-segment contains another vertex,
three consecutive vertices collinear,
two side-segments lie on the same line.

These sets may be combined in any way to produce more elaborate patterns.

1.7 ***N*7** Elements of a simple polygon

G $\mathscr{S}|\mathrm{P}_1 \ldots \mathrm{P}_n$ & $\mathrm{P}_{n+1} = \mathrm{P}_1$ & $\ulcorner \mathrm{P}_i \mathrm{P}_{i+1} \urcorner \cap \ulcorner \mathrm{P}_j \mathrm{P}_{j+1} \urcorner = \emptyset$, $\{i, j\} \subset \{1, \ldots, n\}$

N(i) $\mathrm{P}_1(=\mathrm{P}_{n+1}), \mathrm{P}_2, \ldots, \mathrm{P}_n$: vertices

N(ii) $\mathrm{P}_r \mathrm{P}_{r+1}$: side-lines

N(iii) $\ulcorner \mathrm{P}_r \mathrm{P}_{r+1} \urcorner$: side-segments

N(iv) $\triangle P_1P_2\ldots P_n = \triangle P_2\ldots P_nP_1 = \ldots = P_1 \cup P_2 \cup \ldots \cup P_n \cup \ulcorner P_1P_2\urcorner \cup \ldots \cup \ulcorner P_nP_1\urcorner$: (boundary of the) (simple closed) polygon

N(v) $\mathfrak{R}P_1\ldots P_n = \mathfrak{R}P_2\ldots P_nP_1 = \ldots$: region bounded by the polygon.

The 'region' has yet to be defined. The triangular region was defined initially by

$$\mathfrak{R}ABC = \ulcorner BC(A) \cap \ulcorner CA(B) \cap \ulcorner AB(C),$$

that is, as the following set {H} of points:

$$\mathfrak{R}ABC = \{H : AH \cap BC \mathscr{B} AH \quad \& \quad BH \cap AC \mathscr{B} BH \quad \& \quad CH \cap AB \mathscr{B} CH\}.$$

Theorem *T* 15 enables us to replace this rather clumsy definition by:

$$\mathfrak{R}ABC = \{H : \ulcorner H(P) \cap \triangle ABC = \{1 \text{ point}\}\} \quad \& \quad \{P : \ulcorner H(P) \cap \{A, B, C\} = \emptyset\}.$$

It is on this definition that we are to base the definition of a simple polygonal region; the analogous prescription would be:

$$\mathfrak{R}P_1\ldots P_n = \{H : \ulcorner H(P) \cap \triangle P_1\ldots P_n = \{\text{odd-set of points}\}\} \quad \& \quad \{P : \ulcorner H(P) \cap \{P_i\} = \emptyset\}.$$

It is to be proved first that:

1.7 *T* 17 Any line which does not contain any vertices of $\triangle$ meets $\triangle$ in either no points or an even-set of points.

*G*1 $\triangle P_1P_2\ldots P_n$

*G*2 $\wp|l,\ l \cap \{P_i\} = \emptyset, \quad i \in \{1, \ldots, n\}$

*G*3 (i) $l \cap \triangle = \emptyset$ (i.e. *T* is trivially valid), or

(ii) $l \cap \ulcorner P_nP_1\urcorner = K$†

T $l \cap \triangle =$ even-set of points (including K)

P Divide the sequence of points $P_1, \ldots, P_n$ into a set of subsequences, so that the points of alternate subsequences lie in $\ulcorner l(P_1)$ and $\ulcorner l(\tilde{P}_1)$ ($= \ulcorner l(P_n)$). Each subsequence terminates as l meets $\triangle$ (i.e. as it meets a side-segment $\ulcorner P_rP_{r+1}\urcorner$). Let the subsequences be

$$\{P_1, P_2, \ldots, P_r\} \subset \ulcorner l(P_1)$$
$$\{P_{r+1}, \ldots, P_s\} \subset \ulcorner l(P_n)$$
$$\{P_{s+1}, \ldots, P_t\} \subset \ulcorner l(P_1)$$
$$\ldots\ldots\ldots\ldots\ldots\ldots$$
$$\{P_{u+1}, \ldots, P_n\} \subset \ulcorner l(P_n).$$

† I.e. we assume that l meets at least one side-segment; the segment may be taken as $\ulcorner P_nP_1\urcorner$ by re-numbering the vertices so that P_r becomes P_{r+k} or P_{r+k-n} for the appropriate value of k.

Since the first of these subsequences lies in $\ulcorner l(P_1)$ and the last in $\ulcorner l(P_n)$, the set of subsequences is an even-set.

In consequence of *T* 10 we have

1.7*T*18 Parallelogram region enclosing a polygon

*G*1 $\triangle P_1 \dots P_n$
*G*2 $\# \wp | lm; l \nparallel m$
T There exist $l', l''; l' \parallel l'' \parallel l$
and $m', m''; m' \parallel m'' \parallel m$,
such that $\ulcorner l'l'' \urcorner \cap \ulcorner m'm'' \urcorner \supset \triangle P_1 \dots P_n$.

We clearly wish to define the region $\mathfrak{R}P_1 \dots P_n$ in such a way that this parallelogram region contains every point of it, as well as points which do not belong to it, and that if any segment $\ulcorner XY \urcorner$ meets $\triangle$ in a single point, then one of X, Y lies in $\mathfrak{R}$ and the other does not.

The type of construction we are going to use depends on drawing a parallel l_i to l through each vertex P_i. These lines are not necessarily all distinct, since l might be parallel to side-segments or the joins of non-adjacent vertices.

Suppose that
$$\{l_j'\} = \{l_i\} \ \& \ \mathscr{B} | l' l_1' \dots l_r' l''.$$

In the strips defined by consecutive pairs of these, select lines

$$f_j; \ f_j \parallel l \ \& \ f_j \subset \ulcorner l_j' l_{j+1}' \urcorner.$$

Then, in general, in the strip $\ulcorner f_j f_{j+1} \urcorner$ there is only one vertex of $\triangle$, but there may be a collinear set of vertices (and possibly some side-segments); in any case there is only one of the lines $\{l_k'\}$, namely l_{j+1}', and on this line all these vertices lie.

Suppose that $m \cap \triangle = \{D_1, \dots, D_u\}$, $m \cap l' = L'$, $m \cap l'' = L''$, and that the points have been named so that $\mathscr{B} | L' D_1 \dots D_u L''$. Consider various points $K \in m$:

$K_0 \in \ulcorner L' D_1 \urcorner$: $K \notin \mathfrak{R}P_1 \dots P_n$
$K_1 \in \ulcorner D_1 D_2 \urcorner$: $K \in \mathfrak{R}P_1 \dots P_n$, since $\ulcorner K_0 K_1 \urcorner \supset D_1$ and no other points of $\triangle$
$K_2 \in \ulcorner D_2 D_3 \urcorner$: $K \notin \mathfrak{R}P_1 \dots P_n$, etc.

So that
$$\ulcorner K_r(K_0) \cap \triangle = \{D_r, D_{r-1}, \dots, D_1\}.$$

Thus a point K of m belongs to $\mathfrak{R}$ if the half-line through it along m meets $\triangle$ in an odd-set of points. It does not follow immediately that every half-line from K meets $\triangle$ in an odd-set of points; we have to prove that:

1.7*T*19 The sets of points in which two half-lines with a common end-point, and neither containing a vertex, meet $\triangle$ are either both even-sets or both odd-sets.

*G*1 $\triangle P_1 \ldots P_n$

*G*2 $\mathscr{L}\wp|\mathrm{HKK}'$ & $\{H, K, K'\} \cap \triangle = \emptyset$
& $\ulcorner H(K) \cap \{P_i\} = \emptyset$ & $\ulcorner H(K') \cap \{P_i\} = \emptyset$, $i \in \{1, \ldots, n\}$

T $\ulcorner H(K) \cap \triangle$ and $\ulcorner H(K') \cap \triangle$ are either both even-sets or both odd-sets

N $\pi_i = \ulcorner H(P_i)$, $i \in \{1, \ldots, n\}$,
$\kappa = \ulcorner H(K)$, $\kappa' = \ulcorner H(K')$

P *c*1 Arrange the set of half-lines $\{\pi_i, \kappa\}$ in separation order (*T*13)

*g*1 $\{\pi_r, \kappa\}$, $\{\kappa, \pi_s\}$, $\{\pi_s, \pi_t\}$ are adjacent pairs in this order

*s*1 X: $\ulcorner H(X) \subset \angle H(P_r P_s)$ & $\# | \ulcorner H(X), \kappa$ & $\ulcorner H(X) \cap \{P_i\} = \emptyset$, $i \in \{1, \ldots, n\}$

*s*2 Y: $\ulcorner H(Y) \subset \angle H(P_s P_t)$ & $\ulcorner H(Y) \cap \{P_i\} = \emptyset$, $i \in \{1, \ldots, n\}$

*t*1 There is a one-to-one correspondence between the sets $\{\ulcorner H(K) \cap \triangle\}$ and $\{\ulcorner H(X) \cap \triangle\}$.

p There are no vertices in $\angle H(P_r P_s)$ (*c*1 and *g*1) and therefore any side-segment $\ulcorner P_i P_{i+1} \urcorner$ either meets both π_r and π_s or meets neither, i.e. it meets both κ and $\ulcorner H(X)$ or neither.

*t*2 The sets $\{\ulcorner H(X) \cap \triangle\}$, $\{\ulcorner H(Y) \cap \triangle\}$ either are in one-to-one correspondence or can be put into one-to-one correspondence after removing two members from one of them.

p The only vertex in $\angle H(P_r P_t)$ is P_s, and therefore $\ulcorner H(X)$, $\ulcorner H(Y)$ have points common with the same side-segments, except possibly for $\ulcorner P_{s-1} P_s \urcorner$ and $\ulcorner P_s P_{s+1} \urcorner$. There are three cases:

(i) $\{P_{s-1}, P_{s+1}\} \subset \ulcorner HP_s(K)$: κ meets both segments and $\ulcorner H(Y)$ meets neither, or

(ii) $\pi_s \subset \angle H(P_{s-1} P_{s+1})$: κ meets one segment and $\ulcorner H(Y)$ meets the other, or

(iii) $\{P_{s-1}, P_{s+1}\} \subset \ulcorner HP_s(Y)$: κ meets neither segment and $\ulcorner H(Y)$ meets both.

T *P* There are n or fewer regions $\angle H(P_r P_s)$. For half-lines in regions whose boundaries have a common half-line the sets of intersections with $\triangle$ are both even-sets or both odd-sets. The proof is completed by a simple induction.

We may therefore consistently define the region of which $\triangle P_1 \ldots P_n$ is the boundary, thus:

1.7 *D*9 Region bounded by a simple closed polygon

G $\triangle P_1 \ldots P_n$
$K : K \notin \triangle$
S $\wp|X : \# |XK \;\&\; \ulcorner K(X) \cap \{P_i\} = \emptyset, \quad i \in \{1, \ldots, n\}$
D $K \in \mathfrak{R}P_1 \ldots P_n \Leftrightarrow \ulcorner K(X) \cap \triangle = \text{odd -set of points.}$

We next define and prove the existence of 'cross-segments', leading to the final step of proving that any simple polygonal region can be cut up into triangular regions.

1.7 *D*10 *Cross-segment*: a segment $\ulcorner P_r P_s \urcorner$ whose end-points are vertices of $\triangle P_1 \ldots P_n$ such that (i) it is not a side-segment, (ii) it does not meet $\triangle P_1 \ldots P_n$ and (iii) at least one point of it and therefore every point of it belongs to $\mathfrak{R}P_1 \ldots P_n$.

We are to prove in fact not only that there exists at least one cross-segment, but that if through any vertex there is no cross-segment, then the join of the two vertices adjacent to it is a cross-segment.

The method of proof is to construct a parallelogram region containing $\mathfrak{R}$ and the ordered set of parallel lines $\{l'_j\}$ (*T* 18) and to consider first the vertex which is 'nearest' to l' in the construction.

1.7 *T*20 The existence of cross-segments.

G $\triangle P_1 \ldots P_n;\ \wp|lm : l \nparallel m$
*S*1 $\# |l', l'' : l' \| l'' \| l$
$\# m', m'' : m' \| m'' \| m$
$\ulcorner l' l'' \urcorner \cap \ulcorner m' m'' \urcorner \supset \{P_i\}$ (*T* 18)
C $l_i : l_i \supset P_i,\ l_i \| l$
$\{l'_\rho\} = \{l_i\} \;\&\; \mathscr{B}|l' l'_1 \ldots l'_r l''$
*G*1 $l'_1 \supset P_1$ (i.e., for convenience the vertices have been renumbered in such a way that P_1 is 'nearest' to l')
*G*2 (i) $\mathfrak{R}P_1 P_2 P_n$ contains no vertices
t (i) $\ulcorner P_2 P_n \urcorner$ is a cross-segment, or
*G*2 (ii) $\mathfrak{R}P_1 P_2 P_n \supset P_i, P_j, \ldots$
*G*3 $\{l_i, l_j, \ldots\} = \{l'_{\rho 1}, l'_{\rho j}, \ldots\} \;\&\; \mathscr{B}|l'_{\rho 1} l'_{\rho 2} \ldots$
t (ii) 1 $\ulcorner l'_1 l'_{\rho 1} \urcorner \cap \mathfrak{R}P_1 P_2 P_n \subset \mathfrak{R}P_1 \ldots P_n$
p $l'_1 \cap \mathfrak{R}P_1 P_2 P_n = \emptyset$
t (ii) 2 $\ulcorner P_1 P_{\rho 1} \urcorner$ is a cross-segment.

1.7 *T*21 By the addition of a cross-segment to the set of side-segments, $\mathfrak{R}P_1 \dots P_n$ is split into two simple polygonal regions.

*G*1 $\triangle P_1 \dots P_n$

*G*2 $\ulcorner P_r P_s \urcorner$ is a cross-segment, $r < s-1 < n$

T (i) $\triangle P_r P_{r+1} \dots P_s$, $\triangle P_s P_{s+1} \dots P_n P_1 \dots P_r$ are simple polygons.

(ii) $\mathfrak{R}P_1 \dots P_n = \mathfrak{R}P_r P_{r+1} \dots P_s \cup \mathfrak{R}P_s \dots P_n P_1 \dots P_r \cup \ulcorner P_r P_s \urcorner$.

In each of the derived simple polygons cross-segments can be found, and the process of adding cross-segments can be continued until all the resulting regions are triangular. Since each cross-segment forms part of the boundary of two of these regions, and each side-segment part of the boundary of one, the total number of cross-segments required is $n-3$, and the total number of triangular regions produced is $n-2$. Every vertex P_r must belong to at least one of these triangles; if it belongs to exactly one, then the triangle is $\triangle P_{r-1} P_r P_{r+1}$, and $\ulcorner P_{r-1} P_{r+1} \urcorner$ is a cross-segment; if it belongs to more than one, then the common side-segment of two of them is a cross-segment with end-point P_r.

To sum up we define

1.7 *D*11 *Triangular dissection* of $\mathfrak{R}P_1 P_2 \dots P_n$: any set of $n-2$ triangular regions into which $\mathfrak{R}$ may be split by naming a suitable set of $n-3$ cross-segments.

Finally since every vertex P_r of the polygon belongs to at least one triangle of the dissection, either two side-segments of that triangle are the adjacent side-segments $\ulcorner P_{r-1} P_r \urcorner$, $\ulcorner P_r P_{r+1} \urcorner$ of the polygon, or at least one of them is a cross-segment one of whose end-points is P_r. Analysing the relations among the vertices and cross-segments more closely we find:

1.7 *T*22 Any vertex P_r of $\triangle P_1 \dots P_n$ is the end-point of at least one cross-segment, unless P_r is such that the triangular region defined by the two side-lines $P_r P_{r-1}$, $P_r P_{r+1}$ through P_r and some other side-line $P_s P_{s+1}$ is contained in $\mathfrak{R}P_1 \dots P_n$. In this case $\ulcorner P_{r-1} P_{r+1} \urcorner$ is a cross-segment.

Because of their simplicity we single out one special class of simple polygons, namely,

1.7 *D*12 *Convex polygon*: A simple polygon $\triangle P_1 \dots P_n$ with the additional property that every segment $\ulcorner P_r P_s \urcorner$ which is not a side-segment is a cross-segment.

1.7*T*23*X*

*G*1 $\triangle P_1 \ldots P_n$ a simple polygon

*G*2 For each index i and every pair of indices $\{r, s\}$,

$$(\{r, s\} \cap \{i, i+1\} = \emptyset) \Rightarrow (P_s \in \ulcorner P_i P_{i+1}(P_r))$$

T $\triangle P_1 \ldots P_n$ is convex.

State and prove the converse.

CHAPTER 1.8

SENSE

We return now to the last of the phenomena observed by the F-G in setting markers along his furrow, namely, the distinction between 'northwards' and 'southwards'. When we have devised a method of describing this within the framework of 'ordering by betweenness' we shall have completed the justification for accepting the printing convention in a book on geometry—we shall be able to attach significance to the phrase 'reading from left to right' as well as to recognise pairs of adjacent letters and a first-and-last pair.

In this section again we assume that the ideal line is fixed, and that we are dealing with a geometry that satisfies all four axioms $A\mathscr{S}$.

Suppose we are given $\mathscr{L}\#\wp|$ABPQ; the set of possible ways of ordering these points by betweenness can be split into two complementary subsets Σ_1 and Σ_2, such that Σ_1 is the subset:

betweenness order	adjacent pairs			first-and-last
$\mathscr{B}\|$APBQ	AP	BQ	BP	AQ
PAQB	AP	BQ	AQ	BP
APQB	AP	BQ	PQ	AB
PABQ	AP	BQ	AB	PQ
ABPQ	AB	PQ	BP	AQ
PQAB	AB	PQ	AQ	BP

Σ_2 is obtained from Σ_1 by interchanging either P with Q or A with B. Σ_1 can be defined by the two overlapping conditions: Σ_1 includes (i) all arrangements in which both AP and BQ are adjacent pairs and (ii) all arrangements in which either AQ or BP are the first-and-last. These are then the arrangements which associate A with P and B with Q.

The statement of the pattern of arrangements can be simplified by introducing as in 1.7C1 a new point O as 'origin', thus

G $\mathscr{L}\#|$ABPQ
S O : $\mathscr{B}|$OAB, $\ulcorner$O(A) $\supset$ {B, P, Q}
D Σ_1 is the set of betweenness orderings of A, B, P, Q such that $\mathscr{B}|$OPQ.

We are now able, when we wish to do so, to ascribe distinct meanings to the symbol printed as 'PQ' and that printed as 'QP'. We may assume that each line of type abc... (with repeated symbols distinguished as on p. 50) is a conventional way of representing a set of pairs {(ab), (bc), ...}, which are referred to two fixed markers say, ∗ o and a marker α selected so that $\mathscr{B}|\alpha * \text{o}$ and $\ulcorner\alpha(*) = \{\text{o}, \text{a}, \text{b}, \text{c}, \ldots\}$ and $\mathscr{B}|$oabc.... That is, we can now see some mathematical significance in the physical operation of reading from left to right.

We shall however preserve the convention that a set of symbols is not to be read 'in order', unless an ordering is specified by some symbol. In particular we shall continue to write:

{A, B, C, ...} for a set of elements, and
{$^{\wp}$A, B, C, ...} for the same set of elements arranged as a sequence in printing order.

But in this Chapter it is preferable to introduce an alternative way of expressing the sequence relation, which follows the pattern of the $\mathscr{S}|$ and $\mathscr{B}|$ relations.

So far we have defined

ordering by separation: $\mathscr{S}|$ABC...MN (by prescribing a set {AB, BC, ..., MN, NA} of adjacent pairs),
ordering by betweenness: $\mathscr{B}|$ABC...MN (by prescribing a set of adjacent pairs {AB, BC, ..., MN} and a first-and-last pair AN).

To these we now add first:

1.8*D*1 *Conformably ordered pairs: the relation* ∂.

*G*1 $\mathscr{L}\wp|$ABPQ & $\#|$AB & $\#|$PQ
S O : $\ulcorner$O(A) $\supset$ {B, P, Q}
*G*2 $\mathscr{B}|$OAB & $\mathscr{B}|$OPQ
D The pairs {A, B}, {P, Q} are *conformably ordered* or in the *same sense*.

1.8*N*1 (i) $\partial(\text{AB/PQ}) = \partial(\text{PQ/AB}) = \text{I}$,
(ii) $\partial(\text{AB/PQ}) = \text{I} \Leftrightarrow \partial(\text{AB/QP}) = \text{J}$.

(After the symbol '∂' the names of the points are to be read in printed order.) Returning to the description at the beginning of the Chapter we see that $\partial(\text{AB/PQ}) = \text{I}$ or J according as the betweenness order of {A, B, P, Q} belongs to the set Σ_1 of orderings or to the set Σ_2.

We prove next the theorem which determines the rule of composition for the elements I and J, namely,

1.8*T*1 Rule for compounding '∂' relations.

G	$\mathscr{L}\wp\|$ABCDEF & $\#\|$AB & $\#\|$CD & $\#\|$EF
S	O : ⌜O(A) ⊃ {B, C, D, E, F} & $\mathscr{B}\|$OAB
	*t*1 ∂(AB/CD) = I & ∂(CD/EF) = I
	⇒ $\mathscr{B}\|$OCD & $\mathscr{B}\|$OEF
	⇒ ∂(AB/EF) = I
	*t*2 ∂(AB/CD) = I & ∂(CD/EF) = J
	⇒ $\mathscr{B}\|$OCD & $\mathscr{B}\|$OFE
	⇒ ∂(AB/EF) = J
	*t*3 ∂(AB/CD) = J & ∂(CD/EF) = I
	⇒ ∂(AB/EF) = J
	*t*4 ∂(AB/CD) = J & ∂(CD/EF) = J
	⇒ ∂(AB/EF) = I
T	∂(AB/CD) ∂(CD/EF) = ∂(AB/EF)
	⇔ {II = I, IJ = JI = J, JJ = I}.

It is easily verified that on this basis the rule of composition is associative (e.g. (IJ)J = JJ = I = II = I(JJ)) so that, as was to be expected, I, J are the elements of the two-element group in which I is the unit and $J^2 = I$.

We next introduce the notation for a sequence in relation to a set of three markers in an assigned betweenness relation; in practice the markers will not necessarily be specified.

1.8*N*2 *Conformable ordering of a set. Sequence*

*G*1	$\mathscr{L}\wp\|$ABPQR... & $\#\|$AB & $\#\|$PQR...
S	O : ⌜O(A) ⊃ {B, P, Q, R, ...} & $\mathscr{B}\|$OAB
*G*2	$\mathscr{B}\|$OPQR...
N (i)	{P, Q, R, ...} are *conformably ordered* with {A, B}.
(ii)	$\mathscr{O}\|$PQR ... or {$^{\mathscr{O}}$P, Q, R, ...} : P, Q, R, ... form a *sequence* ordered in relation to the ordered pair $\mathscr{O}\|$AB.
(iii)	$\mathscr{O}\|$AB & $\mathscr{O}\|$PQR ... : {A, B} and {P, Q, R, ...} are (ordered by betweenness) in the *same sense.*

It may be noted that $\mathscr{O}|$AB & $\mathscr{B}|$PQR ⇔ ∂(AB/PQ) ∂(AB/QR) = I.

Before we can apply a relation of 'sense' to the plane, we have to devise a relation of *conformable cyclic ordering* (by betweenness). This is a relation between two sets of three elements, all six ordered by betweenness in relation to a selected marker. The definition is similar to *D*1.

1.8*D*2 *Conformable cyclic ordering.*

*G*1 $\mathscr{L}\wp$|LMNPQR & # |LMN & # |PQR
*G*2 $\mathscr{B}$|LMN
S O : ⌜O(L) ⊃ {M, N, P, Q, R} & $\mathscr{B}$|OLMN
D {L, M, N} and {P, Q, R) are conformably cyclically ordered if either $\mathscr{B}$|OPQR or $\mathscr{B}$|ORPQ or $\mathscr{B}$|OQRP.

The two types of conformable ordering are linked by the theorem:

1.8*T*2 Conformable ordering and conformable cyclic ordering.

*G*1 $\mathscr{L}\wp$|LMNPQR & $\mathscr{B}$|LMN & # |PQR
T ∂(LM/PQ) ∂(LN/PR) ∂(MN/QR) = I
⇔ {L, M, N} and {P, Q, R} are conformably cyclically ordered
P *s* O : ⌜O(L) ⊃ {M, N, P, Q, R} & $\mathscr{B}$|OLMN
p (table of corresponding values).

∂(LM/PQ) =	∂(LN/PR) =	∂(MN/QR) =	⇒
I	I	I	$\mathscr{B}$|OPQR
I	J	J	$\mathscr{B}$|ORPQ
J	I	J	inconsistent
J	J	I	$\mathscr{B}$|OQRP

Against this background we can attack the problem of sense in the plane. From physical observation we see that we require the relation of sense between two sets of points to have the following properties:

(a) *G* $\mathscr{L}\wp$|ABC.
The sequences {$^{\mathscr{O}}$A, B, C} and {$^{\mathscr{O}}$B, C, A} have the same sense, {$^{\mathscr{O}}$A, B, C} and {$^{\mathscr{O}}$A, C, B} have opposite senses.

(b) *G* #$\wp$|ABB′C & $\mathscr{L}$|ABB′ & $\not\mathscr{L}$|ABC.
(i) A$\not\mathscr{B}$BB′ ⇒ {$^{\mathscr{O}}$A, B, C} and {$^{\mathscr{O}}$A, B′, C} have the same sense.
(ii) A$\mathscr{B}$BB′ ⇒ {$^{\mathscr{O}}$A, B, C} and {$^{\mathscr{O}}$A, B′, C} have opposite senses.

(c) *G* #$\wp$|ABCC′ & C∉AB & C′∉AB.
(i) C′∈⌜AB(C) ⇒ {$^{\mathscr{O}}$A, B, C} and {$^{\mathscr{O}}$A, B, C′} have the same sense.
(ii) C′∈⌜AB($\tilde{\text{C}}$) ⇒ {$^{\mathscr{O}}$A, B, C} and {$^{\mathscr{O}}$A, B, C′} have opposite senses.

(d) The relation is consistent, i.e., if two sets of three points each have the same sense as a third set, then they have the same sense as each other.

1.8*N*3 Notation for relative sense of two triads.

G $\wp|\mathrm{ABCDEF}$ & $\mathscr{L}|\mathrm{ABC}$ & $\mathscr{L}|\mathrm{DEF}$

N $\partial(\mathrm{ABC/DEF}) = \begin{cases}\mathrm{I}\\ \mathrm{J}\end{cases} : \{^c\mathrm{A, B, C}\}$ and

$\{^c\mathrm{D, E, F}\}$ have $\begin{cases}\text{the same sense.}\\ \text{opposite senses.}\end{cases}$

1.8*D*3 Definition of $\partial(\mathrm{ABC/DEF})$ in terms of cyclically ordered sets of three points on a line.

G $\mathscr{L}|\mathrm{ABC}$ & $\mathscr{L}|\mathrm{DEF}$

S $\mathrm{g} : \ulcorner\mathrm{g(A)} \supset \{\mathrm{B, C, D, E, F}\}$ (i.e. g is such that all six points lie on the same side of it)

C $\{^c\mathrm{L, M, N, P, Q, R}\} = \mathrm{g} \cap \{^c\mathrm{BC, CA, AB, EF, FD, DE}\}$

D $\partial(\mathrm{ABC/DEF}) = \partial(\mathrm{LM/PQ})\,\partial(\mathrm{LN/PR})\,\partial(\mathrm{MN/QR})$ (i.e. $\partial(\mathrm{ABC/DEF}) = \mathrm{I}$ or J according as $\{^c\mathrm{L, M, N}\}$ and $\{^c\mathrm{P, Q, R}\}$ are conformably or non-conformably cyclically ordered).

This definition is framed explicitly in terms of the line g selected and we have to show that it is in fact independent of the choice of g, and then to show that it leads to properties (*a*), (*b*), (*c*), (*d*) listed above. We need first an adaptation of 1.7*T* 16, namely,

1.8*T*3 *G*1 : *G*, *S*, *C*1, *S*2, *S*3, *C*2 as in 1.7*T* 16

*G*2 $\mathscr{B}|\mathrm{GLMN}$

T either $\mathscr{B}|\mathrm{GL'M'N'}$ or $\mathscr{B}|\mathrm{N'GL'M'}$ or $\mathscr{B}|\mathrm{M'N'GL'}$ or $\mathscr{B}|\mathrm{L'M'N'G}$

P $\mathscr{S}|\mathrm{GL'M'N'}$ (1.7*T* 16).

Next we prove that having selected g and $\mathrm{G} \in \mathrm{g}$, such that $\{\mathrm{L, M, N, P, Q, R}\}$ are all on the same side of G, we obtain the same value for $\partial(\mathrm{L'M'/P'Q'})\,\partial(\mathrm{L'N'/P'R'})\,\partial(\mathrm{M'N'/Q'R'})$ for the points on any line g′ through G and such that $\{\mathrm{L', M', N', P', Q', R'}\}$ are in the same half-plane in relation to it. Explicitly:

1.8*T*4 *G*1 $\{\mathrm{A}, \ldots, \mathrm{F}\}$, g, $\{\mathrm{L}, \ldots, \mathrm{R}\}$ as in 1.8*D*3

*S*1 $\mathrm{G} : \mathrm{G} \in \mathrm{g}$ & $\ulcorner\mathrm{G(L)} \supset \{\mathrm{M, N, P, Q, R}\}$

*G*2 $\mathscr{B}|\mathrm{GLMN}$ & $\mathscr{B}|\mathrm{GPQR}$
(so that $\partial(\mathrm{LM/PQ})\,\partial(\mathrm{LN/PR})\,\partial(\mathrm{MN/QR}) = \mathrm{I}$)

*S*2 $\mathrm{g'} : \mathrm{g'} \supset \mathrm{G}$ & $\ulcorner\mathrm{g'(A)} \supset \{\mathrm{B, C, D, E, F}\}$

*C*2 $\{^c\mathrm{L}, \ldots, \mathrm{R'}\} = \mathrm{g'} \cap \{^c\mathrm{BC}, \ldots, \mathrm{DE}\}$

T $\partial(\mathrm{L'M'/P'Q'})\,\partial(\mathrm{L'N'/P'R'})\,\partial(\mathrm{M'N'/Q'R'}) = \mathrm{I}$

$$P\quad t1\quad \left.\begin{array}{ll}\text{either} & \mathscr{B}|\mathrm{GL'M'N'}\\ \text{or} & \mathscr{B}|\mathrm{N'GL'M'}\\ \text{or} & \mathscr{B}|\mathrm{M'N'GL'}\\ \text{or} & \mathscr{B}|\mathrm{L'M'N'G}\end{array}\right\}\ \&\ \left.\begin{array}{ll}\text{either} & \mathscr{B}|\mathrm{GP'Q'R'}\\ \text{or} & \mathscr{B}|\mathrm{R'GP'Q'}\\ \text{or} & \mathscr{B}|\mathrm{Q'R'GP'}\\ \text{or} & \mathscr{B}|\mathrm{P'Q'R'G}\end{array}\right\}$$

$$s\quad \mathrm{H}' : \ulcorner\mathrm{H}'(\mathrm{G}) \supset \{\mathrm{L}', \ldots, \mathrm{R}'\}$$

$$t2\quad \left.\begin{array}{ll}\text{either} & \mathscr{B}|\mathrm{H'L'M'N'}\\ \text{or} & \mathscr{B}|\mathrm{H'N'L'M'}\\ \text{or} & \mathscr{B}|\mathrm{H'M'N'L'}\end{array}\right\}\ \&\ \left\{\begin{array}{ll}\text{either} & \mathscr{B}|\mathrm{H'P'Q'R'}\\ \text{or} & \mathscr{B}|\mathrm{H'R'P'Q'}\\ \text{or} & \mathscr{B}|\mathrm{H'Q'R'P'}\end{array}\right.$$

$t3$ $\{^{c}\mathrm{L}', \mathrm{M}', \mathrm{N}\}$ and $\{^{c}\mathrm{P}', \mathrm{Q}', \mathrm{R}'\}$ are conformably cyclically ordered

$$T\quad \partial(\mathrm{L'M'/P'Q'})\,\partial(\mathrm{L'N'/P'R'})\,\partial(\mathrm{M'N'/Q'R'}) = \mathrm{I}.$$

The final theorem is a simple consequence of this theorem, which covers all lines g′ which put {A, ..., F} in the same half-plane and are such that $\mathrm{g} \cap \mathrm{g}'$ puts {L, ..., R} in the same half-line on g; given any line h which puts {A, ..., F} in the same half-plane we can link it with g by suitably choosing g′.

1.8*T*5 The definition 1.8*D*3 is independent of the choice of g.

The properties (*a*), (*b*), (*c*), (*d*) listed above, which are required of the sense relation, are stated in the following theorems:

1.8*T*6*X*

(*a*) $\partial(\mathrm{ABC/ABC}) = \mathrm{I}$, $\partial(\mathrm{ABC/ACB}) = \mathrm{J}$, etc.

$\partial(\mathrm{ABC/DEF}) = \partial(\mathrm{DEF/ABC}) = \mathrm{J}\,\partial(\mathrm{ABC/DFE})$ etc.

(*b*) $G\quad \#\wp|\mathrm{ABB'C}\ \&\ \mathscr{L}|\mathrm{ABB'}\ \&\ \mathscr{L}/\mathrm{ABC}$

$T\quad \partial(\mathrm{ABC/AB'C}) = \partial(\mathrm{AB/AB'})$

(*c*) $G\quad \#\wp|\mathrm{ABCC'}\ \&\ \mathscr{L}|\mathrm{ABC}\ \&\ \mathscr{L}|\mathrm{ABC'}$

$$T\quad \partial(\mathrm{ABC/ABC'})\ \left\{\begin{array}{l} = \mathrm{I} \Leftrightarrow \mathrm{C}' \in \ulcorner\mathrm{AB(C)}\\ = \mathrm{J} \Leftrightarrow \mathrm{C}' \in \ulcorner\mathrm{AB}(\tilde{\mathrm{C}})\end{array}\right.$$

(*d*) $\partial(\mathrm{ABC/DEF})\,\partial(\mathrm{DEF/GHK}) = \partial(\mathrm{ABC/GHK})$.

1.8*T*7*X* A test which distinguishes convex quadrilaterals from non convex quadrilaterals.

$G1\quad \mathscr{L}_3\wp|\mathrm{ABCD}$

$G2\quad \partial(\mathrm{ABC/ABD})\,\partial(\mathrm{ABC/ADC})\,\partial(\mathrm{ABC/DBC})$

$$= \left\{\begin{array}{l}\mathrm{I}\\ \mathrm{J}\end{array}\right. \Leftrightarrow \text{the quadrilateral ABCD is} \left\{\begin{array}{l}\text{non-convex.}\\ \text{convex}\end{array}\right.$$

EXCURSUS ON BOOK 1

EXCURSUS 1.1

SOME FINITE GEOMETRIES

A geometrical system which satisfies the four conditions of Chapter 1.1 or the axioms $A\mathscr{I}$ 1 and 2 of Chapter 1.2 needs to have no apparent relation to the physical system which was used as a basis for discovering what was required of a set of axioms. The axioms make no reference to drawing lines and the physical interpretations of the axioms may differ very widely. The selection of interpretations given here has among other objectives that of warning against too readily drawing dots for points, and straightish strokes for lines.

In the following finite sets of symbols, the symbols themselves are the points, and the designated subsets of symbols are the collinear sets. There is no need to write 'let A, B, C, ... stand for the points'; the points are the visible marks A, B, C, ... and the collinear sets are the visible collections of marks A, B, C, ...

Exc. 1.1.1 The geometry $\Gamma^{(2)}$

In this section we construct explicitly a geometry which has three points in a collinear set, and therefore altogether seven points and seven collinear sets. Let v = {A, B, C} be a collinear set, V a point not in v, and s a collinear set which contains A but not V. The third points of {V, B, ...} and {V, C, ...} must be on s. Let them be B′ and C′ respectively. Then, if a geometry with three points on a line exists, six of its points may be taken to be V, A, B, C, B′, C′ and the seventh point, A′ say, must be the third point in each of the sets {B′, C, ...}, {B, C′, ...} and {V, A, ...}. That is, if the geometry exists, then the collinear sets can be exhibited as the columns in the table:

V	A	B	A′	C	C′	B′
A	B	A′	C	C′	B′	V
A′	C	C′	B′	V	A	B

The pattern is displayed more clearly if we use numerals instead of letters, thus

0	1	2	3	4	5	6
1	2	3	4	5	6	0
3	4	5	6	0	1	2

It is now easily verified that each pair of columns (subsets) has exactly one common member, and each pair of elements belongs to exactly one subset.

This system may be used also to exhibit duality by the simple expedient of naming the collinear sets 0, 1, ..., 6 in such a way that each set of three collinear sets with a common member corresponds to a set of three collinear points. Thus, we could name as

	0	1	2	3	4	5	6
the collinear sets	045	516	602	253	364	421	130;

a more striking arrangement is that in which points and collinear sets are named in exactly the same way, as, e.g.

0	1	2	3	4	5	6
405	516	643	352	260	031	124

Then, the three points 0, 1, 3 form the collinear set 5, and the three collinear sets 0, 1, 3 have the common member 5, and, generally, the three points x, y, z form the collinear set s, while the three collinear sets x, y, z have the common member s.

The simplest way of numbering the collinear sets is that in which the collinear set r consists of the points r, r + 1, r + 3 (reduced modulo 7).

Exc. 1.1.2 The geometry $\Gamma^{(3)}$

The smallest geometry (in number of points) in which we can fully test the axioms 1 and 2 is one with at least four points in a collinear set, since we need this number for a proper testing of $A\mathscr{I}1$ (iii). If therefore we can construct a geometry with four points on a line and altogether 13 points and 13 collinear sets, then we can assert that the axioms are consistent and that none of them is nugatory.

Proceeding as in the previous case we take a collinear set v = {A, B, C, D}, a point V not in v, and a collinear set s containing A but distinct from v and {A, V, ...}. Denote the collinear sets {V, A, ...}, {V, B, ...}, {V, C, ...}, {V, D, ...} by a, b, c, d, and name s ∩ b, s ∩ c, s ∩ d respectively as B′, C′, D′, and further points thus

$$\{D', B, \ldots\} \cap \{a, c\} = \{A'', C''\},$$
$$\{D', C, \ldots\} \cap \{a, b\} = \{A', B''\}.$$

We have now named 12 of the points; the thirteenth, D″, is the fourth point of {V, D, ...}. The fourth collinear sets that contain respectively B and C must be {B, A′, C′, D″} and {C, A″, B′, D″}. There remain only three collinear sets to be named, namely, the third set containing D″, and the other two containing D, which contain between them the six points A′, B′, C′, A″, B″, C″. {D, A′, ...} cannot contain B″ (since

{A′, B″, ...} contains D′) and therefore contains B′, and it cannot contain C′, and therefore contains C″. Thus the remaining three collinear sets are {A, B″, C″, D″}, {A′, B′, C″, D} and {A″, B″, C′, D}.

The thirteen collinear sets can be arranged to form a table in the same pattern as that for $\Gamma^{(2)}$, thus:

A	B	A″	B″	C	C″	D	D′	C′	A′	V	B′	D″
B	A″	B″	C	C″	D	D′	C′	A′	V	B′	D″	A
C	C″	D	D′	C′	A′	V	B′	D″	A	B	A″	B″
D	D′	C′	A′	V	B′	D″	A	B	A″	B″	C	C″

where the collinear sets are written as columns. The thirteen-point geometry $\Gamma^{(3)}$ is therefore realizable, and the axioms proposed are consistent.

Again the pattern is clearer if we write numerals (with u, v, w for 10, 11, 12) instead of letters, thus

0	1	2	3	4	5	6	7	8	9	u	v	w
1	2	3	4	5	6	7	8	9	u	v	w	0
4	5	6	7	8	9	u	v	w	0	1	2	3
6	7	8	9	u	v	w	0	1	2	3	4	5

Exercise 1 Name these thirteen sets as 0, 1, ..., w in such a way as to exhibit the duality of the system.

Exc. 1.1.3 A general pattern: difference tables

Let us now look at the numerical patterns for $\Gamma^{(2)}$ and $\Gamma^{(3)}$ to discover the structure that underlies them. The columns for $\Gamma^{(2)}$, $\Gamma^{(3)}$ are respectively of the forms

$$\begin{array}{ll} r & \text{reduced modulo 7} \\ r+1 & \\ r+3 & \end{array}$$

and

$$\begin{array}{ll} r & \text{reduced modulo 13} \\ r+1 & \\ r+4 & \\ r+6 & \end{array}$$

That is, for $\Gamma^{(2)}$, the point collinear with

r and r + 1	is r + 3
r and r + 2	is r + 6
r and r + 3	is r + 1
r and r + 4	is r + 5
r and r + 5	is r + 4
r and r + 6	is r + 2

We may express this in the form of a 'difference' table, in which the entries are $x-y$ reduced modulo 7.

$x-y$ (mod. 7)

y \ x	0	1	3
0	.	1	3
1	6	.	2
3	4	5	.

The essential feature of this table is that the six differences modulo 7 of the three chosen numbers, 0, 1, 3, are the six numbers 1, 2, ..., 6.

In the same way, the thirteen-entry table can be built on the four numbers, 0, 1, 4, 6, since these generate the difference table

$x-y$ (mod. 13)

y \ x	0	1	4	6
0	.	1	4	6
1	12	.	3	5
4	9	10	.	2
6	7	8	11	.

The points collinear with two given points, r, $r+d$ can be found from the following table of collinear sets

r	r	r	r
$r+1$	$r+3$	$r+2$	$r+7$
$r+4$	$r+5$	$r+9$	$r+8$
$r+6$	$r+12$	$r+10$	$r+11$

This method is clearly of general application, so that, provided that for a given value m^2+m+1 we can construct a difference table, we can construct a geometry with $m+1$ points on the line. In Excursus 1.1.4 we find a difference table for $m = 4$, namely, that of which the first line is

0 1 6 8 18

To provide the reader with examples of finite geometries in which he may experiment, we list the first lines of the difference tables for values of m up to 9 that are to be constructed later:

$m = 5$, 31 points (Excursus 2.3.4)

0 1 11 19 26 28

$m = 6$, no geometry exists

$m = 7$, 57 points (Excursus 2.3.5)

0 1 5 27 34 37 43 45

m = 8, 73 points (Excursus 3.2.2)

0 1 5 12 18 21 49 51 59

m = 9, 91 points (Excursus 3.2.3)

0 1 27 33 49 63 72 74 84 87

These are of course in each case only one among many forms, the result of the selection of a particular method of generating the geometry. Geometries which can be specified by a difference-table are known as 'Cyclic Geometries'. If we can form a difference table, then we can apply it to generate a geometry, but if we cannot find a difference table of order m^2+m+1, it does not follow that we cannot find a geometry with $m+1$ points in a collinear set.

Exc. 1.1.4 The geometry $\Gamma^{(4)}$

This geometry is of considerable importance in the later development of the subject and has an intrinsic interest at this stage. It may be constructed by a method similar to that used for $\Gamma^{(3)}$. Take a collinear set v = {A, B, C, D, U}, a point V not belonging to this set, and a collinear set w, which contains A and whose intersections with {V, B, ...}, {V, C, ...}, {V, D, ...}, {V, U, ...} are B_1, C_1, D_1, W respectively. Then construct the other points A_1, B_1, C_1, D_1 as in the diagram at the end of the book (p. 329).

We have now 19 points, the remaining two, J and J′ say, being required to complete the set {U, V, W, J, J′}, and 10 collinear sets. The remaining collinear sets consist of four pairs, each pair meeting in one of the points A, B, C, D, and three containing U. The symmetry of the arrangement is improved by renaming A, B, C, D as A_1, B_2, C_3, D_4. When this is done, and specific points are named as J and J′, it can be verified that the only possible choice is that in the following table:

	A_i	B_j	C_k	D_l	
Subscripts ...	1	3	4	2	J
	1	4	2	3	J′
	4	2	1	3	J
	3	2	4	1	J′
	2	4	3	1	J
	4	1	3	2	J′
	3	1	2	4	J
	2	3	1	4	J′
	2	1	4	3	U
	3	4	1	2	U
	4	3	2	1	U

It may be noted as evidence of almost inevitable pattern that the arrangements above are the eleven even permutations of 1 2 3 4.

Exercise 1 Prove that when points have been named as in the diagram the other two collinear sets containing A can only be $\{A_1, B_3, C_4, D_2\}$, and $\{A_1, B_4, C_2, D_3\}$.

Exercise 2 Prove that A_1, A_2, B_2, C_3, W, D_1, A_3 by themselves form a 7-point geometry $\Gamma^{(2)}$, and find another such set.

One of the difference tables for numbers modulo 21 is generated by 0, 1, 6, 8, 18:

x − y (mod. 21)

y \ x	0	1	6	8	18
0	.	1	6	8	18
1	20	.	5	7	17
6	15	16	.	2	12
8	13	14	19	.	10
18	3	4	9	11	.

The relation of this table to the points $A_1, A_2, \ldots,$ W is set out at the foot of the diagram on p. 329.

Exercise 3 Show that each of the three sets of points, $\{3r\}, \{3r+1\}, \{3r+2\}$, $r = 0, \ldots, 6$, forms a geometry $\Gamma^{(2)}$.

Exercise 4 A geometry $\Gamma^{(2)}$ is given; its points are named 0, 3, 6, 9, 12, 15, 18 in such a way that the collinear sets are $\{3r, 3r+3, 3r+9\}$. A point named 4 is added to the collinear set $\{0, 3, 9\}$. It is joined to the point 6 and the intersections of $\langle 4, 6\rangle$ with the lines of $\Gamma^{(2)}$ other than those through 4 and 6 are named in accordance with the difference-table for $\Gamma^{(4)}$. By joining 4 to 12, 15 and 18, and specifying collinear sets, show that $\Gamma^{(4)}$ can be constructed from $\Gamma^{(2)}$ and the one additional point.

Exc. 1.1.5 $\Gamma^{(2)}$: another approach

Although to do so is to anticipate the introduction of algebra into the geometry, there is some advantage to be gained by building up the simple finite geometries on an algebraic basis by assigning specific vectors to the points and particular rules of arithmetic to the components of the vectors. For example, take the algebraic system of 'evens and odds' in which the only elements are 0 and 1 (namely, the remainders on division by 2) which combine according to the addition table

+	0	1
0	0	1
1	1	0

Consider the set Φ of seven vectors (α, β, γ) where $(\alpha, \beta, \gamma) \in \{0, 1\}$ and not all three of α, β, γ are 0. The result of adding two different members of Φ is to produce a third member of Φ, so that we may investigate whether the vectors may be used to represent the points of $\Gamma^{(2)}$. The relation

$$(\alpha, \beta, \gamma)+(\alpha', \beta', \gamma') = (\alpha'', \beta'', \gamma'')$$

is equivalent to

$$\alpha+\alpha'+\alpha'' = 0$$

$$\beta+\beta'+\beta'' = 0$$

$$\gamma+\gamma'+\gamma'' = 0$$

and may be replaced by a relation

$$\mathrm{lx}+\mathrm{my}+\mathrm{nz} = 0$$

among the components of each of the vectors.

For example,

$$(1, 0, 0)+(1, 1, 1) = (0, 1, 1),$$

i.e.

$$(1, 0, 0)+(1, 1, 1)+(0, 1, 1) = (0, 0, 0),$$

may be replaced by

$$0.\mathrm{x}+1.\mathrm{y}+1.\mathrm{z} \equiv \mathrm{y}+\mathrm{z} = 0.$$

There are seven of these sets of three, and the relations corresponding to them are:

(0, 1, 0),	(0, 0, 1),	(0, 1, 1)	$\mathrm{x} = 0$
(1, 0, 0),	(0, 0, 1),	(1, 0, 1)	$\mathrm{y} = 0$
(1, 0, 0),	(0, 1, 0),	(1, 1, 0)	$\mathrm{z} = 0$
(1, 0, 0),	(0, 1, 1),	(1, 1, 1)	$\mathrm{y}+\mathrm{z} = 0$
(0, 1, 0),	(1, 0, 1),	(1, 1, 1)	$\mathrm{x}+\mathrm{z} = 0$
(0, 0, 1),	(1, 1, 0),	(1, 1, 1)	$\mathrm{x}+\mathrm{y} = 0$
(0, 1, 1),	(1, 0, 1),	(1, 1, 0)	$\mathrm{x}+\mathrm{y}+\mathrm{z} = 0$

We can establish then a complete correspondence between the points of $\Gamma^{(2)}$ and the vectors of Φ, and between the collinear sets of $\Gamma^{(2)}$ and the linear equations in Φ. One such correspondence is

points	0	1	2	3	4	5	6
vectors	(1, 0, 0)	(1, 1, 0)	(0, 1, 1)	(0, 1, 0)	(1, 0, 1)	(0, 0, 1)	(1, 1, 1)

We may use the vectors also to exhibit the duality property by defining the one-to-one correspondence to be that in which the point (α, β, γ) corresponds to the collinear set

$$\alpha\mathrm{x}+\beta\mathrm{y}+\gamma\mathrm{z} = 0.$$

With the system we have used, this gives

Point	Vector for point	Condition satisfied by collinear set	Collinear set
0	(1, 0, 0)	x = 0	2, 3, 5
1	(1, 1, 0)	x+y = 0	5, 1, 6
2	(0, 1, 1)	y+z = 0	6, 2, 0
3	(0, 1, 0)	y = 0	0, 5, 4
4	(1, 0, 1)	x+z = 0	4, 6, 3
5	(0, 0, 1)	z = 0	3, 0, 1
6	(1, 1, 1)	x+y+z = 0	1, 4, 2

Exc. 1.1.6 An algebraic formulation for $T^{(3)}$

We may extend the idea introduced in the previous section, by taking remainders modulo 3 as the basis of the algebra. For the elements of the vectors we use 1, 0, -1 and have two combination tables corresponding to addition and its inverse, namely:

x+y	y \ x	1	0	−1
	1	−1	1	0
	0	1	0	−1
	−1	0	−1	1

x−y	y \ x	1	0	−1
	1	0	−1	1
	0	1	0	−1
	−1	−1	1	0

from which we deduce $\alpha-(1)=\alpha+(-1)$ and $\alpha-(-1)=\alpha+1$.

Omitting the vector (0, 0, 0) there are 26 vectors (α, β, γ) from which 13 have to be selected to represent the points of $\Gamma^{(3)}$. One way to do this is to prescribe that the last non-zero element in the vector (α, β, γ) shall be 1 (and not -1). If, after an algebraic operation the last non-zero element in (α, β, γ) is -1, then this vector is replaced by $(-\alpha, -\beta, -\gamma)$.

From any two vectors we may derive two others by the operations + and −, namely, $(\alpha,\beta,\gamma)\pm(\alpha',\beta',\gamma')$, and the 13 sets of four so obtainable correspond to the collinear sets. E.g.

$$(-1, 0, 1)+(0, -1, 1) = (-1, -1, -1) = -(1, 1, 1)$$
$$(-1, 0, 1)-(0, -1, 1) = (-1, 1, 0)$$

and the components of the four vectors satisfy the relation

$$x+y+z = 0.$$

Exercise 1 Make a list of the 13 vectors and of the 13 equations satisfied by the components together with the four vectors satisfying the equation.

Establish a correspondence between these vectors and the points of the geometry discussed in Excursus 1.1.2.

This process of building up a geometry from vectors whose elements belong to some algebraic field can clearly be extended indefinitely; in the cases above the fields are those of integers reduced modulo 2 and 3 respectively. But the reader is warned that the process is not a simple one. There are, for example, obvious difficulties in attempting to construct in this way a geometry based on integers reduced modulo 4, since 4 is not a prime, and in consequence there is no simple representation of $\Gamma^{(4)}$ by 'vectors'. But if p is a prime number we may construct a geometry $\Gamma^{(p)}$ directly analogous to $\Gamma^{(2)}$ and $\Gamma^{(3)}$.

The points are the 'vectors': $(\alpha, \beta, \gamma) : (\alpha, \beta, \gamma) \in \{0, 1, \ldots, p-1\}$, at least one component being non-zero, and the last non-zero component being 1.

The line determined by (α, β, γ), $(\alpha', \beta', \gamma')$ contains the set of $p+1$ points $(\lambda\alpha + \lambda'\alpha', \lambda\beta + \lambda'\beta', \lambda\gamma + \lambda'\gamma')$, where λ' is determined from λ by the condition that the last non-zero component shall be 1. That is, if $\gamma' \neq 0$, p of points are given by

$$\lambda' = (1 - \lambda\gamma)/\gamma',$$

and the remaining point is

$$[(\gamma'\alpha - \gamma\alpha')/(\gamma'\beta - \gamma\beta'), 1, 0] \quad \text{if} \quad \gamma'\beta - \gamma\beta' \neq 0,$$

and is (1, 0, 0) otherwise. The case $\gamma' = 0$ may be analysed similarly.

Exercise 2 Prove that every pair of lines in this geometry has a single common point.

EXCURSUS 1.2

A GEOMETRY IN WHICH COLLINEAR SETS ARE CIRCLES

For this geometry we need to know something of the geometrical properties of Euclidean circles, and this must raise some doubts as to the propriety of discussing such figures before any property by which a circle can be defined has been introduced. But these geometries are to be regarded merely as illustrations of interpretations of the axioms 1, 2 of which the diagrammatic representation is quite different from that of the Euclidean geometry (e-geometry) based on the relations of 'objects in a paddock' with which the main text is concerned. Properties of these geometries can be logically established after circles have been defined, but for the present purpose (which is essentially only the drawing of diagrams) it is sufficient to accept our intuitively based ideas of what circles look like.

We need effectively only two properties of (real) circles:

(i) Two circles in e-geometry which have one common point, P say, have a second common point, say Q. If Q 'coincides' with P (a phrase which can be made precise in a logically arranged context), then the two circles 'touch' at P. A circle and a line which are specified so as to have one common point, P, have a second common point, Q. If Q 'coincides' with P then the line is the 'tangent' to the circle at P.

(ii) There is a single circle containing any three non-collinear proper points. If the points are collinear the circle is replaced by a line.

We have also to anticipate Chapter 1.5 and adjoin to the e-geometry the collinear set of 'ideal points' (or, colloquially, 'points at infinity') by which parallelism is defined.

On this basis we can relate a *c-geometry* to *e-geometry* in which 'e-lines' and 'e-circles' are the ordinary lines and circles of Euclidean geometry. With the exceptions to be noted, every point of the e-geometry is a point of the c-geometry, but the sub-sets of points in the c-geometry which are described as 'c-collinear' form e-circles in the e-geometry.

The correspondence between the two systems may be tabulated thus:

c-geometry		*e-geometry*
General c-point	↔	general e-point
one special point J	→	no e-point

no c-point	$\leftarrow$	one special point Ω
point J	$\leftarrow$	any point of the ideal line
general c-collinear set, $c_2 = \{P, Q, \ldots\}$	$\leftrightarrow$	e-circle through Ω, P, Q, …
special c-collinear set, $c_1 = \{P' Q', \ldots\}$	$\leftrightarrow$	e-line P′ Q′ through Ω
any point of one special c-collinear set, ω	$\rightarrow$	the single e-point Ω
intersection $c_2 \cap c_2$	$\leftrightarrow$	residual intersection of two circles through Ω
intersection $c_2 \cap c_1$	$\leftrightarrow$	residual intersection of e-circle and e-line through Ω
intersection $c_1 \cap c_1 = J$	$\leftrightarrow$	no single corresponding point (J corresponds to every point of the ideal line).
intersection $c_2 \cap \omega$ or $c_1 \cap \omega$	$\rightarrow$	e-point Ω
$c_2 \cap c_2$ lies on ω	$\leftrightarrow$	two e-circles touch at Ω
$c_2 \cap c_1$ lies on ω	$\leftrightarrow$	e-circle touches e-line at Ω.

The whole correspondence may be summarised by using Cartesian equations.

c-point (X, Y)	$\leftrightarrow$	e-point (x, y)
$X = \dfrac{x}{x^2+y^2}, \quad Y = \dfrac{y}{x^2+y^2}$		$x = \dfrac{X}{X^2+Y^2}, \quad y = \dfrac{Y}{X^2+Y^2}$
origin, J	$\rightarrow$	all points of the ideal line
all points of c-line ω	$\leftarrow$	origin Ω
c-collinear set $\alpha x + \beta y + \gamma = 0$	$\leftrightarrow$	e-locus $\gamma(x^2+y^2) + \alpha x + \beta y = 0$
c_2-set: $\gamma \neq 0$, not both $\alpha = 0$ and $\beta = 0$	$\rightarrow$	e-circle through Ω
c_1-set $\gamma = 0$, not both $\alpha = 0$ and $\beta = 0$	$\rightarrow$	e-line through Ω
set ω $\alpha = 0, \beta = 0, \gamma \neq 0$	$\leftarrow$	point Ω.

We may note also the following systems which conform to the axioms $A\mathscr{I}$ 1, 2.

s-geometry

s-points. h-points of Euclidean hemisphere: each pair of diametrically opposite h-points on the rim of the hemisphere correspond to the same s-point.

s-lines. h-semi-great-circles (semicircles cut by planes through the centre of the base of the hemisphere); also the rim with each pair of diametrically opposite points treated as the same point.

Except for the h-points on the rim all points in this representation of the geometry are equivalent in the e-geometry (on the hemisphere).

t-geometry

Any t-system (transformed system) whose elements are in continuous one-to-one correspondence with a geometry whose elements satisfy conditions 1 and 2 is itself a geometry whose elements satisfy these conditions.

Thus, for example, we may derive from the s-geometry, by orthogonal projection on to the base of the hemisphere, a geometry in the Euclidean plane in which the lines are the e-semi-ellipses whose major axes are diameters of a fixed e-circle (together with the circle itself with pairs of diametrically opposite points identified).

EXCURSUS 1.3

TWO 91-POINT GEOMETRIES

(A finite non-Desarguesian geometry)

The two geometries to be described in this Excursus are both derived from $\Gamma^{(3)}$ in the same way as $\Gamma^{(4)}$ was shown in Excursus 1.1.4, Exercise 4, to be derived from $\Gamma^{(2)}$. But while the pattern of collinearities in $\Gamma^{(4)}$ was shown to be essentially unique, we show in this excursus that we can construct two geometries, $\Gamma^{(9)}$ and $\Omega^{(9)}$, both based on $\Gamma^{(3)}$ and both containing 91 points, yet differing completely in structure.

The basic geometry $\Gamma^{(3)}$, say [K], consists of 13 points $K_0, \ldots, K_{12}$, and 13 collinear sets $k_0, \ldots, k_{12}$, incident according to the table (cf. Excursus 1.1.2)

index of line k	0	1	2	3	4	5	6	7	8	9	10	11	12
index of point k	0	1	2	3	4	5	6	7	8	9	10	11	12
	1	2	3	4	5	6	7	8	9	10	11	12	0
	4	5	6	7	8	9	10	11	12	0	1	2	3
	6	7	8	9	10	11	12	0	1	2	3	4	5

Adjoin to the collinear set $\{K_0, K_1, K_4, K_6\}$ the point L_0. The collinear set K_2L_0 must contain a point of each of the eight lines other than k_0 which does not pass through K_2, namely, the lines k_3, k_4, k_5, k_6, k_7, k_8, k_{10}, k_{12}. None of these new points can belong to [K], since, if one of them were K_j, then the line K_2K_j would meet k_0 in a point, namely L_0, which does not belong to [K]. Thus, by joining L_0 in turn to the nine points of [K] which do not lie on k_0, we shall derive 72 new points and shall then have 86 points (the remaining five are on k_0) and 22 collinear sets. So far we have had no choice of structure, but the next step is to assign relations of collinearity, and in doing this we determine whether the geometry shall be $\Gamma^{(9)}$ or $\Omega^{(9)}$ or something other. Theoretically the problem of assigning incidences is a combinatorial problem, soluble by trial and error; in fact, of course, both geometries have an algebraic background and were first constructed by algebraic processes. We discuss these processes in Excursus 3.3.

Exc. 1.3.1 The geometry $\Gamma^{(9)}$

The geometry $\Gamma^{(9)}$ is compounded from seven geometries $\Gamma^{(3)}$, no two having any common point; each line of $\Gamma^{(9)}$ contains at least one point of each of the geometries $\Gamma^{(3)}$. We designate the geometries [K], [L], [M], [N], [L′], [M′], [N′], with points $K_1, L_1, \ldots$ and lines $k_1, l_1, \ldots$. The whole pattern may be exhibited by assigning names to the points on the lines through K_0. Take these to be:

Table 1 (i) $\Gamma^{(9)}$. *Indices of points on lines* k_r, *where* $k_r \supset K_0$ & $k_r \subset [K]$

line	K				L	L′	M	M′	N	N′
	indices of points									
k_0	0	1	4	6			0			
k_{12}	12	0	3	5			12			
k_9	9	10	0	2			9			
k_7	7	8	11	0			7			

(ii) $\Gamma^{(9)}$. *Indices of points on lines* x_0, *where* $x_0 \supset K_0$ & $x_0 \not\subset [K]$

line	K	L	L′	M	M′	N	N′
	indices of points						
l_0	0	2, 3, 6, 8	10	1	11	4	5
l_0'	0	10	2, 3, 6, 8	11	1	5	4
m_0	0	1	11	4, 5, 8, 10	3	6	2
m_0'	0	11	1	3	4, 5, 8, 10	2	6
n_0	0	4	5	6	2	10, 11, 1, 3	8
n_0'	0	5	4	2	6	8	10, 11, 1, 3

In this arrangement the lines $l_r, m_r, \ldots, n_r'$ contain sets of points typified by:

$$l_r : K_r, L_{r+2}, L_{r+3}, L_{r+6}, L_{r+8}, L'_{r+10}, M_{r+1}, M'_{r+11}, N_{r+4}, N'_{r+5}.$$

We can verify that points and collinear sets defined in this way satisfy axioms $A\mathcal{I}$ 1 and 2 by showing explicitly that there is one and only one collinear set containing any two points. For reasons which appear shortly we do not need to carry out the analysis in detail (as we shall have to do when considering the corresponding problem for $\Omega^{(9)}$), but we can indicate the method by determining the line through a pair of points. Take for example L_4 and M'_{10}. We have first to find the lines among the set $l_0, m_0, n_0, l_0', m_0', n_0'$ which contains the two points L_r and M'_{r+6} for some value of r. The only such line is m_0' which contains L_{11} and M_4'. The line containing L_4 and M'_{10} is therefore m_6'.

Alternatively we can relate the points named $K_r, L_r, \ldots, N_r'$ to the points numbered 0, ..., 90 in a difference table. Using the final table constructed in Excursus 3.2.3, we may take the collinear sets k_r to be:

k_0			k_r	
K_0	0	$= 0$	K_r	$= 7r$
K_1	7×1	$= 7$	K_{r+1}	$= 7r+7$
K_4	7×4	$= 28$	K_{r+4}	$= 7r+28$
K_6	7×6	$= 42$	K_{r+6}	$= 7r+42$
L_0	$7\times 0+6$	$= 6$	L_r	$= 7r+6$
L_0'	$7\times 10+1$	$= 71$	L_r'	$= 7r+71$
M_0	$7\times 5+3$	$= 38$	M_r	$= 7r+38$
M_0'	$7\times 3+4$	$= 25$	M_r'	$= 7r+25$
N_0	$7\times 4+2$	$= 30$	N_r	$= 7r+30$
N_0'	$7\times 11+5$	$= 82$	N_r'	$= 7r+82$

With this numbering, the line l_0 in the table above contains:

$$l_0 : 0, 20, 27, 48, 62, 50, 45, 11, 58, 26.$$

The lines represented on the difference table (that is, the lines through $K_0 = 0$) are:

k_0	0	7	28	42	71	6	25	38	82	30
k_1	84	0	21	35	64	90	18	31	75	23
k_4	63	70	0	14	43	69	88	10	54	2
k_6	49	56	77	0	29	55	74	87	40	79
l_0	20	27	48	62	0	26	45	58	11	50
l_0'	85	1	22	36	65	0	19	32	76	24
m_0	66	73	3	17	46	72	0	13	57	5
m_0'	53	60	81	4	33	59	78	0	44	83
n_0	9	16	37	51	80	15	34	47	0	39
n_0'	61	68	89	12	41	67	86	8	52	0

It would be a formidable task to verify that, for every set of points selected to be the points K, A, A′, B, B′, C, C′ of a Desargues figure, the condition of the axiom $A\mathscr{I}$ 3 is satisfied, but the reader is recommended to experiment with one or two cases. We shall find later that $\Gamma^{(9)}$ belongs to the main stream of geometries whose structure depends on $A\mathscr{I}$ 1, 2 and 3.

Exc. 1.3.2 $\Omega^{(9)}$, the Veblen–Wedderburn geometry

We use for the core of the geometry $\Omega^{(9)}$ the same system [K] as for $\Gamma^{(9)}$ and again construct six new sets of thirteen points, say, $\{A_i\}$, $\{A_i'\}$, $\{B_i\}$, $\{B_i'\}$, $\{C_i\}$, $\{C_i'\}$, but in $\Omega^{(9)}$ these sets do not form geometries $\Gamma^{(3)}$ isomorphic with [K]. The pattern of indices to be adopted is such that we can easily verify that every pair of points determines a unique collinear set, and hence that every two collinear sets have a single common member. The indices of the points on the thirteen lines through K_0 are given by the following tables:

Table 1 (i) $\Omega^{(9)}$. *Indices of points on lines* k_r, *where* $k_r \supset K_0$ & $k_r \subset [K]$

line	points									
	K				A	A′	B	B′	C	C′
k_0	0	1	4	6				0		
k_{12}	12	0	3	5				12		
k_9	9	10	0	2				9		
k_7	7	8	11	0				7		

(ii) $\Omega^{(9)}$. *Indices of points on lines* x_0, *where* $x_0 \supset K_0$ & $x_0 \not\subset [K]$

line	points						
	K	A	A′	B	B′	C	C′
a_0	0	3, 8, 10	2, 6		1, 11		4, 5
a_0'	0	2, 6	3, 8, 10	1, 11		4, 5	
b_0	0		1, 11	3, 8, 10	4, 5		2, 6
b_0'	0	1, 11		4, 5	3, 8, 10	2, 6	
c_0	0		4, 5		2, 6	3, 8, 10	1, 11
c_0'	0	4, 5		2, 6		1, 11	3, 8, 10

If x_0 is one of the lines in table 1 (i) or 1 (ii) and x_0 contains a point Y_r, then x_s contains Y_{r+s} (all indices being reduced modulo 13).

The arrangements of indices in table 1 (i) are identical for $\Gamma^{(9)}$ and $\Omega^{(9)}$, while those in table 2 (ii) are strikingly similar. If the pairs of sets {A}, {A′} and {L}, {L′}, etc. are taken together the arrangements of indices are identical, but the indices are somewhat differently distributed between the members of a pair.

Returning to table 1 (ii) $\Omega^{(9)}$ we can make the pattern clearer by rewriting the table in the form:

Table 2 $\Omega^{(9)}$. *Arrangement of points by indices*

set of indices	δ 3, 8, 10	α 2, 6	β 1, 11	γ 4, 5
line a_0	A	A′	B′	C′
a_0'	A′	A	B	C
b_0	B	C′	A′	B′
b_0'	B′	C	A	B
c_0	C	B′	C′	A′
c_0'	C′	B	C	A

Table 3 $\Omega^{(9)}$. *Differences*, y − x, *modulo* 13, *between pairs of indices*

x \ y	δ	α	β	γ
δ	2, 5, 6, 7, 8, 11	3, 5, 7, 9, 11, 12	1, 3, 4, 6, 8, 11	1, 2, 7, 8, 9, 10
α	1, 2, 4, 6, 8, 10	4, 9	5, 8, 9, 12	2, 3, 11, 12
β	2, 5, 7, 9, 10, 12	1, 4, 5, 8	10, 3	3, 4, 6, 7
γ	3, 4, 5, 6, 11, 12	1, 2, 10, 11	6, 7, 9, 10	1, 12

From tables 2 and 3 we can verify that any pair of the 91 points belongs to exactly one collinear set. Consider, for example, the pairs $\{K_1, C'_{11}\}$, $\{B_1, C'_{11}\}$ and $\{B'_1, C'_{11}\}$.

K_1 and C'_{11} belong to a line in the same set of 13 lines as $K_0C'_{10}$. From table 1 (ii), $K_0C'_{10}$ is c'_0, so that

$$K_1C'_{11} = c'_1.$$

B_1 *and* C'_{11}. From table 2, B and C′ occur together on b_0 (from table 3, the possible forward differences from B to C′ are $\{\alpha-\delta\} = \{3, 5, 7, 9, 11, 12\}$) and on c'_0 (forward differences $\{\delta-\alpha\} = \{1, 2, 4, 6, 8, 10\}$). Thus $B_1C'_{11}$ (forward difference 10) is a line c'_i for some i. $c'_0 = B_6C'_3$, so that

$$B_1C'_{11} = c'_8.$$

B'_1 *and* C'_{11}. B′ and C′ occur together on a_0 (forward differences $\{\gamma-\beta\} = \{3, 4, 6, 7\}$), on b_0 (forward differences $\{\alpha-\gamma\} = \{1, 2, 10, 11\}$) and on c_0 (forward differences $\{\beta-\alpha\} = \{5, 8, 9, 12\}$). Thus $B'_1C'_{11}$ (forward difference 10) is a line b_i for some i. $b_0 = B'_5C'_2$, so that

$$B'_1C'_{11} = b_9.$$

A full proof that every pair of points determines a unique line would be tedious, but it should be clear from the examples above that the patterns of table 2, which picks out the possible sets of lines, and table 3, which identifies the set giving the required difference, ensure that the system conforms to this condition.

We can now show, by selecting points to be K, A, A′, B, B′, C, C′, in the notation of *A𝒮* 3, and verifying that the required set of three points {L, M, N} are not collinear, that the geometry satisfies neither the Desargues condition nor the triply-special Desargues condition. This does not mean that there are no Desargues configurations in the system; there are, for example, some quadruply-special figures in [K]. In the tables below there are listed the points and collinear sets of:

(i) a set of points not satisfying the general Desargues condition,

(ii) a set of points not satisfying the triply-special Desargues condition,

(iii) a set of points which satisfy the general Desargues condition. All three sets include the three collinear points K_0, K_2, B_9, and points of the collinear set k_0.

(i) *A non-special non-Desargues figure*

Points in $\mathcal{AI}$ 3	Points in $\Omega^{(9)}$	Collinear sets	
K	K_0	K_0, K_1, C_0	k_0
A	K_1	K_0, K_2, B_9	k_9
B	K_2	K_0, A'_2, A'_6	a_0
C	A'_2	K_2, A'_2, C_2	k_2
A′	C_0	B_9, A'_6, C_2	a'_{11}
B′	B_9	A'_2, K_1, A'_{12}	b_1
C′	A'_6	A'_6, C_0, A'_{12}	a'_9
L	C_2	K_1, K_2, A_1	k_1
M	A'_{12}	C_0, B_9, A_1	a'_8
N	A_1	(MN) $A'_{12}A_1 = a_6$	
		(NL) $A_1C_2 = b'_0$	
		(LM) $C_2A'_{12} = c_7$	

It may be noted that, in the notation of $\mathcal{AI}$ 3, $\mathcal{L}$|LMN, but $\mathcal{L}$|NLK.

(ii) *A triply-special non-Desargues figure*

Points in $\mathcal{AI}$ 3	Points in $\Omega^{(9)}$	Collinear sets	
K	C'_{10}	C'_{10}, K_0, A_5	c'_1
A	K_0	C'_{10}, B_0, B_9	c'_7
B	B_0	C'_{10}, A'_9, B'_0	b_8
C	A'_9	B_0, A'_9, A_5, B_{10}	a'_{12}
A′	A_5	B_9, B'_0, B_{10}	b'_5
B′	B_9	A'_9, K_0, B_9, K_2	k_9
C′	B'_0	B'_0, A_5, K_2	a_2
L	B_{10}	K_0, B_0, B'_0, K_4	k_0
M	K_2	A_5, B_9, K_4	b'_4
N	K_4	$K_2B_{10} = b_2$	
		$B_{10}K_4 = c'_4$	
		$K_4K_2 = k_{11}$	

(iii) *A non-special complete Desargues figure*

Points in $A\mathscr{I}3$	Points in $\Omega^{(9)}$	Collinear sets	
K	K_0	K_0, K_1, A_0	k_0
A	K_1	K_0, K_2, B_9	k_9
B	K_2	K_0, K_3, C_{12}	k_{12}
C	K_3	K_2, K_3, K_8	k_2
A′	A_0	B_9, C_{12}, K_8	a_8'
B′	B_9	K_3, K_1, A_{10}'	k_{10}
C′	C_{12}	C_{12}, A_0, A_{10}'	a_7'
L	K_8	K_1, K_2, A_1'	k_1
M	A_{10}'	A_0, B_9, A_1'	a_{11}'
N	A_1'	K_8, A_{10}', A_1'	a_8

The most remarkable features of $\Omega^{(2)}$ are displayed in the 'harmonic constructions'. As an example, let us attempt to construct, as in 1.4*C*1,

$$\mathscr{H}_{K_1K_2}B_1',$$

by taking, as A B C L M N K D of 1.4*C*1, the points

$$K_1 \quad K_2 \quad B_1' \quad K_0 \quad X \quad Y \quad Z \quad D.$$

The fixed lines in the construction are

$$K_1K_2 = k_1, \quad K_0K_1 = k_0, \quad K_0K_2 = k_9, \quad B_1'K_0 = a_0.$$

X may be any point, other than B_1' and K_0, on a_0, and the other points are defined by

$$Y = K_1X \cap k_9, \quad Z = K_2X \cap k_0, \quad D = YZ \cap k_1.$$

The eight constructions are set out below, and it appears that for the four points X, $X \in \{A_3, C_4', C_5', A_6'\}$, $D = B_1'$; for the four points X, $X \in \{A_2', A_8, A_{10}, B_{11}'\}$, $D = B_1$. That is, in four cases the figure collapses into a set of seven points forming a system $\Gamma^{(2)}$, while in the other four it forms a standard harmonic figure, and, as we shall see, the points of this figure in fact belong to a set of thirteen forming a system $\Gamma^{(3)}$.

$X \in a_0$	$XK_1 = y$	$y \cap k_9 = Y$	$XK_2 = z$	$z \cap k_0 = Z$	$x = YZ$	$x \cap k_1 = D$
A_2'	b_1	B_9	k_2	K_6	b_6	B_1
A_3	a_1'	A_9'	b_2'	A_0	a_3	B_1'
C_4'	c_1'	C_9'	b_2	A_0'	c_8	B_1'
C_5'	a_1	A_9	c_2'	C_0	b_{11}'	B_1'
A_6'	c_1	C_9	c_2	C_0'	c_{12}	B_1'
A_8	k_8	K_9	a_2'	B_0	b_9'	B_1
A_{10}	k_{10}	K_{10}	a_2	B_0'	b_{10}'	B_1
B_{11}'	b_1'	B_1'	k_{11}	K_4	b_4	B_1

The systems $\Gamma^{(2)}$ and $\Gamma^{(3)}$ formed from or containing some of these points are

$$\Gamma^{(2)}\quad \begin{array}{ccccccc} K_0 & K_1 & K_2 & Z & B_1' & X & Y \\ K_1 & K_2 & Z & B_1' & X & Y & K_0 \\ Z & B_1' & X & Y & K_0 & K_1 & K_2 \end{array}$$

$$\Gamma^{(3)}\quad \begin{array}{ccccccccccccc} K_0 & . & B_1' & K_1 & . & . & B_1 & Y & K_2 & X & . & . & Z \\ . & B_1' & K_1 & . & . & B_1 & Y & K_2 & X & . & . & Z & K_0 \\ . & . & B_1 & Y & K_2 & X & . & . & Z & K_0 & . & B_1' & K_1 \\ B_1 & Y & K_2 & X & . & . & Z & K_0 & . & B_1' & K_1 & . & . \end{array}$$

In the case $X = A_8$, the complete system $\Gamma^{(3)}$ is:

Points

$$\begin{array}{ccccccccccccc} K_0 & A_{10}' & B_1' & K_1 & A_8' & B_0' & B_1 & K_9 & K_2 & A_8 & A_{10} & K_{10} & B_0 \\ A_{10}' & B_1' & K_1 & A_8' & B_0' & B_1 & K_9 & K_2 & A_8 & A_{10} & K_{10} & B_0 & K_0 \\ A_8' & B_0' & B_1 & K_9 & K_2 & A_8 & A_{10} & K_{10} & B_0 & K_0 & A_{10}' & B_1' & K_1 \\ B_1 & K_9 & K_2 & A_8 & A_{10} & K_{10} & B_0 & K_0 & A_{10}' & B_1' & K_1 & A_8' & B_0' \end{array}$$

Lines

$$\begin{array}{ccccccccccccc} a_0' & b_9 & k_1 & k_8 & a_2 & b_{10}' & b_9' & k_9 & a_2' & a_0 & k_{10} & b_{10} & k_0 \end{array}$$

The reader must remember that $\Omega^{(9)}$ has been deliberately constructed as a non-Desarguesian geometry. The whole of the main part of the book is concerned with geometries which satisfy the Desargues condition, and it should be particularly noted that we are to prove later that such a geometry cannot contain both a set of seven points forming $\Gamma^{(2)}$ and a set of thirteen points forming $\Gamma^{(3)}$. A finite geometry which conforms to $A\mathscr{I}$ 1, 2, 3 (Desargues) can contain only systems $\Gamma^{(p)}$, for a fixed prime p, and derived systems $\Gamma^{(p^2)}$, $\Gamma^{(p^3)}$, ...

EXCURSUS 1.4

GEOMETRY OF THREE DIMENSIONS AND THE PLANE DESARGUES FIGURE

In formulating the axioms $A\mathscr{I}$ we confined our attention to the positions of objects 'in' the paddock, without regard to the extension above or below the paddock. If we accept axiom $A\mathscr{I}\,3$ which lays down the existence of the Desargues configuration, then we may develop the geometry corresponding to objects in the paddock without ever taking into consideration the existence of points, in the mathematical sense, which do not belong to the set of points prescribed by axioms $A\mathscr{I}\,1, 2$. But the surprising result is that if axioms corresponding to $A\mathscr{I}\,1, 2$ are formulated for three-dimensional space, then we do not need further axioms to establish the existence of the Desargues configuration in any plane in the space. Thus we may replace axiom $A\mathscr{I}\,3$ by a set of axioms which embed the plane in a three-dimensional space.

$A\mathscr{I}_3 1$ $= A\mathscr{I}\,1$ (p. 22)

Exc. 1.4*N*1 First definition of a coplanar set.

G $\mathscr{L}|\mathrm{ABC}$
S $\mathrm{D} : \#|\mathrm{CD}\ \&\ \mathscr{P}|\mathrm{AB} \cap \mathrm{CD}$
N Plane $\langle \mathrm{AB}; \mathrm{C}\rangle = \{\mathrm{D} : \mathscr{P}|\mathrm{AB} \cap \mathrm{CD}\} \cup \mathrm{C}$.

$A\mathscr{I}_3 2$ *G* $\mathscr{L}|\mathrm{ABC}$
A $\langle \mathrm{AB}; \mathrm{C}\rangle = \langle \mathrm{AC}; \mathrm{B}\rangle$.

Exc. 1.4*T*1

G $\mathscr{L}|\mathrm{ABC}\ \&\ \mathscr{L}|\mathrm{ABP}$
T $\langle \mathrm{AB}; \mathrm{C}\rangle = \langle \mathrm{AP}; \mathrm{C}\rangle$

Exc. 1.4*T*2

G $\mathscr{L}_3|\mathrm{ABCD}\ \&\ \mathscr{P}|\mathrm{AB} \cap \mathrm{CD}$
T (i) $\mathscr{P}|\mathrm{AD} \cap \mathrm{BC}$
(ii) $\langle \mathrm{AB}; \mathrm{C}\rangle = \langle \mathrm{AB}; \mathrm{D}\rangle = \langle \mathrm{CD}; \mathrm{A}\rangle$.

Exc. 1.4*N*2 Definition of coplanar set.

G $\mathscr{L}|\mathrm{ABC}$.
N Plane $\mathrm{ABC} = \{\mathrm{D} : \mathscr{P}|\mathrm{AB} \cap \mathrm{CD}\ \&\ \#|\mathrm{CD}\} \cup \mathrm{C}$.

From $A\mathscr{I}_3 2$ we have ABC = {E : $\mathscr{P}$|AC ∩ BE}, etc. and from $T1$ that:

Exc. 1.4*T*3

G $\mathscr{L}_3$|ABCD & $\mathscr{P}$|AB ∩ CD
T ABC = ABD = ACD = BCD,

and thence

Exc. 1.4*T*4*X* A plane is defined by any three non-collinear points of itself.

*G*1 $\mathscr{L}$|ABC
*G*2 $\mathscr{L}$|PQR & {P, Q, R} ⊂ ABC
T ABC = PQR

Exc. 1.4*N*3 Coplanar points.

G $\mathscr{L}_3$|ABCD & $\mathscr{P}$|AB ∩ CD
N (i) ϖ|ABCD : the points ABCD are coplanar
(ii) α = ABC : α is the plane ABC.

$\alpha, \beta, \lambda, \ldots$ will be used to denote planes.

Exc. 1.4*T*5*X* If two distinct points lie in a plane then every point of the collinear set determined by them lies in the plane.

*G*1 $\mathscr{L}$|ABC
*G*2 {P, Q} ⊂ ABC & ≠|PQ
T R ∈ PQ ⇒ R ∈ ABC.

Exc. 1.4*T*6*X* Any two coplanar collinear sets have a common point.

*G*1 $\mathscr{L}$|ABC
*G*2 {P, Q, R, S} ⊂ ABC & ≠|PQ, RS
T $\mathscr{P}$|PQ ∩ RS.

***A*$\mathscr{I}_3$3** Not all points are coplanar.

G $\mathscr{L}$|ABC
A There exists D, D ∉ ABC.

Exc. 1.4*N*4 Non-coplanar points.

G $\mathscr{L}$|ABC & D : D ∉ ABC
T *A* ∉ BCD etc.
N not-ϖ|ABCD : A, B, C, D are non-coplanar.

Exc. 1.4*N*5 Non-coplanar lines: skew lines.

G a, b : $\emptyset$|a $\cap$ b, i.e. not-ϖ|a, b : a, b are *skew*.

Exc. 1.4*T*7 not-ϖ|ABCD $\Rightarrow$ $\mathscr{L}_3$|ABCD
not-ϖ|ABCD $\Rightarrow$ $\#$|ABCD.

***A*$\mathscr{I}_3$4** Axiom restricting space to three dimensions.

G not-ϖ|ABCD
S E : $\#$|ED
A $\mathscr{P}$|DE $\cap$ ABC.

Exc. 1.4*T*8*X* Any line and any plane either have a single common point or the plane contains the line.

Exc. 1.4*T*9*X* Any two distinct planes have a common line.

Exc. 1.4*T*10*X* Any three planes either have a single common point or have a common line or coincide.

Exc. 1.4*T*11*X* Test for a point not to lie in a plane.

$G1$ α, A, B : A $\in \alpha$, B $\notin \alpha$
$G2$ C : C $\in$ AB & $\#$|CA
T C $\notin \alpha$.

Exc. 1.4*T*12*X* Test for collinearity of three points.

G ϖ|ABCD & ϖ|ABCE & E $\notin$ ABCD
T $\mathscr{L}$|ABC.

Exc. 1.4*T*13 Through a point not on either of two skew lines a single line can be drawn to meet the two lines.

$G1$ a, b : $\emptyset$|a $\cap$ b
$G2$ P : P $\notin$ a, P $\notin$ b
$C1$ l : l = Pa $\cap$ Pb
T (i) l $\supset$ P & $\mathscr{P}$|l $\cap$ a & $\mathscr{P}$|l $\cap$ b
$C2$ m : m $\supset$ P & $\mathscr{P}$|m $\cap$ a & $\mathscr{P}$|m $\cap$ b
T (ii) l = m
P assume $\#$|lm
l $\cap$ m = P $\Rightarrow$ ϖ|lm
t lm $\supset$ {a, b} $\Rightarrow$ ab = lm $\Rightarrow$ $\mathscr{P}$|a $\cap$ b contradicting $G1$.

Exc. 1.4*T*14*X* If three lines are such that each pair is concurrent then the three lines are either all coplanar or all concurrent.

G a, b, c : $\mathscr{P}|a \cap b$ & $\mathscr{P}|a \cap c$ & $\mathscr{P}|b \cap c$
T either $\varpi|a, b, c$
or $\mathscr{P}|a \cap b \cap c$.

Desargues theorem

We adopt the symmetric notation for the figure introduced on p. 28, so that the theorem which is to replace $A\mathscr{I}3$ in virtue of the adoption of the axioms $A\mathscr{I}_3$ may be expressed thus:

Exc. 1.4*T*15 Desargues theorem

*G*1 α
$\#|\{P_{ij}\}$ & $P_{ij} \in \alpha$; $\{i, j\} \subset \{1, \ldots, 5\}$ & $P_{ij} = P_{ji}$
*G*2 $\mathscr{L}|$each set $\{P_{ij}, P_{ik}, P_{jk}\}$ except $\{P_{12}, P_{13}, P_{23}\}$
T $\mathscr{L}|P_{12}P_{13}P_{23}$
P *s* $P_4 : P_4 \notin \alpha$
$P_5 : P_5 \in P_4P_{45}$ & $\#|P_5P_4P_{45}$
*t*1 $\varpi|P_4P_5P_{45}P_{r4}P_{r5}$, each r, $r \in \{1, 2, 3\}$
c $P_r : P_r = P_4P_{r4} \cap P_5P_{r5}$, $r \in \{1, 2, 3\}$
*t*2 $P_r \notin \alpha$
p $P_4 \notin \alpha$, $P_{r4} \in \alpha$
*t*3 $\varpi|P_rP_sP_{rs}P_{rk}P_{sk} : \{r, s\} \subset \{1, 2, 3\}$; $k \in \{4, 5\}$
*t*4 $\mathscr{L}|P_rP_sP_{rs} : \{r, s\} \subset \{1, 2, 3\}$
p *t*3, *T*12
*t*5 $\varpi|P_1P_2P_3P_{13}P_{23}P_{12}$
T $\mathscr{L}|P_{23}P_{31}P_{12}$
p $P_1 \notin \alpha$, *T*12, *t*5, *G*1.

We may now reorganise the figure using the five points P_i as the primary elements. These points, which are supposed to be distinct, no three collinear, and no four coplanar (this last condition ensures the other two), define

10 lines P_iP_j, 10 planes $P_iP_jP_k$.

Let α be any plane not through any of the ten lines, then from *T*8, α meets each of the lines P_iP_j, say in the point P_{ij}. We have

$$\varpi|P_iP_jP_kP_{ij}P_{jk}P_{ki} \text{ \& } \{P_{ij}, P_{jk}, P_{ki}\} \subset \alpha \text{ \& } P_i \notin \alpha.$$

So that from *T*12 $\mathscr{L}|P_{ij}P_{jk}P_{ki}$.

Thus the plane section of these ten lines and planes is the complete Desargues figure. We obtain the special Desargues figures (1.3*T*5–7), by fixing the five points P_i (no four coplanar) in the space and selecting special positions of the plane α in relation to them. Construct the points Q_{ij},

$$Q_{ij} = P_iP_j \cap P_kP_lP_m, \quad \{i, j, k, l, m\} = \{1, 2, 3, 4, 5\}.$$

The various special cases are typified by:

$\alpha \supset Q_{45} \Rightarrow \mathscr{L}|P_{45}P_{23}P_{31}P_{12}$. Simply-special.

$\alpha \supset Q_{45}Q_{23} \Rightarrow \alpha \supset P_1P_4P_5 \cap P_1P_2P_3$.

$\Rightarrow \alpha \supset P_1$. Degenerate configuration.

$\alpha \supset Q_{45}Q_{53} \Rightarrow \mathscr{L}|P_{45}P_{23}P_{31}P_{12}$ & $\mathscr{L}|P_{53}P_{24}P_{41}P_{12}$. Doubly-special.

$\alpha = Q_{45}Q_{53}Q_{34} \Rightarrow$ Triply-special.

No other non-degenerate specialization is possible (except in a plane section which includes $\Gamma^{(3)}$).

EXCURSUS 1.5

SOME SPECIAL THREE-DIMENSIONAL GEOMETRIES

As we are concerned almost entirely with geometry in the plane, we confine our attention to a few special three-dimensional geometries, which are direct extensions of geometries investigated in Exc. 1.1.

Exc. 1.5.1 Fano's geometry

This geometry is the direct analogue of $\Gamma^{(2)}$: the points, of which there are 15, are the 'vectors' $(\alpha, \beta, \gamma, \delta)$ in which each component is either 0 or 1, and not all four of them are zero. If u and v are any two of these, they determine a collinear set of which the only other member is u+v (i.e. the vector whose components are the sums reduced modulo 2 of the components of u and v). If w is not u+v, then u, v and w determine seven points which are collinear by sets of three, namely,

u	v	w	u+v	v+w	u+v+w	w+u
v	w	u+v	v+w	u+v+w	w+u	u
u+v	v+w	u+v+w	w+u	u	v	w

These seven are coplanar, since the plane of u, v, w contains, e.g. u+v (which is collinear with u and v), and u+v+w (which is collinear with w and u+v). These seven points correspond to the points in the geometry $\Gamma^{(2)}$ described in Excursus 1.1.1 (the arrangement above corresponds exactly to the table on p. 85) or to those described in Excursus 1.1.5, where u, v and w are taken to be (1, 0, 0, 0), (0, 1, 0, 0), (0, 0, 1, 0) and the last zero is then omitted.

If t is any point not in this coplanar set, then t+u, t+v, ... are seven points each collinear with t and one point of the coplanar set. In particular if u, v, w are taken to be (1, 0, 0, 0), (0, 1, 0, 0), and (0, 0, 1, 0), we may take t to be (0, 0, 0, 1). In this way all fifteen points of the geometry are accounted for.

There are three points on each line, and seven lines through each point, and therefore altogether 35 lines. There are seven points in each plane and seven planes through each point (joining the point to the lines in any plane that does not contain the point), and consequently altogether 15 planes.

An alternative notation for the points is this: take

1000 0100 0010 0001 1111

to correspond to the pairs of digits

16 26 36 46 56

If u and v, any two of these, correspond to $\alpha 6$, $\beta 6$, then $u+v$ corresponds to $\alpha\beta$ (the same as $\beta\alpha$). In this way the fifteen points correspond to the 15 digit pairs ij ($=$ji), where $\{i, j\} \subset \{1, 2, \ldots, 6\}$. The collinear sets of three are typified by

23, 31, 12 (pairs from three digits, 20 sets)
12, 34, 56 (all digits different, 15 sets);

the coplanar sets of seven are typified by

23, 31, 12; 14, 24, 34; 56.

The simple method which was available for $\Gamma^{(2)}$ of determining the arrangement of the set of seven points from a difference table modulo 7 cannot of course be applied to a three-dimensional geometry, but we may adapt the cyclic table provided we allow for more than one cycle. One set is the following:

Table 1

0	1	2	3	4
5	6	7	8	9
10	11	12	13	14

Table 2

0	1	2	3	4	5	6	7	8	9	10	11	12	13	14
1	2	3	4	5	6	7	8	9	10	11	12	13	14	0
4	5	6	7	8	9	10	11	12	13	14	0	1	2	3

Table 3

0	1	2	3	4	5	6	7	8	9	10	11	12	13	14
2	3	4	5	6	7	8	9	10	11	12	13	14	0	1
8	9	10	11	12	13	14	0	1	2	3	4	5	6	7

The thirty-five collinear sets are displayed in the columns. One result which stands out immediately is that there exist sets of five mutually skew lines $a_i, i = 0, \ldots, 4$ say, typified by the columns of table 1, which contain all 15 points.

The first two and the last columns of table 1 form the scheme

$$\begin{array}{ccc} 0 & 1 & 4 \\ 5 & 6 & 9 \\ 10 & 11 & 14 \end{array}$$

in which the rows are collinear sets from table 2. Since all 15 points lie on the five lines a_i, each of the remaining 30 lines must meet three lines a_i, and the 30 form ten sets of three transversals to sets of three lines a_i.

Exercise Show that there are three planes through a line and that there are 16 lines skew to any given line (find the lines skew to 0, 5, 10), that each line belongs to twenty-four sets of five mutually skew lines, and that there are 168 sets of five mutually skew lines altogether.

Exc. 1.5.2 The extension of $\Gamma^{(3)}$

In the same way a three-dimensional geometry may be constructed to follow the pattern of the plane geometry $\Gamma^{(3)}$. There are forty points, the 'vectors' $(\alpha, \beta, \gamma, \delta)$, $(\alpha, \beta, \gamma, \delta) \in \{0, 1, -1\}$, where not all components are zero, and the last non-zero component is $+1$. Collinear with two points u, v, there are the two points $\pm(u+v)$, $\pm(u-v)$, 2 being replaced by -1 whenever it occurs, and the sign being selected so that the last non-zero component is $+1$.

The 13 points $(\alpha, \beta, \gamma, 0)$ form the plane geometry $\Gamma^{(3)}$. If these are joined to $(0, 0, 0, 1)$ we obtain thirteen collinear sets, each containing two more points $(\alpha, \beta, \gamma, 1)$ and $(-\alpha, -\beta, -\gamma, 1)$. These complete the total of 40 points.

Exercise 1 Prove that there are 130 lines.

Exercise 2 (i) Complete the plane Desargues' figure based on the two triads

P_{14}	P_{24}	P_{34}	P_{15}	P_{25}	P_{35}
1000	0100	0010	0110	1010	1100

and show that $P_{23}P_{31}P_{12}$ passes through P_{45}.

(ii) Complete the three-dimensional figure, of which this is the plane section, taking P_4 to be $(0, 0, 0, 1)$, P_5 to be $(1, 1, 1, 1)$, and show that the plane of the figure passes through the point $P_1P_2P_3 \cap P_4P_5$.

(iii) By considering sections of the figure determined by $P_1(1000)$, $P_2(0100)$, $P_3(0010)$, $P_4(0001)$, $P_5(1111)$, by planes which do not pass through any of the points P_1, show that in any plane Desargues' figure of this geometry in which the ten points are distinct there are four sets of four collinear points.

Exercise 3 The forty points can be arranged as a cycle in the same way as the points of Fano's 15-point geometry were. Verify by constructing difference tables that the columns of the following tables (the first containing 10 points and the others 40 each) give the 130 collinear sets of four points.

I		II		III		IV	
0	1 ...	0	1 ...	0	1 ...	0	1 ...
10	11 ...	1	2 ...	2	3 ...	7	8 ...
20	21 ...	26	27 ...	5	6 ...	19	20 ...
30	31 ...	32	33 ...	18	19 ...	36	37 ...

The method of generating these and the Fano tables will be found in Exercise 9 at the end of Excursus 2.3.

Exercise 4 Assuming that a finite geometry of three dimensions has the same number $m+1$ of points on every line prove that it contains m^3+m^2+m+1 points, $m^4+m^3+2m^2+m+1$ lines and m^3+m^2+m+1 planes.

Exc. 1.5.3 Dual geometry in three dimensions

In the same way as a dual system was established in the plane, a dual system may be set up in space of three dimensions by identifying 'd-points' and 'p-planes', thus

d-system	*p-system*
d-point	p-plane
two distinct d-points define a d-line, their join	two distinct p-planes define a p-line, their meet
three non-collinear d-points define a d-plane	three p-planes which do not have a common p-line, meet in a p-point.

Let Φ be any prescribed set of points forming given collinear sets and given coplanar sets, and Φ_d its representation in system 'd'. Then Φ_d, when d-points are interpreted as p-planes in the p-geometry, provides the dual of Φ.

Exercise Write down the theorems dual to Exc. 1.4*T*12 and *T*13.

Exc. 1.5.4 Γ-geometry

The analogue of the c-geometry is a geometry in which γ-planes are e-spheres through a point Ω. Most simply we may represent the system algebraically thus:

γ-points rectangular cartesian points (x, y, z)

γ-planes $\delta(x^2+y^2+z^2)+\alpha x+\beta y+\gamma z = 0.$

With conventions such as those adopted in establishing the plane c-geometry, this system can be proved to satisfy the axioms $A\mathscr{I}_3$.

BOOK 2

GEOMETRY AND COUNTING

With possibly some lingering doubts about the 'inaccessible' points, the F-G is now satisfied that he has a mathematical model which corresponds at least qualitatively to his observations of the relations among the objects in the paddock. But he has become aware that there is another aspect of his operation of 'counting', an operation which he has so far applied only to the individual objects.

His observation is this: if he makes a mark at the end of each pace as he proceeds along a furrow, and then counts these marks, then each time he walks from ⁑A to ⁑B he records about the same number of marks. That is, he realises that he should be able to associate a number with every pair of objects. The underlying idea is that the separate steps taken in walking along a furrow should correspond to elements in some sense mathematically equivalent. Or again, in ploughing a second furrow by 'visual reference' (cf. Chapter 1.5) to the first there is a recognition of some sort of equivalence among the pairs of 'corresponding places' in the two furrows (where a footmark in one furrow is 'nearly beside' one in the other).

This last observation in fact gives the lead to the path of development that we are to adopt, namely, that the basic figure in setting up a model for these physically equivalent paces is this: a pair of parallel lines, corresponding to the furrows, with a sequence of parallel lines intersecting them, corresponding to the ⁑collinear sets determined by pairs of footmarks more or less side by side in the two furrows.

We have to devise a mathematical model which enables us to make the transition from 'counting' to 'measurement'.

We use the full printing convention now so that the printed order of symbols may have geometric significance, i.e., 'A, B' can have a different significance from 'B, A' when used in a specific combination of symbols. For example, we have already introduced $\partial(\mathrm{AB}/\mathrm{CD})$ and $\mathcal{O}|\mathrm{AB}$, which bear different meanings when A and B are interchanged (1.8D1).

8-2

CHAPTER 2.1

DISPLACEMENT: THE GEOMETRIC TRANSFORMER $\mathscr{D}$

We assume initially only the incidence axioms $A\mathscr{I}1$ and 2, and as the ideas develop we shall see the part that has to be played by the Desargues and Separation Axioms.

Effectively what is to be done is this: point-pairs in the plane are to be assigned to 'equivalence classes', in somewhat the same way as lines were assigned to equivalence classes by the relation of parallelism (at 1.5*N*1). Then with each equivalence class is associated an operation or 'transformer' $\mathscr{D}$ ('displacement'), with the property that when we are given a pair A, B and a point P, then P and $\mathscr{D}$P are a pair in the equivalence class to which A, B belongs. The transformer $\mathscr{D}$ transforms each point P into a point $\mathscr{D}$P; it 'maps' the plane onto itself.

2.1*D*1 A displacement transformer, $\mathscr{D}$, defined by any two distinct proper points.

G $\mathscr{L}\wp|\mathrm{ABP}$
C $\mathrm{Q} : \mathrm{PQ} \parallel \mathrm{AB},\ \mathrm{BQ} \parallel \mathrm{AP}$
D $\mathscr{D}_{\mathrm{AB}}\mathrm{P} = \mathrm{Q}.$

Note that we must at present have $\mathrm{P} \notin \mathrm{AB}$, a restriction we shall remove shortly. From the symmetry of the relations among the four points, we have at once

2.1*T*1 Elementary properties of the transformers $\mathscr{D}$.

G $\mathscr{D}_{\mathrm{AB}}\mathrm{P} = \mathrm{Q}$
T (i) $\mathscr{D}_{\mathrm{BA}}\mathrm{Q} = \mathrm{P}$
(ii) $\mathscr{D}_{\mathrm{PQ}}\mathrm{A} = \mathrm{B}$
(iii) $\mathscr{D}_{\mathrm{AP}}\mathrm{B} = \mathscr{D}_{\mathrm{AB}}\mathrm{P} = \mathrm{Q}.$

We have also to verify that the transformer is uniquely defined by a point-pair, i.e.

2.1*T*2 $\mathscr{D}_{\mathrm{AB}}\mathrm{P}$ is uniquely determined by $\mathscr{L}\wp|\mathrm{ABP}$.

*G*1 $\mathscr{D}_{\mathrm{AB}}\mathrm{P} = \mathrm{Q}$
*G*2 $\mathscr{D}_{\mathrm{AB}}\mathrm{P} = \mathrm{Q}'$
T $\mathrm{Q} = \mathrm{Q}'$

P $t1$ $Q' \in BQ, \quad Q' \in PQ$

p $BQ \parallel AP \parallel BQ', \quad PQ \parallel AB \parallel PQ'$

$t2$ $\# | BQ, PQ$

p $Q \notin AB$

T $Q' = BQ \cap PQ = Q.$

If the transformer $\mathscr{D}$ is to define an equivalence class, then, when $\mathscr{D}_{AB} P = Q$, the transformers $\mathscr{D}_{AB}$ and $\mathscr{D}_{PQ}$ must be identical. This we can prove provided we accept the 'simply-special Desargues' axiom, $A\mathscr{I}3^*$.

2.1*T*3 $\mathscr{D}_{AB} P = Q \Rightarrow \mathscr{D}_{PQ} = \mathscr{D}_{AB}.$

G $\mathscr{D}_{AB} P = Q$

S $\wp | X : X \notin AB, \quad X \notin PQ$

C $Y : \mathscr{D}_{AB} X = Y$

T $\mathscr{D}_{PQ} X = Y.$

Let us restate the data in terms of parallelisms, namely,

G $\mathscr{L} \# \wp | ABPQ \ \& \ AB \parallel PQ \ \& \ AP \parallel BQ$

S $\wp | X : X \notin AB, \quad X \notin PQ$

C $Y : AB \parallel XY \ \& \ AX \parallel BY$

T $PX \parallel QY.$

c ideal$| UVW : U \in AB, V \in AP, W \in AX$

P cf. simply-special Desargues, $A\mathscr{I}3^*$, with points

	K	A	B	C	A′	B′	C′	L	M	N
renamed:	U	A	P	X	B	Q	Y	.	W	V

t $PX \cap QY \in VW.$

Note that we may now write unambiguously

$$\mathscr{D}_{AB} = \mathscr{D}_{PQ} = \mathscr{D}_{XY},$$

with the meaning that, for all points Z,

$$\mathscr{D}_{AB} Z = \mathscr{D}_{PQ} Z = \mathscr{D}_{XY} Z.$$

Moreover, from $T2$ and $T3$ it follows that if for any particular proper point G

$$\mathscr{D}_{AB} G = \mathscr{D}_{PQ} G,$$

then for all proper points Z

$$\mathscr{D}_{AB} Z = \mathscr{D}_{PQ} Z$$

i.e.

$$\mathscr{D}_{AB} = \mathscr{D}_{PQ}.$$

It should be noticed also that, if $\# | AB$, there can be no proper point J such that $\mathscr{D}_{AB} J = J$.

2.1 *T*4*X* The displacement of a collinear set is a parallel collinear set

*G*1 $\# \wp | AB$
*G*2 $\mathscr{L}\# \wp | PQR$ & $PQ \nparallel AB$
C $\{{}^{\mathscr{O}}P', Q', R'\} = \mathscr{D}_{AB}\{{}^{\mathscr{O}}P, Q, R\}$
T (i) $\mathscr{L}|P'Q'R'$
(ii) $P'Q' \parallel PQ$.

We can express this theorem and an earlier remark together in the form:

2.1 *T*5 ideal$|U \Leftrightarrow \mathscr{D}_{AB} U = U$, for any pair $\#|AB$.

In *T* 3, X was required not to lie on AB or PQ, and we have next to remove this restriction.

2.1 *D*2 Operation of $\mathscr{D}_{AB}$ on X when $X \in AB$.

G $\mathscr{L}\# \wp | ABX$
S $\wp | P : P \notin AB$
*C*1 $Q : Q = \mathscr{D}_{AB} P$
*C*2 $Y : Y = \mathscr{D}_{PQ} X$
D $Y = \mathscr{D}_{AB} X$.

For this definition to be valid we require that Y should be independent of the choice of P. This follows at once from *T* 3. As a minor consequence we have also: for all point-pairs A, B

$$\mathscr{D}_{AB} A = B.$$

Also since, if $\mathscr{D}_{AB} P = Q$, then $\mathscr{D}_{BA} Q = P$,

it follows that $\mathscr{D}_{BA} (\mathscr{D}_{AB} P) = P$ for all P,

a relation which we may write as

$$\mathscr{D}_{BA} \mathscr{D}_{AB} = \mathscr{I},$$

where

2.1 *N*1 The identity transformer: $\mathscr{I}$

$$\mathscr{I} X = X, \quad \text{for all X.}$$

It is convenient also to use a conventional extension of the definition of $\mathscr{D}$, to cover the case in which $A = B$, namely,

2.1 *N*2 $$\mathscr{D}_{AA} = \mathscr{I}.$$

T 1, *T* 2, *T* 3 have been proved only for non-collinear sets, and we have to verify them for collinear sets; the proofs of all are immediate, except for *T* 1 (iii), namely $\mathscr{D}_{AB}C = \mathscr{D}_{AC}B$. This appears shortly as a simple consequence of *T* 8.

We have next to discuss the compounding of a general pair of displacements, namely, if $\mathscr{D}_{AB}P = Q$ and $\mathscr{D}_{LM}Q = R$, what is the relation between P and R?

2.1 *T*6 The resultant of two displacements is a displacement.

G $\mathscr{D}_{AB}, \mathscr{D}_{LM} \quad AB \nparallel LM$
C 1 $C : \mathscr{D}_{LM}B = C$
S $\wp | P$
C 2 $Q : \mathscr{D}_{AB}P = Q$
C 3 $R : \mathscr{D}_{LM}Q = R$
T $R = \mathscr{D}_{AC}P$
P *t* 1 $\mathscr{D}_{LM} = \mathscr{D}_{BC}$
c ideal|UVT : $U \in AB$, $V \in BC$, $T \in AP$
p compare with the simply-special Desargues,
the points K, A, B, C, A′, B′, C′, L, M N
being renamed T, A, B, C, P, Q, R, V, ., U
t 2 $W = AC \cap PR \in UV$
T $\mathscr{D}_{AC}P = R$
P $AC \parallel PR$ (*t* 2)
$AP \parallel BQ \parallel CR$ (*C* 2, *C* 3, *t* 1).

We have proved the theorem under the assumption $LM \nparallel AB$; proof in the case of parallel displacements is best deferred until after *T* 8.

2.1 *T*7 The transformers $\mathscr{D}$ commute.

G $\mathscr{D}_{AB}, \mathscr{D}_{LM}$; $AB \nparallel LM$
T $\mathscr{D}_{AB}\mathscr{D}_{LM} = \mathscr{D}_{LM}\mathscr{D}_{AB}$
C $C : C = \mathscr{D}_{LM}B$
$B' : B' = \mathscr{D}_{LM}A$
P $\mathscr{D}_{LM}\mathscr{D}_{AB} = \mathscr{D}_{BC}\mathscr{D}_{AB} = \mathscr{D}_{AC}$
$\mathscr{D}_{AB}\mathscr{D}_{LM} = \mathscr{D}_{B'C}\mathscr{D}_{AB'} = \mathscr{D}_{AC}$.

Again the proof for $AB \parallel LM$ is deferred until after *T* 8.

2.1 *T*8 The transformers $\mathscr{D}$ are associative.

G $\mathscr{D}_{AB}, \mathscr{D}_{LM}, \mathscr{D}_{PQ}$; no two of AB, LM, PQ parallel
T $(\mathscr{D}_{PQ}\mathscr{D}_{LM})\mathscr{D}_{AB} = \mathscr{D}_{PQ}(\mathscr{D}_{LM}\mathscr{D}_{AB})$
C $C : \mathscr{D}_{LM}B = C$
$D : \mathscr{D}_{PQ}C = D$
P $(\mathscr{D}_{PQ}\mathscr{D}_{LM})\mathscr{D}_{AB} = (\mathscr{D}_{CD}\mathscr{D}_{BC})\mathscr{D}_{AB}$
$= \mathscr{D}_{BD}\mathscr{D}_{AB} = \mathscr{D}_{AD}$
$= \mathscr{D}_{CD}\mathscr{D}_{AC} = \mathscr{D}_{PQ}(\mathscr{D}_{LM}\mathscr{D}_{AB})$.

We turn now to the proofs of T6, 7, 8 in the cases of parallel displacements.

2.1 T6, 7(||) Case of parallel displacements.

G $\mathscr{L}\#\wp|\mathrm{ABC}$

T $\mathscr{D}_{\mathrm{AB}}\mathscr{D}_{\mathrm{BC}} = \mathscr{D}_{\mathrm{BC}}\mathscr{D}_{\mathrm{AB}} = \mathscr{D}_{\mathrm{AC}}$

P s $\wp|\mathrm{X}:\mathrm{X}\notin\mathrm{AB}$

c $\mathrm{Z}:\mathrm{Z} = \mathscr{D}_{\mathrm{AC}}\mathrm{X}$

$$p \quad \mathscr{D}_{\mathrm{AC}} = \mathscr{D}_{\mathrm{XZ}} = \mathscr{D}_{\mathrm{BZ}}\mathscr{D}_{\mathrm{XB}}$$
$$= (\mathscr{D}_{\mathrm{BC}}\mathscr{D}_{\mathrm{CZ}})\mathscr{D}_{\mathrm{XB}}$$
$$= \mathscr{D}_{\mathrm{BC}}(\mathscr{D}_{\mathrm{AX}}\mathscr{D}_{\mathrm{XB}}) = (\mathscr{D}_{\mathrm{AX}}\mathscr{D}_{\mathrm{XB}})\mathscr{D}_{\mathrm{BC}}$$
$$= \mathscr{D}_{\mathrm{BC}}\mathscr{D}_{\mathrm{AB}} = \mathscr{D}_{\mathrm{AB}}\mathscr{D}_{\mathrm{BC}}.$$

2.1 T8(||)X

G $\mathscr{L}\#\wp|\mathrm{ABCD}$

T $\mathscr{D}_{\mathrm{CD}}(\mathscr{D}_{\mathrm{BC}}\mathscr{D}_{\mathrm{AB}}) = (\mathscr{D}_{\mathrm{CD}}\mathscr{D}_{\mathrm{BC}})\mathscr{D}_{\mathrm{AB}}.$

2.1 T9(i) = T1 (iii) in the case of collinear points.

G $\mathscr{L}\#\wp|\mathrm{ABC}$

T $\mathscr{D}_{\mathrm{AB}}\mathrm{C} = \mathscr{D}_{\mathrm{AC}}\mathrm{B}$

$$P \quad \mathscr{D}_{\mathrm{AC}}\mathrm{B} = (\mathscr{D}_{\mathrm{AB}}\mathscr{D}_{\mathrm{BC}})\mathrm{B}$$
$$= \mathscr{D}_{\mathrm{AB}}(\mathscr{D}_{\mathrm{BC}}\mathrm{B})$$
$$= \mathscr{D}_{\mathrm{AB}}\mathrm{C}.$$

We can express this theorem in a slightly different form which is of use later, since it corresponds to the following property of numbers (not restricted to 'rational numbers'), namely, 'If $\mathrm{a} = \mathrm{b}$ then $\mathrm{a}+\mathrm{x} = \mathrm{x}+\mathrm{b}$', which is equivalent to '$\mathrm{a}+\mathrm{x} = \mathrm{x}+\mathrm{a}$', i.e. addition is commutative.

2.1 T9(ii) Geometric form of commutativity of addition.

G $\mathscr{L}\#\wp|\mathrm{ABC}$

C $\mathrm{D}:\mathscr{D}_{\mathrm{AB}} = \mathscr{D}_{\mathrm{CD}}$

T $\mathscr{D}_{\mathrm{AC}} = \mathscr{D}_{\mathrm{BD}}.$

We have now proved for the system, $\mathfrak{D}$ say, of transformers $\mathscr{D}$ that

(i) $\mathfrak{D}$ contains the identity transformer $\mathscr{I}$;

(ii) every transformer in $\mathfrak{D}$ has a unique inverse, since we may rewrite the relation

$$\mathscr{D}_{\mathrm{AB}}\mathscr{D}_{\mathrm{BA}} = \mathscr{I},$$

as

$$\mathscr{D}_{\mathrm{BA}} = \mathscr{D}_{\mathrm{AB}}^{-1};$$

(iii) the resultant of any pair of transformers in $\mathfrak{D}$ is a transformer in $\mathfrak{D}$ (T6);

(iv) the transformers are associative (T8).

The transformers therefore form a GROUP; in addition

(v) the transformers commute (T7),

so that $\mathfrak{D}$ is an ABELIAN GROUP or COMMUTATIVE GROUP.

We have assumed so far that differently named points are in general distinct; since we have not so far excluded geometries like the seven-point geometry $\Gamma^{(2)}$, this assumption is by no means necessarily valid. Consider first a theorem which is, in its first form, one about 'mid-points', and essentially one about harmonic pairs.

2.1*T*10 The resultant of two equal displacements.

G $\mathscr{L}\#|$ABU & ideal$|$U
C C : C = $\mathscr{D}_{AB}$B
T BU $\mathscr{H}$ AC
P *s* $\wp|$H : H $\notin$ AB
*c*1 ideal$|$VW : V $\in$ AH, W $\in$ BH
*c*2 K : K = VB $\cap$ HU
C′ : C′ = WK $\cap$ AB
*t*1 $\mathscr{D}_{AB}$H = K
$\mathscr{D}_{HK}$B = C′ = C
*t*2 AC $\mathscr{H}$ BU
p compare the figure of 1.4*C*1,

the points	A	B	C	D	K	L	M	N
being renamed	U	B	A	C	K	H	V	W

T BU $\mathscr{H}$ AC.

2.1*N*3 The 'mid-point' determined by two proper points.

G $\mathscr{L}\#|$ACU & ideal$|$U
C B : BU $\mathscr{H}$ AC
N B = mid-⌜AC⌝.

We use this notation even when, because we have not introduced the separation axioms, 'segments' cannot or have not been defined.

If this construction is applied to $\Gamma^{(2)}$, we find, on naming the points, an arrangement such as the following

A	B	H	U	V	W	K	C
0	1	4	3	5	2	6	0

i.e. $\mathscr{D}_{AB}$B = A.

To avoid complications of this sort we could reintroduce the axiom of the separate harmonic conjugates ($A\mathscr{S}$4).

2.1 *C*1 Construction of mid-$\ulcorner AB \urcorner$, i.e., the point M, $M \in AB$, such that $\mathscr{D}_{AM} = \mathscr{D}_{MB}$.

G $\# \wp | AB$
S $\wp | H : H \notin AB$
*C*1 $K : K = \mathscr{D}_{AH} H \quad (\Rightarrow \mathscr{D}_{AH} = \mathscr{D}_{HK})$
*C*2 $M : M \in AB, \quad HM \parallel BK$
T $M = \text{mid-}\ulcorner AB \urcorner$
P *c* $G : G \in KB, \quad MG \parallel AK$
*t*1 $\mathscr{D}_{HK} M = G \quad (c, C2)$
$\Rightarrow \mathscr{D}_{AH} M = G \Rightarrow HG \parallel AM$
$\Rightarrow \mathscr{D}_{AM} H = G \ \& \ \mathscr{D}_{HG} M = B$
T $\mathscr{D}_{AM} M = \mathscr{D}_{HG} M = B.$

The converse of this theorem, which is easily deducible, may be expressed as

2.1 *T*11*X* The join of the mid-points of two of the sides of a triangle is parallel to the third.

G $\mathscr{L} \wp | ABK$
C $M = \text{mid-}\ulcorner AB \urcorner$
$H = \text{mid-}\ulcorner AK \urcorner$
T $MH \parallel BK.$

We shall need another theorem on mid-points, which can be expressed as

2.1 *T*12*X* The diagonals of a parallelogram meet at their mid-points.

G $\mathscr{L} \# \wp | ABA'B' \ \& \ AB \parallel A'B' \ \& \ AA' \parallel BB'$
C $K : K = AB' \cap A'B$
T $K = \text{mid-}\ulcorner AB' \urcorner = \text{mid-}\ulcorner A'B \urcorner$
i.e. $\mathscr{D}_{AK} K = B', \quad \mathscr{D}_{BK} K = A'$
P (Introduce the ideal points.)

2.1 *T*13*X* Converse of *T* 12.

G $\mathscr{L} \wp | ABK$
C $A' = \mathscr{D}_{BK} K$
$B' = \mathscr{D}_{AK} K$
T $B' = \mathscr{D}_{AB} A'.$

CHAPTER 2.2

THE PRIMARY NET ON A LINE, $\mathfrak{N}_R$ AND $\mathfrak{N}_p$

In the transformer $\mathscr{D}$ we clearly have a model of the Farmer's paces along a furrow, since the sequence of points $\mathscr{D}_{OA}O = A$, $\mathscr{D}_{OA}A = B$, $\mathscr{D}_{OA}B = C, \ldots$ corresponds exactly to the succession of footprints. Also $\mathscr{D}_{AO}C = B$, $\mathscr{D}_{AO}B = A$, etc. We are able therefore to introduce the counting numbers and the negative integers on the basis of the description of repetitions of a geometric operation. When this has been done we may follow, in geometric terms, the usual route through the successive amplifications of the integers to the rationals, and, in later chapters, the reals, the complexes and other fields based on the reals.

We are however already aware of the possibility of interpreting almost all the axioms by geometric systems which contain only finite numbers of points, so that as the system develops we have to allow for these interpretations.

In this Chapter we are concerned mainly with the net on a single line, but of necessity make some use of constructions on other lines. While in general we regard the ideal point on the line as being fixed, we can simplify some of the explanations by naming first one point and then another as ideal. In many theorems however, the ideal point and line do not need to be specified; $\mathscr{L}\#|OEU$ & ideal$|U$, and $\#\wp|OE$ are then effectively equivalent statements.

In the following constructions O and E_0, and E and E_1 are alternative names for the same points.

2.2*C*1 Construction of the 'integer' points.

G $\quad \#\wp|OE \quad (O = E_0,\ E = E_1,\ \mathscr{D}_{OE}E_0 = E_1)$
*C*1 $\quad \{E_r : r = 2, 3, 4, \ldots,\ E_r = \mathscr{D}_{OE}E_{r-1}\}$
*C*2 $\quad \{E'_s : s = 1, 2, 3, \ldots,\ E'_1 = \mathscr{D}_{EO}E_0,\ E'_s = \mathscr{D}_{EO}E'_{s-1}\}$.

2.2*T*1 (i) $E_r = \mathscr{D}_{OE}\mathscr{D}_{OE}$ (altogether r factors) $\mathscr{D}_{OE}E_0$

(ii) $E'_s = \mathscr{D}_{EO}\mathscr{D}_{EO}$ (altogether s factors) $\mathscr{D}_{EO}E_0$.

2.2*N*1 $\quad (\mathscr{D}_{OE})^r$ or $\mathscr{D}^r_{OE} = \mathscr{D}_{OE}$ (altogether r factors) $\mathscr{D}_{OE}$,

i.e. $\mathscr{D}^r_{OE}$ is the resultant of r displacements each $\mathscr{D}_{OE}$.

From 2.1*T*6, 7, 8($\|$) for displacements on the same line, we see that we can adopt the usual rules for the addition and subtraction of integers, and shall write therefore

$$\mathscr{D}_{EO} = \mathscr{D}_{OE}^{-1}$$
$$\mathscr{D}_{EO}^{s} = \mathscr{D}_{OE}^{-s}, \quad \text{i.e. } E'_s = E_{-s}$$
$$\mathscr{D}_{OE}^{r} E_s = \mathscr{D}_{OE}^{r}\mathscr{D}_{OE}^{s} E_0 = \mathscr{D}_{OE}^{r+s} E_0$$
$$\mathscr{D}_{E_r E_s} = \mathscr{D}_{OE_{s-r}} = \mathscr{D}_{OE}^{s-r}.$$

These results may be summarised in the theorem

2.2*T*2 The set of points $\{E_r\} = \{D_{OE}^{r} E_0\}$, $r = \ldots -2, -1, 0, 1, 2, \ldots$ form a closed system under operations of the transformer $\mathscr{D}_{OE}$.

For every set of three integers r, s, t

$$\mathscr{D}_{E_r E_s} E_t = E_{t+s-r}.$$

An important property of the set $\{E_r\}$ is

2.2*T*3 $\{E_r\}$ is determined by O, E and the ideal point U on OE, and is independent of the choice of ideal line, except that the ideal line must pass through U.

*G*1	$\mathscr{L}\#\|OEU$ & ideal$\|U$ $(O = E_0, E = E_1)$	
*G*2	$V, V' : V, V' \notin OE$	
C	$\{E_r\}$ from O, E with ideal$\|UV$	
	$\{E'_r\}$ from O, E with ideal$\|UV'$	
T	$E_r = E'_r$	
P(i)	*g*(i)	$r > 0$
	*t*1	$E'_2 = E_2$
	p	$E_2 : E_0E_2 \;\mathscr{H}\; EU$
		$E'_2 : E_0E'_2 \;\mathscr{H}\; EU$
	*t*2	$E'_r = E_r$
	g	(inductive hypothesis) $E'_i = E_i$, $i = 2, \ldots, r-1$
	p	$E_rE_{r-2} \;\mathscr{H}\; E_{r-1}U$
		$E'_rE'_{r-2} \;\mathscr{H}\; E'_{r-1}U$, i.e. $E'_rE_{r-2} \;\mathscr{H}\; E_{r-1}U$
P(ii) *X*	*g*(ii)	$r < 0$.

We make frequent use of this freedom to vary the ideal line; we do so, for example, in the course of the proof of the next theorem, which plays a key part in the investigation of the relation between geometry and algebra.

2.2*T*4 First ⁑parallels-and-proportions theorem.

G	$\mathscr{L}\#\|OEU$ & ideal$\|U$ $O = E_0, E = E_1$.
	$\mathscr{L}\#\|O'E'U$ (i.e. $OE \parallel_U O'E'$) & $\#\|OE, O'E'$
*C*1	$E_r : E_r = \mathscr{D}_{OE}^{r} E_0$
*C*2	$V : V = OO' \cap EE'$

*C*3 $E'_r : E'_r = VE_r \cap O'E'$

T $\mathscr{D}^r_{O'E'} E'_0 = E'_r$

P $\mathscr{D}^r_{O'E'} E'_0$ depends only on O′E′U. Take UV to be the ideal line

t $\mathscr{D}_{OE} = \mathscr{D}_{O'E'}$ (ideal|UV)

$\mathscr{D}_{OE_r} = \mathscr{D}_{O'E'_r}$

T $\mathscr{D}^r_{O'E'} E'_0 = E'_r$.

2.2*T*5 Construction for multiplying integers.

G $\# \wp | OE$

*C*1 E_r, E_s

S $\wp | H : H \notin OE$

*C*2 $X : X = \mathscr{D}_{OE_r} H$

*C*3 $M : M = EX \cap OH$

*C*4 $Z : Z = HX \cap ME_s$

*C*5 $E_* : E_* \in OE \;\&\; E_* Z \parallel OH$

T $E_* = E_{rs}$

P *t*1 $\mathscr{D}^s_{HX} H = Z$ (*T*4)

*t*2 $\mathscr{D}^s_{OE_r} O = E_*$ (*C*5)

T $E_* = [\mathscr{D}_{OE}$ (r factors) $\mathscr{D}_{OE}]^s E_0 = E_{rs}$.

If the geometry contains only a finite number of points, then clearly for some pairs of counting numbers m and n we shall find $E_m = E_n$ while $m \neq n$; we can prove in fact the stronger theorem:

2.2*T*6 If, and only if, for some pair of unequal integers m, n, $E_m = \mathscr{D}^m_{OE} E_0 \;\&\; E_n = \mathscr{D}^n_{OE} E_0 \;\&\; E_m = E_n$ then there are only a finite number of points in the collinear set $\{E_r\}$.

*G*1 $\# \wp | OE$

*G*2 $E_m = \mathscr{D}^m_{OE} E_0 = \mathscr{D}^n_{OE} E_0 = E_n, \quad m < n$

S $E_r = \mathscr{D}^r_{OE} E_0$

T $E_r \in \{E_m, E_{m+1}, \ldots, E_{n-1}\}$

P *t*1 $\mathscr{D}^{n-m}_{OE} E_0 = E_0$

p $E_0 = \mathscr{D}^{-m}_{OE} \mathscr{D}^m_{OE} E_0$

$= \mathscr{D}^{-m}_{OE} \mathscr{D}^n_{OE} E_0$

c $k : 0 \leqslant r - m + k(n-m) < n - m$

i.e. $m \leqslant r + k(n-m) < n$†

*t*2 $(\mathscr{D}_{OE})^{r+k(n-m)} = (\mathscr{D}_{OE})^{k(n-m)} \mathscr{D}^r_{OE}$

$= \mathscr{D}^r_{OE}$

T $E_r = E_{r+k(n-m)} \;\&\; E_{r+k(n-m)} \in \{E_m, E_{m+1}, \ldots, E_{n-1}\}$.

† This is a counting operation: If $r < m$, the sequence of numbers $\mathscr{O}|\{r-m, r-m+(n-m), r-m+2(n-m), \ldots\}$ is formed, and if $r > m$ the sequence $\mathscr{O}|\{r-m, r-m-(n-m), r-m-2(n-m), \ldots\}$ is formed. From the appropriate sequence the member is selected which satisfies the required condition.

Effectively this theorem shows that the proper points of the line in a finite simple net are $E_0, E_1, E_2, \dots E_{p-1}$, and that $\mathscr{D}^n_{OE} E_0 = E_r$, where r is the remainder on dividing n by p. In the next four theorems we are to prove that p is prime.

2.2*T*7 If $\mathscr{D}^{rs}_{OE} = \mathscr{I}$, then either $\mathscr{D}^r_{OE} = \mathscr{I}$ or $\mathscr{D}^s_{OE} = \mathscr{I}$ or both.

*G*1 $\# \wp | OE$
*G*2 $\mathscr{D}^n_{OE} E_0 = E_n = E_0$
& $n = rs$ (r, s, n are positive integers)
& $\# | E_0 E_r$ (i.e. $\mathscr{D}^r_{OE} \neq \mathscr{I}$)
T $\mathscr{D}^s_{OE} = \mathscr{I}$
C as in *T*5 (construction for multiplying integers, which would not be possible if $E_r = E_0$)
P *t*1 $Z = H$
p $E_{rs} Z \parallel E_0 H$ & $E_{rs} = E_0$
T $E_s = E_0$
p $E_s = MZ \cap E_0 E_1$ & $MZ = ME_0$.

2.2*T*8 If $E_r = E_0$ and $E_s = E_0$ then $E_t = E_0$, where t is the H.C.F. of r and s.

G $E_r = E_0, E_s = E_0$
& $r = tu, s = tv$, & u, v relatively prime
T $E_t = E_0$.
P *t*1 There exist integers a, b, such that $ar + bs = t$. (This arithmetical theorem is assumed.)
T $\mathscr{D}^r_{OE} E_0 = E_0$ & $\mathscr{D}^s_{OE} E_0 = E_0 \Rightarrow \mathscr{D}^t_{OE} E_0 = E_0$
p $(\mathscr{D}^{ar}_{OE} \mathscr{D}^{bs}_{OE}) E_0 = \mathscr{D}^t_{OE} E_0 = E_0$.

Corollary: (i) If r and s are relatively prime and $E_r = E_s = E_0$, then $\mathscr{D}_{OE} E_0 = E_0$, contradicting the statement '$G \# | OE$' in the construction 2.2*C*1.

(ii) If r and s are relatively prime and the geometry is non-trivial, then either $E_r = E_0$ and $\# | E_s E_0$ or $E_s = E_0$ and $\# | E_r E_0$.

2.2*T*9 If $E_n = E_0$, then $E_{p^\alpha} = E_0$, where p is a prime and p^α is a factor of n.

*G*1 $n = p_1^{\alpha_1} p_2^{\alpha_2} \dots p_h^{\alpha_h}$ where each p_i is prime & $\# | p_i p_j$
*G*2 $E_n = E_0$ & $\# | E_0 E_1$
T For one particular p_i, which may be assumed to be p_1, $E_{p_1^{\alpha_1}} = E_0$, while, for $j \in \{2, 3, \dots, h\}$, $\# | E_{p_j^{\alpha_j}} E_0$

P   *s* (i) $r = p_1^{\alpha_1}$, $s = n/r$
   t $E_r = E_0$ & $\# | E_s E_0$
   s (ii) $r' = p_j^{\alpha_j}$, $s' = n/r' = mp_1^{\alpha_1}$ for some m
   t $E_{s'} = E_0$ & $\# | E_0 E_{r'}$.

2.2*T*10 If $E_{p^\alpha} = E_0$ then $E_p = E_0$.

G $E_{p^\alpha} = E_0$
T $E_p = E_0$
P $\mathscr{D}_{OE}^{p^\beta p^{\alpha-\beta}} = \mathscr{I}$ $(0 < \beta < \alpha)$
  $\Rightarrow$ either $\mathscr{D}_{OE}^{p^\beta} = \mathscr{I}$ or $\mathscr{D}_{OE}^{p^{\alpha-\beta}} = \mathscr{I}$.
In either case, successive reduction gives
T $\mathscr{D}_{OE}^{p} = \mathscr{I}$.

2.2*T*11 If $E_n = E_0$, then $E_p = E_0$, where p is some prime factor of n.

P   *T* 9 and *T* 10.

As a consequence of *T* 6 and *T* 11:

2.2*T*12

*G*1 $\mathscr{L}\wp | OEP$, $\# | OE$
*G*2 $(\mathscr{D}_{OE})^m P = Q$
   $(\mathscr{D}_{OE})^n P = Q$
T (i)   Net non-finite : $m = n$
  (ii)   Net finite : $m \equiv n$ (mod. p), where p is a prime (and independent of m).

2.2*N*2 $\mathfrak{F}_p$: In a finite geometry, the 'algebraic field' $\mathfrak{F}_p$ of the primary net of proper points on a line is GF(p) for a given prime p, i.e., there are in the net p proper points $E_0, E_1, \ldots, E_{p-1}$, and all arithmetical operations are carried out modulo p.

With this notation, we may replace *T* 12 by the simpler form

2.2*T*13

G $\{m, n\} \subset \{\text{integers}\}$ or $\{m, n\} \subset \mathfrak{F}_p$
T $(\mathscr{D}_{OE})^m = (\mathscr{D}_{OE})^n \Leftrightarrow m = n$.

It should be noticed that, as a consequence of theorem *T* 11 the net on a line of $\Gamma^{(4)}$ (Excursus 1.1.3) is not a *primary* net. In fact if we take ideal$|P_{18}P_5$ and consider the collinear points $P_0P_1P_6P_8$ and ideal$|P_{18}$, we find

$$(\mathscr{D}_{P_0P_1})^2 P_0 = P_0, \quad (\mathscr{D}_{P_0P_6})^2 P_0 = P_0, \quad (\mathscr{D}_{P_0P_8})^2 P_0 = P_0 \quad \text{etc.}$$

Every pair of the proper points P_0, P_1, P_6, P_8 forms a complete primary net. We consider the compounding of primary nets in Book 3.

In the case of the primary net in a finite geometry there are no further points to consider; the field is complete and closed under the operations of addition and multiplication and their inverses. On the other hand the non-finite primary net is not complete: we have yet to construct, for example, a point P to satisfy the condition $\mathscr{D}^r_{OP} E_O = E_1$ for an integer r.

2.2*C*2 Construction of the point $E_{1/r}$.

G $\#\wp|OE \quad (O = E_0, E = E_1)$
*C*1 $E_r : E_r = (\mathscr{D}^r_{OE}) E_0,\ r \in \{\text{integers}\}$
S $\wp|H,\ H \notin OE$
*C*2 $K : K = \mathscr{D}_{OE} H$
$L : L = E_r H \cap EK$
$M : M \in OE \quad \text{and} \quad ML \parallel OK$
T $\mathscr{D}^r_{OM} E_0 = E_1 \quad (\text{i.e. } M = E_{1/r})$
P *c* $P' : P' = LM \cap HK$
g (assumption) $\mathscr{D}^s_{OM} E_0 = E_1$
*t*1 $\mathscr{D}^s_{P'K} K = H \quad (\mathscr{D}_{MO} = \mathscr{D}_{P'K}, \mathscr{D}_{KH} = \mathscr{D}_{EO})$
*t*2 $\mathscr{D}^s_{ME} E_1 = E_r \quad (T4)$
*t*3 $\mathscr{D}^s_{ME} \mathscr{D}^s_{OM} E_0 = E_r \quad (g)$
*t*4 $(\mathscr{D}_{OM} \mathscr{D}_{ME})^s E_0 = E_r \quad (\mathscr{D} \text{ commutative})$
*t*5 $\mathscr{D}^s_{OE} E_0 = E_r$
*t*6 $s = r \quad (T13)$
T $\mathscr{D}^r_{OM} E_0 = E_1$.

While we require this construction explicitly for the non-finite net, it is a valid construction also for the $\mathfrak{F}_p$-nets; if $M = E_t$, then $tr = 1$ (mod. p).

The proof of construction *C*2 depends on the assumption (g) that there is an integer s such that $\mathscr{D}^s_{OM} E_O = E_1$, and justifies the assumption by showing that r is such an integer. This method of proof does not exclude the possibility that there may be other points which satisfy the condition $D^r_{OM} E_0 = E_1$. Assume then that there is a point M′ such that $\mathscr{D}^r_{OM'} = \mathscr{D}^r_{OM}$. We may use *C*2 to construct the point E_r from a point M′ said to be such that $\mathscr{D}^r_{OM'} E_0 = E_1$. From M′ construct: $L' : L' \in EK$ and $M'L' \parallel OK$, $E' : E' = OE \cap HL'$. $\#|MM' \Rightarrow \#|LL' \Rightarrow \#|E_r E'$. So that $\mathscr{D}^r_{OM'} = \mathscr{D}^r_{OM} \Rightarrow M = M'$.

In the non-finite net we now have points corresponding to the integers and to the reciprocals of the integers; by combining *C*2 with *C*1 we shall obtain the point corresponding to any rational number.

2.2*C*3 Construction of the point $E_{s/r}$.

G $\# \wp | OE$
*C*1 $E_{1/r}$ as in 2.2*C*2
*C*2 $E_{s/r} : E_{s/r} = \mathscr{D}^{s}_{OE_{1/r}} E_0$ as in 2.2*C*1.

2.2*N*3 $E_{s/r}$ is the unique point such that

$$\mathscr{D}^{r}_{OE_{s/r}} = \mathscr{D}^{s}_{OE}.$$

The finite and non-finite nets so far constructed will be called the PRIMARY NETS.

2.2*N*4 *Types of primary nets.*

(i) $\mathfrak{N}_p(O, E)$: the finite primary net of a prime number p of proper points on a line, defined, in relation to an ideal point on the line, by

$$\mathscr{D}^{r}_{OE} E_0 = E_r \quad (r = 0, \ldots, p-1),$$
$$E_{r+p} = E_r.$$

(ii) $\mathfrak{N}_R(O, E)$: the primary net of points P on the line, defined, in relation to a given ideal point, by $(\mathscr{D}_{OP})^m = (\mathscr{D}_{OE})^n$, $m \neq 0$, where m and n are integers. $P = E_{n/m}$.

(iii) $\mathfrak{N}(O, E)$: any primary net, $\mathfrak{N}_p(O, E)$ or $\mathfrak{N}_R(O, E)$.

I.e.

$\mathfrak{N}_R(O, E) = [P : \mathscr{D}^{n}_{OP} = \mathscr{D}^{m}_{OE}, m \in \{\text{integers}\} \mathbin{\&} n \in \{\text{counting numbers}\}]$.

For $\mathfrak{N}_p(O, E)$, $n = 1$, and the integers are replaced by the remainders modulo p.

The corresponding sets of numbers which combine under the operations of addition and multiplication (that is, repeated addition) and their inverses will be called PRIMARY FIELDS.

2.2*N*5 $\mathfrak{F}_R$: the field of the rationals. Compare $\mathfrak{F}_p$, the field of remainders modulo p introduced in *N*2.

The theorems that follow are valid over any primary net, but they may require slightly different interpretations for the finite nets $\mathfrak{N}_p$ and the rational net $\mathfrak{N}_R$. The numbers p, q used are different from the characteristic, p, of $\mathfrak{N}_p$.

2.2*T*14 Given any two points $\{P, Q\} \subset \mathfrak{N}(O, E)$ we can find a point L and numbers p, q such that $(\mathscr{D}_{OL})^p E_0 = P$ and $(\mathscr{D}_{OL})^q E_0 = Q$.

*G*1 $\{P, Q\} \subset \mathfrak{N}(O, E)$
*G*2 $P : \mathscr{D}^{a'}_{OP} = \mathscr{D}^{a}_{OE}$
$Q : \mathscr{D}^{b'}_{OQ} = \mathscr{D}^{b}_{OE}$

C $\quad L:(\mathscr{D}^{a'}_{OL})^{b'} = \mathscr{D}^{a'b'}_{OL} = \mathscr{D}_{OE}$ (2.2*C*2)

T $\quad P = \mathscr{D}^{b'a}_{OL}E_0$

$Q = \mathscr{D}^{a'b}_{OL}E_0$

P $\quad \mathscr{D}^{a'}_{OP} = \mathscr{D}^{a}_{OE} = \mathscr{D}^{a'b'a}_{OL}$

$\Rightarrow \mathscr{D}_{OP} = \mathscr{D}^{b'a}_{OL}$

similarly $\mathscr{D}_{OQ} = \mathscr{D}^{a'b}_{OL}$

T $\quad$ L is given by $\mathscr{D}^{b'a}_{OL} = \mathscr{D}_{OP}$ (*C*2)

$p = b'a, \quad q = a'b.$

So far we have assumed that a primary net is determined (in relation to a given ideal point) by two given base points O, E, but we prove next:

2.2*T*15 A primary net is determined by any two distinct points of itself (in relation to a given ideal point).

*G*1 $\quad \mathscr{L}\# \wp|PQOE \;\&\; \{P, Q\} \subset \mathfrak{N}(O, E)$

T $\quad \mathfrak{N}(P, Q) = \mathfrak{N}(O, E)$

*G*2 $\quad \mathscr{D}^{p'}_{OP} = \mathscr{D}^{p}_{OE}$

$\mathscr{D}^{q'}_{OQ} = \mathscr{D}^{q}_{OE}$

$n = p'q, \quad m = pq'$

C $\quad E':\mathscr{D}^{p'q'}_{OE'}E_0 = E_1$ (2.2*C*2)

*t*1 $\quad \mathscr{D}^{m}_{OE'}E_0 = P$

p $\quad \mathscr{D}^{p'q'p}_{OE'} = \mathscr{D}^{p}_{OE} = \mathscr{D}^{p'}_{OP}$

$\mathscr{D}^{n}_{OE'}E_0 = Q$

*t*2 $\quad \mathscr{D}_{PQ} = \mathscr{D}^{n-m}_{OE'}$

S $\quad X:X\in\mathfrak{N}(O, E) \;\&\; \mathscr{D}^{z}_{OX} = \mathscr{D}^{y}_{OE} = \mathscr{D}^{yp'q'}_{OE'} = \mathscr{D}^{y'}_{OE'}$

*t*3 $\quad \mathscr{D}^{(n-m)z}_{PX} = \mathscr{D}^{y'-mz}_{PQ}$

p $\quad \mathscr{D}^{z}_{PX} = \mathscr{D}^{z}_{PO}\mathscr{D}^{z}_{OX} = \mathscr{D}^{mz}_{E'O}\mathscr{D}^{y'}_{OE'} = \mathscr{D}^{y'-mz}_{OE'}$

$\mathscr{D}^{(n-m)z}_{PX} = (\mathscr{D}^{y'-mz}_{OE'})^{(n-m)} = (\mathscr{D}^{n-m}_{OE'})^{(y'-mz)} = \mathscr{D}^{y'-mz}_{PQ}$

T $\quad X\in\mathfrak{N}(P, Q).$

2.2*T*16 The explicit relation between the systems $\mathfrak{N}(O, E)$, $\mathfrak{N}(P, Q)$.

*G*1, 2 as in *T*15

S $\quad X:X\in\mathfrak{N}(O,E) \;\&\; X = E_x$

(i.e. $\mathscr{D}^{z}_{OX} = \mathscr{D}^{y}_{OE}$ & $y/z = x$)

T $\quad X\in\mathfrak{N}(P, Q) \;\&\; X = Q_{x'}$

(i.e. $\mathscr{D}^{y'}_{PX} = \mathscr{D}^{z'}_{PQ}$ & $z'/y' = x'$)

$$\& \; x' = \frac{(p'x-p)\,q'}{p'q-pq'}$$

P $$x' = \frac{y'-mz}{(n-m)z} \quad (T15\;(t3))$$

$$= \frac{xzp'q'-pq'z}{(p'q-pq')z} \quad (G2).$$

For arithmetical use the notation which expresses the numerical relations among the points in terms of powers of a transformer $\mathscr{D}$ is not very convenient, and we introduce therefore a notation which expresses the relation more simply. We define

2.2 *D*1 *Span-ratio* of a point-pair $\{P, Q\} \subset \mathfrak{N}(A, B)$ in relation to the base points A, B (with an assigned ideal point U) : s(PQ/AB), or, where explicit reference to the ideal point is required, $s_U(PQ/AB)$:

$$D \quad s(PQ/AB) = m/n \overset{\text{def}}{\Leftrightarrow} (\mathscr{D}_{PQ})^n = (\mathscr{D}_{AB})^m.$$

2.2*T*17*X* $s(AB/PQ)\, s(PQ/AB) = 1.$

2.2*T*18 $s(AB/PQ)\, s(PQ/XY) = s(AB/XY).$

$$G \quad \mathscr{D}^m_{AB} = \mathscr{D}^n_{PQ} \ \&\ \mathscr{D}^{m'}_{PQ} = \mathscr{D}^{n'}_{XY}$$
$$t \quad \mathscr{D}^{mm'}_{AB} = \mathscr{D}^{nm'}_{PQ} = \mathscr{D}^{nn'}_{XY}$$
$$T \quad s(AB/XY) = \frac{nn'}{mm'} = s(AB/PQ)\, s(PQ/XY).$$

The numbers s(AB/OE) behave in exactly the same way as either the rational numbers or the elements of GF(p) so long as all the points concerned are drawn from the same primary net. At a later stage we shall have to investigate the rules of combination when the points are not all in the same net, that is, effectively, how we can define and operate with s(OF/OE) when $F \in OE$ but $F \notin \mathfrak{N}(O, E)$. For example, in the geometry $F^{(4)}$ consider the collinear set $\mathscr{L}|P_0P_1P_6P_8P_{18}$ and take ideal$|P_{18}$. Then

$$\mathscr{D}_{P_0P_1}P_1 = P_0$$

so that $\mathfrak{N}(P_0, P_1) = \{P_0, P_1\}$, and the net is $\mathfrak{N}_2$. The question to be answered is: in what sense can we usefully introduce a symbol $s(P_0P_1/P_0P_6)$? We consider this question in Book 3.

The rest of this Chapter is concerned with properties of primary nets and span-ratios, which depend largely on adaptations of 2.2*T*4, *T*5 and *C*2. We re-state these therefore in terms of span-ratios.

2.2*T*19 (*T*4) The intercepts on two parallel lines by three concurrent lines give equal span-ratios.

$$G1 \quad \mathscr{L}\#|ABCU,\ \text{ideal}|U \ \&\ C \in \mathfrak{N}(A, B)$$
$$G2 \quad \mathscr{L}\#A'B'C'U \ \&\ \mathscr{P}|AA' \cap BB' \cap CC'$$
$$T \quad s(AB/AC) = s(A'B'/A'C').$$

2.2*C*4 (*T*5) Given $\{P, Q\} \subset \mathfrak{N}(O, E)$, to construct R:

$$s(OR/OE) = s(OP/OE)\, s(OQ/OE).$$

G $\{P, Q\} \subset \mathfrak{N}(O, E)$
S $\wp | H, H \notin OE$
C $X : \mathcal{D}_{OP} H = X$
$M : M = OH \cap EX$
$Z : Z = HX \cap MQ$
$R : R \in OE \ \&\ RZ \parallel OH$
T $s(OR/OE) = s(OP/OE)\, s(OQ/OE)$.

2.2*C*5 (*C*2) Construction of the 'inverse point', i.e. the point M such that $s(OM/OE)\, s(OP/OE) = 1$.

G $P \in \mathfrak{N}(O, E)$
S $\wp | H, H \notin OE$
C $K : K = \mathcal{D}_{OE} H$
$L : L = KE \cap HP$
$M : M \in OE \ \&\ ML \parallel OK$
T $s(OP/OE)\, s(OM/OE) = 1$.

So long as points are restricted to belonging to a primary net, the constructions 2.2*C*4 and *C*5 do no more than provide methods of naming the points corresponding to the product mm′/nn′ and the reciprocal n/m shorter than the methods of iteration applied to the points with span-ratios m/n and m′/n′. Later, when we consider points outside the primary net, no iterative process is available, and we use *C*4 and *C*5 rather as *definitions* of products and inverses.

We have seen in 2.2*T*15 that any two distinct points of the primary net on a line may be used as base points, but so far we have treated the ideal point differently from other points on the line. We prove next that the construction *C*5 can be interpreted as the operation of interchanging O and U, so that we could reorganise the points on the line as $\mathfrak{N}(U, E)$ with ideal|O. The construction in the following theorem is identical with *C*5 but the ideal line is explicitly named and some points have been renamed to emphasise the symmetry.

Wherever necessary, when naming the base points of the net, we shall name also the ideal point:

2.2*N*6 $\mathfrak{N}(O, E/U)$: net with base points O, E and ideal point U.

2.2*T*20 Construction of $\mathfrak{N}(U, E)$ with ideal O.

*G*1	$\mathscr{L}_3$|OUVK	
*G*2	ideal|UV	ideal|OV
*C*1	$E : E \in OU$ & $EK \parallel OV$	$E : E \in UO$ & $EK \parallel UV$
	$H : H \in OV$ & $HK \parallel OU$	$H' : H' \in UV$ & $H'K \parallel UO$
	$H' : H' \in OK$ & ideal|H′	$H : H \in UK$ & ideal|H
S	$X : X \in \mathfrak{N}(O, E)$	$X' : X' \in \mathfrak{N}(U, E)$
*C*2	$L : L = XH \cap EK$	$L : L = X'H' \cap EK$
*C*3	$X' : X' \in OE$ & $X'L \parallel OK$	$X : X \in UE$ & $XL \parallel UK$

(either X determines L determines X′ or X′ determines L determines X)

T† $\mathscr{D}^{r}_{OE} = \mathscr{D}^{s}_{OX} \Rightarrow \mathscr{D}^{r}_{OX'} = \mathscr{D}^{s}_{OE}$ $\qquad$ $\mathscr{D}^{r}_{UE} = \mathscr{D}^{s}_{UX'} \Rightarrow \mathscr{D}^{r}_{UX} = \mathscr{D}^{s}_{UE}$.

The symmetry of this construction together with *C*5 enable us to state explicitly

2.2*T*21 The relation between $\mathfrak{N}(O, E)$ with ideal|U and $\mathfrak{N}(U, E)$ with ideal|O.

G $s_U(OX/OE) = x$
$s_O(UX/UE) = x'$

T $xx' = 1.$

This relation may also be expressed as

2.2*T*22 $s_U(OX/OE) = s_O(UE/UX)$.

The next step is to combine the operations of interchanging O and U with that of replacing O, E by another pair {P, Q} of $\mathfrak{N}(O, E)$, so that we may obtain a relation for any point $P \in \mathfrak{N}(O, E)$ between $s_U(OP/OE)$ and $s_V(OP/OE)$, where $V \in \mathfrak{N}$ (O, E/U).

2.2*T*23 The relation between $s_V(OU/OE)$ and $s_U(OV/OE)$

G $s_U(OV/OE) = v$
$s_V(OU/OE) = u$

T $u + v = 1$

P $u = s_V(OU/OE) = s_O(VE/VU)$ $\quad$ (*T*22)
$v = s_U(OV/OE) = s_O(UE/UV)$
$u + v = s_O(UE/UV) + s_O(EV/UV)$
$= s_O(UV/UV) = 1.$

† See note on p. 216.

2.2*T*24 The relation between $s_U(OP/OE)$ and $s_V(OP/OE)$.

G $s_U(OP/OE) = x \ \& \ s_U(OV/OE) = v$
$s_V(OP/OE) = x' \ \& \ s_V(OU/OE) = u$

T $x' = x(1-v)/(x-v)$
$x = x'(1-u)/(x'-u) = vx'/(v+x'-1)$

P *t*1 $s_U(VP/VE) = (x-v)/(1-v)$

$$p \quad s_U(VP/VE) = \frac{s_U(OP/OE)-s_U(OV/OE)}{s_U(OE/OE)-s_U(OV/OE)}$$

T *p* $s_U(VP/VE) = s_V(UE/UP)$ (*T*22)

$$\Rightarrow \frac{x-v}{1-v} = \frac{1-u}{x'-u} = \frac{v}{x'+v-1} \quad (t1, T23)$$

$$\Rightarrow x' = x(1-v)/(x-v).$$

2.2*T*25 A number, determined by four points in a net, which is independent of the choice of ideal point.

G $\mathscr{L} \# \wp | OEUP \ \& \ P \in \mathfrak{N}(O, E/U)$

S $V : V \in \mathfrak{N}(O, E)$

T $s_V(OP/OE) \, s_V(UE/UP) = s_U(OP/OE)$
which is independent of the choice of V.

P *t* (Notation of *T*24)

$$\begin{cases} x = s_U(OP/OE) \\ x' = s_V(OP/OE) \\ (1-u)/(x'-u) = s_V(UE/UP) \quad (T24, t1) \end{cases}$$

$$x = x'(1-u)/(x'-u) \quad (T24).$$

From *T*23 we have

$$x = s_U(OP/OE) = 1 - s_P(OU/OE)$$
$$= s_P(UE/OE),$$

so that, combining *T*22 and *T*23:

$$x = s_U(OP/OE) = s_O(UE/UP)$$
$$= s_P(EU/EO) = s_E(PO/PU).$$

The symbols O, U, E, P separate into two relatively ordered pairs, {O, U}, {P, E}, or say, {A, B}, {A′, B′}, with $\partial(AB/A'B') = I$ (Section 1.8*D*1), such that

$$x = s_A(BA'/BB') = s_B(AB'/AA').$$

The span-ratio of a point P in relation to given zero, ideal and unit points therefore provides a number which depends only on the four points themselves and an arrangement of these points as two ordered pairs (namely, {$^{\mathcal{O}}$zero, ideal}, {$^{\mathcal{O}}$P, unit}). This number is called the CROSS RATIO:

2.2*D*2 Cross-ratio, R, of two ordered pairs:

$$\mathrm{R}(AB/A'B') = s_A(BB'/BA') = s_B(AA'/AB') = s_{A'}(B'B/B'A) = s_{B'}(A'A/A'B).$$

Given four objects we may arrange them as ordered pairs in six ways, so that given four collinear points they will determine in general six (related) numbers. From relations among the span-ratios $s_A(BB'/BA')$ etc. we see that these are:

2.2*T*26 *G* $s_A(BB'/BA') = x \;\&\; x \notin \{0, 1\}$

T $\mathrm{R}(AB/A'B') = x, \quad \mathrm{R}(AB/B'A') = 1/x$

$\mathrm{R}(AA'/BB') = 1-x, \quad \mathrm{R}(AA'/B'B) = 1/(1-x)$

$\mathrm{R}(AB'/BA') = 1-\frac{1}{x}, \quad \mathrm{R}(AB'/A'B) = \frac{x}{x-1}.$

If two of the points coincide some values of the cross-ratio are still determinate, and it is easily verified that

2.2*T*27 $\mathrm{R}(AB/A'B') = 0 \Leftrightarrow A = A'$ or $B = B'$

$\mathrm{R}(AB/A'B') = 1 \Leftrightarrow A = B$ or $A' = B'$

$\mathrm{R}(AB/A'B')$ is undefined if $A = B'$ or $B = A'$.

2.2*T*28*X* Given $\mathscr{L}\#|PQR$ and $\mathrm{R}(PQ/RX)$, the point X is uniquely determined.

2.2*X*1 If $x_1+x_2+x_3 = 0$, prove that the six values of the cross-ratio of a set of four points one of the values of which is $-x/_1x_2$ are $-x_i/x_j$ $(i, j = 1, 2, 3, i \neq j)$.

2.2*X*2 Prove that of the six values R for four distinct points (i) in $\mathfrak{N}_R$ four are positive and two negative, (ii) in $\mathfrak{N}_p$ for various values of p, examples can be found in which none, two, four or all six are squares.

2.2*T*29 The cross-ratio of two harmonic pairs

G $\mathscr{L}\#|ABC \;\&\; C \in \mathfrak{N}(A, B)$

C $D : AB \;\mathscr{H}\; CD$

T $\mathrm{R}(AB/CD) = -1$

P *s* ideal|A

*t*1 $\mathscr{D}_{CB}B = D$ (ideal|A, 2.1*T*10)

*t*2 $s_A(BD/BC) = -1$ (*t*1)

T $\mathrm{R}(AB/CD) = -1.$

2.2 *T* 30 *X* Converse of *T* 29

G $\mathscr{L}\# | \mathrm{ABCD}$ & $\mathrm{R}(\mathrm{AB/CD}) = -1$
T AB $\mathscr{H}$ CD.

2.2 *X* 3 *G* $\mathscr{L}\# | \mathrm{ABCD}$ & $\mathrm{R}(\mathrm{AB/CD}) = \mathrm{R}(\mathrm{AB/DC})$
T $\mathrm{R}(\mathrm{AB/CD}) = -1$ and there are only three different values of R for the four points namely -1, 2, $\frac{1}{2}$.

2.2 *X* 4 *G* 1 $\{\mathrm{A_i}\} \subset \mathfrak{N}(\mathrm{O, E})$ $(\mathrm{i} = 1, 2, 3, 4)$
G 2 $\mathrm{s}(\mathrm{OA_i/OE}) = \mathrm{a_i}$
T $\mathrm{R}(\mathrm{A_1A_2/A_3A_4}) = \dfrac{(\mathrm{a_1 - a_3})(\mathrm{a_2 - a_4})}{(\mathrm{a_1 - a_4})(\mathrm{a_2 - a_3})}$.

2.2 *X* 5 *G* $\mathscr{L}\# | \{\mathrm{A_i}\}$, $\mathrm{i} \in \{1, \ldots, 5\}$ & $\mathrm{A_i} \in \mathfrak{N}(\mathrm{O, E})$
T $\mathrm{R}(\mathrm{A_1A_2/A_3A_4})\,\mathrm{R}(\mathrm{A_1A_2/A_4A_5})\,\mathrm{R}(\mathrm{A_1A_2/A_5A_3}) = 1$.

We conclude this Chapter by expressing betweenness and separation relations for $\mathfrak{N}_\mathrm{R}$ in terms of span-ratios and cross-ratios.

2.2 *T* 31 *X* (i) *G* $\mathscr{L}\# | \mathrm{ABP}$ & $\mathrm{P} \in \mathfrak{N}_\mathrm{R}(\mathrm{A, B})$
T $\mathrm{P} \in \ulcorner\mathrm{AB}\urcorner \Leftrightarrow \mathrm{s}(\mathrm{AP/PB}) > 0$
$\mathrm{P} \in \ulcorner\mathrm{B}(\tilde{\mathrm{A}}) \Leftrightarrow \mathrm{s}(\mathrm{AP/PB}) < -1$
$\mathrm{P} \in \ulcorner\mathrm{A}(\tilde{\mathrm{B}}) \Leftrightarrow 0 > \mathrm{s}(\mathrm{AP/PB}) > -1$

(ii) *G* $\mathrm{s}(\mathrm{OA/OE}) = \mathrm{a}$, $\mathrm{s}(\mathrm{OB/OE}) = \mathrm{b}$, $\mathrm{s}(\mathrm{OP/OE}) = \mathrm{x}$
T $\mathrm{P} \in \ulcorner\mathrm{AB}\urcorner \Leftrightarrow (\mathrm{x-a})(\mathrm{x-b}) < 0$
$\mathrm{P} \in \ulcorner\mathrm{B}(\tilde{\mathrm{A}}) \Leftrightarrow 0 < \dfrac{\mathrm{x-b}}{\mathrm{x-a}} < 1$
$\mathrm{P} \in \ulcorner\mathrm{A}(\tilde{\mathrm{B}}) \Leftrightarrow 0 < \dfrac{\mathrm{x-a}}{\mathrm{x-b}} < 1$
or P $\mathscr{B}$ AB $\Leftrightarrow (\mathrm{x-a})(\mathrm{x-b}) < 0$
B $\mathscr{B}$ AP $\Leftrightarrow 0 < \dfrac{\mathrm{x-b}}{\mathrm{x-a}} < 1$

(iii) *G* $\mathscr{L}\# | \mathrm{ABCP} \subset \mathfrak{N}_\mathrm{R}(\mathrm{A, B})$
T $\partial(\mathrm{AB/PQ}) = \mathrm{I} \Leftrightarrow \mathrm{s}(\mathrm{AB/PQ}) > 0$

(iv) *G* $\mathscr{L}\# | \mathrm{ABCD} \subset \mathfrak{N}_\mathrm{R}(\mathrm{A, B})$
T AB $\mathscr{S}$ CD $\Leftrightarrow \mathrm{R}(\mathrm{AB/CD}) < 0$.

2.2 *X* 6 Prove, using *T* 31 (iv) and *A*$\mathscr{S}$ 1, 2, 3 (and not the theory of quadratic equations) that in $\mathfrak{N}_\mathrm{R}$ no set of distinct points can be found such that $\mathrm{R}(\mathrm{AB/A'B'}) = \mathrm{R}(\mathrm{AA'/B'B})$.

CHAPTER 2.3

THE PRIMARY NET IN THE PLANE

From two given points O, E in the plane we have derived a primary net on the line OE, and in relation to those points have shown how to assign a span-ratio to every point-pair A, B of $\mathfrak{N}(O, E)$. Further the span-ratio of any point-pair A, B of $\mathfrak{N}(O, E)$ in relation to any other such point-pair C, D, namely s(AB/CD), is independent of the choice of base points O and E in the net. We wish now to cover the plane with a primary net, defined by three non-collinear proper points in the plane, and consisting of points such that any pair determines on their join a primary net all of whose points belong to the net in the plane.

To accomplish this we need to investigate the relations between primary nets on two non-parallel lines: effectively these are the properties of ⁑parallels-and-proportions for a triangle and a second parallel to one of the sides. In the first two theorems we need, in the course of the proof, to designate different points as ideal, but for the rest of the Chapter the ideal line is to be regarded as fixed.

We require first:

2.3*T*1 The converse of 2.2*T* 19. If two sets of three collinear points, one on each of two parallel lines, give the same span-ratios, then the joins of the corresponding points in the two sets are concurrent.

*G*1 $\mathscr{L}\#|\text{LMNU}$
$\mathscr{L}\#|\text{L'M'N'U} \;\&\; \#|\text{LM, L'M'}$
*G*2 $s_U(\text{LM/LN}) = s_U(\text{L'M'/L'N'})$
T $\mathscr{P}|\text{LL'} \cap \text{MM'} \cap \text{NN'}$
P *c* $\text{H} : \text{H} = \text{LL'} \cap \text{MM'}$
s ideal|H, i.e. $\text{LL'} \parallel_{\text{H}} \text{MM'}$
*t*1 $\mathscr{D}_{\text{LM}} = \mathscr{D}_{\text{L'M'}}$ (ideal|UH)
*t*2 $\mathscr{D}_{\text{LN}} = \mathscr{D}_{\text{L'N'}}$ (*t*1, *G*2)
*t*3 $\text{NN'} \parallel_{\text{H}} \text{LL'}$
T $\mathscr{P}|\text{LL'} \cap \text{MM'} \cap \text{NN'} = \text{H}$.

2.3*T*2 P, Q are points on the ⁑sides AB, AC respectively of a ⁑triangle. The necessary and sufficient condition for s(AP/AB) = s(AQ/AC) is BC$\parallel$PQ.

G $\mathscr{L}\#|\text{APBU}$
$\mathscr{L}\#|\text{AQCV} \;\&\; \not{\mathscr{L}}|\text{ABC}$

Necessary

G $PQ\|_{UV}BC$ & $P\in\mathfrak{N}(A, B/U)$

T (nec.) $Q\in\mathfrak{N}(A, C/V)$ & $s_U(AP/AB) = s_V(AQ/AC)$

P c $R : R = BC\cap PQ$

$t1$ $PQ\|_{UV}BC \Rightarrow UV\supset R$

s ideal$|$AR

$t2$ $\mathscr{D}_{PB} = \mathscr{D}_{QC}, \mathscr{D}_{PU} = \mathscr{D}_{QV}$ (ideal$|$AR)

$\Rightarrow s_A(UP/UB) = s_A(VQ/VC)$ $(P\in\mathfrak{N}(A, B/U))$

$t3$ $s_U(AB/AP) = s_A(UP/UB)$ (2.2T22)

$= s_A(VQ/VC)$ (t2)

$= s_V(AC/AQ)$ (2.2T22)

T (necessary) $s_U(AB/AP) = s_V(AC/AQ)$

Sufficient

G $s_U(AB/AP) = s_V(AC/AQ)$

T (suff.) $PQ\|_{UV}BC$

P $t1$ $s_A(UB/UP) = s_U(AP/AB)$ (2.2T22)

$= s_V(AQ/AC)$ (G)

$= s_A(VC/VQ)$ (2.2T22)

$t2$ $\mathscr{P}|UV\cap BC\cap PQ$ (2.3T1, t1)

T (sufficient) $BC\|_{UV}PQ$.

2.3T3 If a line parallel to BC cuts AB, AC in respectively P, Q then s(PQ/BC) = s(AP/AB).

$G1$ $\mathscr{L}\#|APBU$, ideal$|$U, $P\in\mathfrak{N}(A, B/U)$

$\mathscr{L}\#|AQCV$, ideal$|$V & $\mathscr{L}|ABC$

$G2$ $BC\|PQ$ $(\Rightarrow Q\in\mathfrak{N}(A, C))$

T $s(PQ/BC) = s(AP/AB)$

P c $D : D\in BC$ & $QD\|AB$

$t1$ $\mathscr{D}_{PQ} = \mathscr{D}_{BD}$

$t2$ $s(PQ/BC) = s(BD/BC)$

$= s(AQ/AC)$ (2.3T2)

$= s(AP/AB)$ (2.3T2).

We could rewrite theorem T2 in the form:

$G1$ $\mathscr{L}\wp|OEF$ (ideal line fixed)

$G2$ $P : \mathscr{D}^m_{OP} = \mathscr{D}^n_{OE}$ & $\#|OP$

$Q : \mathscr{D}^m_{OQ} = \mathscr{D}^n_{OF}$

T $PQ\|EF$.

It follows that if $\mathfrak{N}(O, E)$ is $\mathfrak{N}_p$, then $\mathfrak{N}(O, F)$ is also $\mathfrak{N}_p$, since there is a one-to-one correspondence between points of $\mathfrak{N}(O, E)$ and points of $\mathfrak{N}(O, F)$ established by the parallels to EF through points of $\mathfrak{N}(O, E)$. Thus

2.3*T*4 If $\mathfrak{N}_p(O, E)$ and $\mathfrak{N}_q(O, F)$ belong to the same geometry then $p = q$.

2.3*T*5 Two sets of four collinear points in perspective have equal cross-ratios.

*G*1 $\mathscr{P}\#|\{l_i\}, i \in \{1, 2, 3, 4\}, l_1 \cap l_2 \cap l_3 \cap l_4 = V$
*G*2 $\#|ab \ \& \ V \notin a, V \notin b$
C $\{A_i\} = \{l_i \cap a\}, \{B_i\} = \{l_i \cap b\}$
T $R(A_1A_2/A_3A_4) = R(B_1B_2/B_3B_4)$
P *c* $U : U = a \cap b$
s $\text{ideal}|UV$
t $a \parallel_U b \ \& \ l_i \parallel_V l_j$
$\Rightarrow \mathscr{D}_{A_iA_j} = \mathscr{D}_{B_iB_j}$ (ideal $|UV$)
T $R(A_1A_2/A_3A_4) = R(B_1B_2/B_3B_4)$ (2.2*X*4).

As a consequence of *T*5 we may formulate the definition

2.3*D*1 The cross-ratios of four concurrent lines.

*G*1 $\mathscr{P}|l \cap m \cap l' \cap m', \#|lm', \#|l'm$
*G*2 $p : p \not\ni l \cap m'$
C $p \cap \{^{\mathscr{O}}l, m, l', m'\} = \{^{\mathscr{O}}A, B, A', B'\}$
D $R(lm/l'm') \overset{\text{def}}{=} R(AB/A'B')$.

2.3*X*1 Prove that in the planar net $\mathfrak{N}_7$ there are sets of four collinear points such that $R(AB/CD) = R(BC/AD)$. In particular, in the definition of the geometry $\Gamma^{(p)}$ for $p = 7$ given in Excursus 1.1.6, take

A	B	C	D	A′	B′	C′
(0, 0, 1)	(1, 0, 1)	(3, 0, 1)	(1, 0, 0)	(0, 1, 1)	(5, 1, 1)	(1, 1, 1)

and show that there are points H, K such that

$$ABCD \overset{H}{\barwedge} A'B'C'D \overset{K}{\barwedge} BCAD.$$

From algebraic considerations, what is the next value of p in order of magnitude such that $\mathfrak{N}_p$ contains sets of four points with this property?

So far the value of a cross-ratio has been made to depend on a selection of ideal point and the construction of a net in relation to it, even though, as was proved in 2.2*T*25, the value is in fact independent of this selection. It is not possible therefore on this basis to define the cross-ratio of two ordered pairs of points of the ideal line, but we could expect that the value would be unchanged if some other line were taken as ideal. $T3$ enables us to formalise this.

2.3 $D2$ The cross-ratio of two ordered pairs of ideal points is equal to the cross-ratio of any two ordered pairs of points of a collinear set in perspective with them.

G ideal|UVU′V′, # |UV′, # |U′V
S ℘|K, h, K∉h
C {$^{\mathcal{O}}$A, B, A′, B′} = h ∩ {$^{\mathcal{O}}$KU, KV, KU′, KV′}
D ℞(UV/U′V′) = ℞(AB/A′B′).

From $T5$, the definition is consistent (i.e. it is independent of the choice of K and h).

Since in the enunciation of theorem $T2$, P, Q may be any points in the respective nets $\mathfrak{N}$(A, B), $\mathfrak{N}$(A, C), $T2$ in conjunction with $T3$, shows how, given primary nets on any two non-parallel lines, we may establish a related net on any other line. In order to complete the definition of a net extending over the plane, we have to prove:

2.3 $T6$ The primary net induced on any line by two given primary nets on non-parallel lines.

$G1$ $\mathscr{L}$℘|ABC (ideal points fixed)
$G2$ B′ ∈ $\mathfrak{N}$(A, B), # |B′AB†
C′ ∈ $\mathfrak{N}$(A, C), # |C′AC
C T : T = BC ∩ B′C′
T T ∈ $\mathfrak{N}$(B, C), T ∈ $\mathfrak{N}$(B′, C′)
P c C″ : C″ ∈ AC, B′C″ ∥ BC
$t1$ C″ ∈ $\mathfrak{N}$(A, C) = $\mathfrak{N}$(C′, C)
⇒ C ∈ $\mathfrak{N}$(C′, C″)
$t2$ CT ∥ C″B′ & C ∈ $\mathfrak{N}$(C′, C″)
⇒ T ∈ $\mathfrak{N}$(C′, B′) ($T2$)
$t3$ T ∈ $\mathfrak{N}$(B, C) (cf. $t2$)
T T ∈ $\mathfrak{N}$(B, C), T ∈ $\mathfrak{N}$(B′, C′).

As a preliminary to the definition of the primary net over the plane, we restate the theorem as:

$G1$ $\mathfrak{N}$(O, E), $\mathfrak{N}$(O, F), $\mathscr{L}$℘|OEF
$G2$ {P, P′} ⊂ $\mathfrak{N}$(O, E), {Q, Q′} ⊂ $\mathfrak{N}$(O, F) & # |EPP′, # |FQQ′
C X : X = PQ ∩ P′Q′
R : R = EF ∩ PQ
R′ : R′ = EF ∩ P′Q′
T R ∈ $\mathfrak{N}$(P, Q), R′ ∈ $\mathfrak{N}$(P′, Q′), X ∈ $\mathfrak{N}$(P′, Q′), etc.

† This excludes the case $\mathfrak{N}_2$. For $\mathfrak{N}_3$ some precautions have to be taken in the course of the proof.

2.3*D*3 A PRIMARY NET on the plane is the set of points X which can be named by the following construction in relation to a fixed ideal line:

*G*1 $\mathscr{L}\wp|$OEF
*S*1 P : P $\in \mathfrak{N}$(O, E), Q $\in \mathfrak{N}$(O, F)
*S*2 X : X $\in \mathfrak{N}$(P, Q).

The theorems above show that this definition is unambiguous: thus, if P′ $\in \mathfrak{N}$(O, E) and P′X $\cap$ OF = Q′, then Q′ $\in \mathfrak{N}$(O, F) and X $\in \mathfrak{N}$(P′, Q′).

2.3*N*1 $\mathfrak{N}$(O, E, F): the primary planar net determined by three non-collinear proper points, in relation to a given ideal line. If the ideal line u or UV needs to be designated explicitly, we use the symbol $\mathfrak{N}$(O, E, F/u) or $\mathfrak{N}$(O, E, F/UV). If the field needs to be specified, we write

$$\mathfrak{N}_p(O, E, F) \quad \text{or} \quad \mathfrak{N}_R(O, E, F).$$

2.3*T*7*X* If two points belong to a primary planar net then all points of the primary net determined by them belong to the planar net.

G {X, Y} $\subset \mathfrak{N}$(O, E, F)
T $\mathfrak{N}$(X, Y) $\subset \mathfrak{N}$(O, E, F).

2.3*T*8*X* A primary planar net is determined, in relation to a given ideal line, by any three non-collinear points of itself.

G $\mathscr{L}\wp|$XYZ, {X, Y, Z} $\subset \mathfrak{N}$(O, E, F)
T $\mathfrak{N}$(X, Y, Z) = $\mathfrak{N}$(O, E, F).

2.3*T*9 A primary planar net is completely determined by any (ordered) set of four points, no three collinear.

G $\mathscr{L}_3|$PQRS
S Ideal|QR
C E : E $\in$ PQ & SE $\|_R$ PR
F : F $\in$ PR & SF $\|_Q$ PQ
T {P, Q, R, S} determine the net $\mathfrak{N}$(P, E, F/QR).

The type of the net is of course not determined; the points P, Q, R, S provide a basis for all the operations $\mathscr{D}$ required to generate other points in the net, but cannot prescribe whether the resulting net is $\mathfrak{N}_p$ for some p or $\mathfrak{N}_R$.

We conclude this Chapter with some theorems which are consequences in the planar net of the construction for 'products', 2.2*C*4 (p. 133), when the construction is repeated with two points interchanged.

2.2*C*4 Constructions for products (ideal line fixed).

G $\{P, Q\} \subset \mathfrak{N}(O, E)$

S $\wp|H, H \notin OE$

C $X : X = \mathscr{D}_{OP}H$ $\quad$ $Y : Y = \mathscr{D}_{OQ}H$

$M : M = OH \cap EX$ $\quad$ $N : N = OH \cap EY$

$Z : Z = HX \cap MQ$ $\quad$ $Z' : Z' = HY \cap NP$

$R : R \in OE$ & $RZ \| OH$ $\quad$ $R' : R' \in OE$ & $R'Z' \| OH$

T $s(OQ/OE) = s(OR/OP)$ $\quad$ $s(OP/OE) = s(OR'/OQ)$.

Since in $\mathfrak{F}_p$ and in $\mathfrak{F}_R$, multiplication is the result of repeated additions, in a primary planar net these two constructions must lead to the same point for s(OQ/OE) s(OP/OE) and s(OP/OE) s(OQ/OE). In fact, in the notation we have used, we have already assumed that no distinction is to be made. In consequence $R = R'$ and $Z = Z'$. The two sequences of constructions, one leading to Z and the other to Z′ therefore result in the same point. But this is a pure incidence theorem, not apparently depending on the assigning of span-ratios, except insofar as they are involved in the restriction $\{P, Q\} \subset \mathfrak{N}(O, E/U)$.

The nature of the theorem appears more clearly if we strip the construction of some unessential detail. The statement $Z = Z'$ is equivalent to $N \in PZ$, i.e. $\mathscr{P}|PZ \cap EY \cap OH$. As far as the incidences in the figure are concerned O, H and R are irrelevant. Without them the construction can be expressed as

G $\mathscr{L} \# \wp|PQE$

$\# \wp|XZ$ & $XZ \| PQ$ & $\#|XZ, PQ$

C $Y : Y \in XZ$ & $PX \| QY$

$M : M = XE \cap ZQ$

$N : N = YE \cap ZP$

T $MN \| XP \quad (\| YQ)$.

Now express the conditions of parallelism as incidences. We have only two families of parallel lines in the construction, so that only two ideal points are involved, and the ideal line plays no significant part. Moreover, there is no third parallel to XY and PQ, so that their relation of parallelism is irrelevant. For symmetry let us rename the points.

	X	Y	Z	Q	P	E	YE ∩ ZP = N	ZQ ∩ XE = M	XP ∩ YQ
as	A	B	C	A′	B′	C′	A″	B″	C″

The theorem becomes:

2.3*T*10 The Pappus theorem.

G $\# \wp|ABCA'B'C' \in \mathfrak{N}(O, E, F)$ & $\mathscr{L}|ABC$ & $\mathscr{L}|A'B'C'$ & $\#|AB, A'B'$

C $\{A'', B'', C''\} : A'' = BC' \cap B'C, B'' = CA' \cap C'A, C'' = AB' \cap A'B$

T $\mathscr{L}|A''B''C''$.

The points O, E, F are dummies, and may be replaced by any three non-collinear points in the net; moreover we could assume for the ideal line any line in the net not through any of the three selected points. Let the net be defined as $\mathfrak{N}(A, B', C/A'B)$. Then the construction can be completed when C′ is assigned arbitrarily in the net. Thus:

$$
\begin{aligned}
&A' : A' \in B'C' \ \&\ \text{ideal}|A', \quad B : B \in AC \ \&\ \text{ideal}|B,\\
&A'' : A'' \in CB' \ \&\ A''C' \parallel_B AC,\\
&C'' : C'' \in AB' \ \&\ \text{ideal}|C'',\\
&B'' : B'' \in AC' \ \&\ B''C \parallel_{A'} B'C'.
\end{aligned}
$$

The final incidence in the figure is $A''B'' \parallel_{C''} AB'$.

So long as C′ is selected to lie in $\mathfrak{N}(A, B', C/A'B)$ this final incidence is a consequence of the others. We cannot prove this on the basis of *A𝒥* 1, 2, 3, alone, if we do not assume that C′ belongs to the primary net, and in fact shall construct in Chapter 3.5 geometries satisfying these axioms in which the eight assigned incidences do not imply the ninth.

The following theorem, while valid in any primary planar net, for the special case of $\mathfrak{N}_R$ provides a quantitative form of the descriptive theorem 1.7*T*4 ('a line, not through a vertex of a triangle, meets either two or none of the side segments'), namely,

2.3*T*11 The Menelaus theorem.

*G*1 $\mathscr{L}\wp|ABC$ (ideal points fixed)
*G*2 $I : I \in \mathfrak{N}(B, C)$, $J : J \in \mathfrak{N}(A, C)$
C $K : K = IJ \cap AB$
T $s(BI/IC)\, s(CJ/JA)\, s(AK/KB) = -1$
P *c* $H : H \in BC \ \&\ AH \parallel IJK$
T $s(BI/IC)\, s(CJ/JA)\, s(AK/KB)$
$= s(BI/IC)\, s(CI/IH)\, s(HI/IB)$
$= -1.$

The figure is essentially that of 2.2*C*4 with

	O	P	Q	Z	M	N	H
replaced by	I	J	K	A	B	C	H.

2.3*X*2 *G*1 $\mathscr{L}\wp|ABC$
*G*2 $\{N, W\} \subset \mathfrak{N}(A, B)$
$\{M, V\} \subset \mathfrak{N}(A, C)$
C $L : L = MN \cap BC$
$U : U = VW \cap BC$
T $R(LU/BC)\, R(MV/CA)\, R(NW/AB) = 1.$

CHAPTER 2.4

AFFINE CO-ORDINATES AND INVARIANTS

When we have set up primary nets ('scales') on two non-parallel lines (in relation to a given ideal line) we may establish an 'affine co-ordinate system' very much in the same way as, on the basis of physical Euclidean geometry, we set up a Cartesian co-ordinate system, and may carry out some of the corresponding elementary operations with it. Progress is of course at present limited by there not being yet available for the affine system a 'distance function' corresponding to the function in the Cartesian geometry derived from the Pythagorean relation.

To establish a usable co-ordinate system in a plane, all we need are procedures for assigning a one-to-one correspondence between points and (ordered) number-pairs, and between lines and appropriate functions of these number-pairs. We proceed to do this under the assumption that all points belong to a primary net over the plane in relation to an assigned ideal line. The word 'number' is used to stand for a member of $\mathfrak{F}_p$ or $\mathfrak{F}_R$.

2.4*T*1 In relation to any two non-collinear point-pairs O, E; O, F, a one-to-one correspondence can be established between the proper points of the plane and the set of ordered pairs of numbers.

*T*1 (i)	To determine the number-pair corresponding to a given point
*G*1	$\mathscr{L}\wp$|OEF
*S*1	$\wp$|P, $\quad$ P $\in \mathfrak{N}$(O, E, F)
*C*1	L : L $\in$ OE & PL $\parallel$ OF
*C*2	M : M $\in$ OF & PM $\parallel$ OE
*C*3	s(OL/OE) = l, $\quad$ s(OM/OF) = m
*T*1 (i)	P determines the ordered pair of numbers (l, m) uniquely
*T*1 (ii)	To determine the point corresponding to a given number pair
*G*2	$\mathscr{L}\wp$|OEF
*S*2	number pair (l, m)
*C*4	L : L $\in$ OE & s(OL/OE) = l
*C*5	M : M $\in$ OF & s(OM/OF) = m
*C*6	P : MP $\parallel$ OE & LP $\parallel$ OF
*T*1 (ii)	(l, m) determines P uniquely.

We may use the number-pair (l, m) which determines and is determined by P in relation to O, E, F as an alternative name for the point, and thus avoid the necessity of writing 'the point whose co-ordinates are (l, m)', writing instead (without any ambiguity) 'the point (l, m)'. We use the following notation

2.4*N*1 $\mathfrak{C}$(OE, OF): the system of affine co-ordinates determined by O, E, F (in relation to an assigned ideal line).

2.4*N*2 P = (l, m) ⊂ $\mathfrak{C}$(OE, OF), or P = (l, m) (when the co-ordinate system is unambiguously known): the point whose co-ordinates are (l, m) (relative to $\mathfrak{C}$(OE, OF)).

We require next some algebraic method of recognising 'collinear sets' of points; we shall find: the set $\{(x_i, y_i)\}$ is a collinear set if and only if there exist numbers a, b such that for each point of the set:

either $$ax_i + by_i + 1 = 0,$$

or $$ax_i + y_i = 0,$$

or $$x_i = 0.$$

This is to be deduced from theorems about span-ratios which correspond to the Joachimsthal relations.

2.4*T*2 Proportionality of segments and co-ordinates.

*G*1 $\mathfrak{C}$(OE, OF)
*G*2 $\# \wp | P_1P_2$
*G*3 $\wp | P : P \in P_1P_2$
*C*1 $\{L_1, L_2, L\} \subset OE : OF \parallel P_1L_1 \parallel P_2L_2 \parallel PL$
*C*2 $\{M_1, M_2, M\} \subset OF : OE \parallel P_1M_1 \parallel P_2M_2 \parallel PM$
T $s(P_1P/P_1P_2) = s(L_1L/L_1L_2) = s(M_1M/M_1M_2)$
P 2.2*T*19.

2.4*T*3 Converse of *T*2.

*G*1 $\mathfrak{C}$(OE, OF)
*G*2 $\# \wp | PP_1P_2$
$P_1P_2 \nparallel OE, \quad P_1P_2 \nparallel OF$
*C*1 $L_1, L_2, L \in OE; \quad OF \parallel P_1L_1 \parallel P_2L_2 \parallel PL$
*C*2 $M_1, M_2, M \in OF; \quad OE \parallel P_1M_1 \parallel P_2M_2 \parallel PM$
*G*3 $s(L_1L/L_1L_2) = s(M_1M/M_1M_2)$
*T*3(i) $P \in P_1P_2$
(ii) $s(P_1P/P_1P_2) = s(L_1L/L_1L_2)$

P   *c*1 $P' : P' \in P_1P_2, \quad LP' \parallel OF$
    *c*2 $M' : M' \in OF, \quad P'M' \parallel OE$
    t $M = M'$
    p $s(M_1M/M_1M_2) = s(L_1L/L_1L_2)$
$$= s(P_1P'/P_1P_2) = s(M_1M'/M_1M_2)$$

*T*3 (i)   *P* $M = M'$
*T*3 (ii)   *P*   2.2*T*19.

2.4*T*4 Interpretation of *T*2 and *T*3 in the co-ordinate system.

*G*1 $\mathfrak{C}(OE, OF)$
*G*2 $\wp | P_1P_2P; \quad P_1 \neq P_2$
*G*3 (i) $P_1 = (x_1, y_1), \quad P_2 = (x_2, y_2), \quad P = (x_p, y_p)$
    (ii) $x_1 \neq x_2, \quad y_1 \neq y_2$

T $P \in P_1P_2 \Leftrightarrow \dfrac{x_p - x_1}{x_2 - x_1} = \dfrac{y_p - y_1}{y_2 - y_1}$

P $\dfrac{x_p - x_1}{x_2 - x_1} = s(L_1L/L_1L_2) = s(M_1M/M_1M_2)$
$$= \frac{y_p - y_1}{y_2 - y_1}.$$

If $x_2 = x_1$ the relation reduces to $x_p = x_1$; since $P_1 \neq P_2$ we cannot have both $x_1 = x_2$, and $y_1 = y_2$.

From this relation we deduce the 'equation of the line P_1P_2', i.e. the algebraic relation satisfied by a pair of numbers (x, y) if and only if it is the co-ordinate-pair of a point of the line. The relation of *T*4 reduces to
$$(y_2 - y_1)x - (x_2 - x_1)y - (x_1y_2 - x_2y_1) = 0.$$
Thus, given any quantities a, b, h such that not both a and b are zero,
$$ax + by + h = 0$$
is the equation of a line, and any line (other than the ideal line) may have its equation reduced to this form. For all values of λ other than zero
$$\lambda(ax + by + h) = 0$$
represents the same line. We shall usually choose λ so that the last written coefficient (non-zero) is 1, i.e. $\lambda = 1/h$ when $h \neq 0$, $\lambda = 1/b$ when $h = 0$, $b \neq 0$.

2.4*T*5 Alternative interpretation of *T*4.

*G*1, 2, 3   As *T*4
*G*4 $P : s(P_1P/PP_2) = \lambda/\mu \quad (\wp | P \Rightarrow \lambda + \mu \neq 0)$

T $x_p = \dfrac{\lambda x_1 + \mu x_2}{\lambda + \mu}, \quad y_p = \dfrac{\lambda y_1 + \mu y_2}{\lambda + \mu}.$

This is the 'parametric' (Joachimsthal) form of equation of the line. From 2.2*T*31 we see that in the case of the rational net:

$$\lambda/\mu > 0 \Leftrightarrow \mathrm{P} \in \ulcorner \mathrm{P}_1\mathrm{P}_2 \urcorner$$

$$\lambda/\mu < -1 \Leftrightarrow \mathrm{P} \in \ulcorner \mathrm{P}_2(\tilde{\mathrm{P}}_1)$$

$$0 > \lambda/\mu > -1 \Leftrightarrow \mathrm{P} \in \ulcorner \mathrm{P}_1(\tilde{\mathrm{P}}_2).$$

The point of intersection of two lines (in any primary net)

$$\mathrm{ax+by+h = 0}$$

$$\mathrm{cx+dy+k = 0}$$

is obtained as the number-pair (x, y) satisfying both equations, namely, provided $\mathrm{ad-bc \neq 0}$, $\left(\dfrac{\mathrm{bk-hd,\ ch-ak}}{\mathrm{ad-bc}}\right)$.

If ad = bc the equations are in algebraic language 'inconsistent' (unless a/c = b/d = h/k, in which case they are not independent and represent the same line), but geometrically they represent two lines with a common ideal point. The equations of the lines may be written as

$$\mathrm{ax+by+h = 0}$$

$$\mathrm{ax+by+h' = 0.}$$

For, writing first $\mathrm{b}+\epsilon$ for b in the second equation, we find for the common point

$$\frac{1}{\mathrm{a}\epsilon}[\mathrm{b(h'-h)-h}\epsilon,\quad \mathrm{a(h-h')}].$$

For all values of ϵ, other than $\epsilon = 0$, this is a proper point on the line $\mathrm{ax+by+h} = 0$, and moreover every proper point on the line corresponds to some value of ϵ other than zero. The ideal point may therefore be taken to be that given by $\epsilon = 0$. Correspondingly we could regard the ideal line as that member of the family of lines with equations

$$\epsilon(\mathrm{ax+by})+\mathrm{h} = 0$$

for which $\epsilon = 0$, but a more satisfactory formulation is to be derived shortly.

The ratios of the quantities a, b, h are the span-ratios of certain segments determined by the line and $\mathfrak{C}$(OE, OF), namely,

2.4*T*6 Meaning of coefficients in $\mathrm{ax+by+h} = 0$.

*G*1 $\mathfrak{C}$(OE, OF)
*G*2 $\mathrm{l:ax+by+h = 0}$
C $\mathrm{L:L = l \cap OE,\quad L = (x_L, y_L)}$

T $s(OL/OE) = -h/a \quad (a \neq 0)$

P $L \in OE : y_L = 0 \quad (C)$

$L \in l : ax_L + by_L + h = 0 \quad (G2)$

$L = l \cap OE : L = (-h/a, 0) \quad (C).$

At this stage there is a considerable gain to be made in clarity of exposition by the adoption of matrix notation. Write, for the proper point P, = (x, y), the column-vector

$$\boldsymbol{\xi} = \begin{bmatrix} x \\ y \\ 1 \end{bmatrix}$$

and for the line a, not passing through the origin, the row-vector

$$\boldsymbol{\alpha}^T = [u, v, 1].$$

The condition that $P \in a$, or that $a \supset P$, may then be expressed as

$$\boldsymbol{\alpha}^T \boldsymbol{\xi} = 0, \quad \text{i.e.} \quad ux + vy + 1 = 0.$$

For ideal points we take the vectors

$$\begin{bmatrix} x \\ 1 \\ 0 \end{bmatrix} \text{ and } \begin{bmatrix} 1 \\ 0 \\ 0 \end{bmatrix};$$

for lines through the origin we take the vectors [u, 1, 0] and [1, 0, 0]. The origin is the point

$$\begin{bmatrix} 0 \\ 0 \\ 1 \end{bmatrix},$$

the ideal line is the line [0, 0, 1].

The special incidences are then represented as follows:

(i) The ideal point

$$\begin{bmatrix} p \\ 1 \\ 0 \end{bmatrix}$$

lies on [u, v, 1] if

$$up + v = 0,$$

i.e. the family of parallel lines defined by (p, 1, 0) is the family with equations $ux - upy + 1 = 0$, together with the line $x - py = 0$ through the origin.

The ideal point (1, 0, 0) lies on lines for which $u = 0$, i.e. lines $vy + 1 = 0$, which are parallel to $y = 0$.

(ii) The line [k, 1, 0] through the origin contains the point (x, y, 1) if $kx + y = 0$, i.e. it contains the points (x, $-kx$, 1), together with the ideal point ($-1/k$, 1, 0).

The form $[\xi_1, \xi_2]$ is to be used to represent the matrix of two columns and three rows

$$[\xi_1, \xi_2] = \begin{bmatrix} x_1 & x_2 \\ y_1 & y_2 \\ 1 & 1 \end{bmatrix}.$$

Then, if $P_0 \in P_1P_2$ and $s(P_1P_0/P_0P_2) = \lambda_1/\lambda_2$, we have

$$(\lambda_1+\lambda_2)\,\xi_0 = [\xi_1, \xi_2]\begin{bmatrix}\lambda_1 \\ \lambda_2\end{bmatrix}.$$

In the same way a line a_0 through the point of intersection of a_1 and a_2 is given by

$$(\rho_1+\rho_2)\,\boldsymbol{\alpha}_0^T = [\rho_1, \rho_2]\begin{bmatrix}\boldsymbol{\alpha}_1^T \\ \boldsymbol{\alpha}_2^T\end{bmatrix},$$

where ρ_1/ρ_2 is essentially only a parameter but has in fact a geometric derivation, namely,

$$-\rho_1/\rho_2 = \mathfrak{R}(a_1a_2/a_0a_*),$$

where a_* is the line through the point of intersection and parallel to $y = 0$.

2.4*X*1 If the geometry is confined to a primary net, then the statement as an axiom of the simply-special Desargues construction $A\mathscr{I}3^*$ is sufficient to ensure the existence of the general Desargues figure.

In the Desargues figure $A\mathscr{I}3$, take

K	A	B	C
(1, 1, 1)	(1, 0, 0)	(0, 1, 0)	(0, 0, 1)

(i.e. take ideal|AB and $\mathfrak{C}$(CA″, CB″) where A″ = KB ∩ AC, B″ = KA ∩ BC). A′ ∈ KA, so may be taken as

$$\alpha\left\{\alpha^{-1}\begin{bmatrix}1\\1\\1\end{bmatrix}+(1-\alpha^{-1})\begin{bmatrix}1\\0\\0\end{bmatrix}\right\} = \begin{bmatrix}\alpha\\1\\1\end{bmatrix}.$$

Define B′ and C′ similarly and prove $\mathscr{L}$|LMN. The theorem follows since the operations $\mathscr{D}$ on which the coordinate system is based depend only on the axiom $A\mathscr{I}3^*$.

2.4*X*2 Select suitable co-ordinate-systems, and prove the Pappus theorem and the Menelaus theorem for points in a primary planar net.

As a preamble to the discussion of the equations representing changes in the co-ordinate-system, we consider:

2.4*T*7 The equations representing the geometric transformer $\mathscr{D}$.

G	$\mathfrak{C}(\mathrm{OE}, \mathrm{OF})$	
	$\mathrm{K}:\mathrm{K} = (\mathrm{h}, \mathrm{k})$	
S	$\mathrm{P}:\mathrm{P} = (\mathrm{x}_0, \mathrm{y}_0)$	
C	$\mathrm{Q}:\mathrm{Q} = \mathscr{D}_{\mathrm{OK}}\mathrm{P} = (\mathrm{x}_1, \mathrm{y}_1)$	
T	$\mathrm{x}_1 = \mathrm{x}_0+\mathrm{h}, \quad \mathrm{y}_1 = \mathrm{y}_0+\mathrm{k}$	
P	*c*	$\{\mathrm{L}, \mathrm{G}, \mathrm{M}\} \subset \mathrm{OE}\ \&\ \mathrm{LP}\parallel\mathrm{MQ}\parallel\mathrm{GK}\parallel\mathrm{OF}$
	*t*1	$\mathscr{D}_{\mathrm{OG}}\mathrm{L} = \mathrm{M}$
	*t*2	$\mathrm{s(OM/OE)} = \mathrm{s(OL/OE)}+\mathrm{s(OG/OE)}$
T		$\Rightarrow \mathrm{x}_1 = \mathrm{x}_0+\mathrm{h}$
	similarly	$\mathrm{y}_1 = \mathrm{y}_0+\mathrm{k}.$

These equations may be written

$$\xi_1 = \mathbf{D}_{\mathrm{OK}}\xi_0,$$

where

$$\mathbf{D}_{\mathrm{OK}} = \begin{bmatrix} 1 & 0 & \mathrm{h} \\ 0 & 1 & \mathrm{k} \\ 0 & 0 & 1 \end{bmatrix}.$$

The inverse is

$$\mathbf{D}_{\mathrm{KO}} = \begin{bmatrix} 1 & 0 & -\mathrm{h} \\ 0 & 1 & -\mathrm{k} \\ 0 & 0 & 1 \end{bmatrix}.$$

Corresponding to the resultant $\mathscr{D}_{\mathrm{OK}'}\mathscr{D}_{\mathrm{OK}}$ of two displacements we have the product of the matrices $\mathbf{D}_{\mathrm{OK}'}\mathbf{D}_{\mathrm{OK}}$, which is again a matrix of the pattern

$$\begin{bmatrix} 1 & 0 & \lambda \\ 0 & 1 & \mu \\ 0 & 0 & 1 \end{bmatrix}.$$

This of course must be the case since the matrices (under multiplication) form a group isomorphic to (i.e. of structure identical with) the group of displacements.

There are three different types of changes of co-ordinate system to be considered and then combinations of these; the types are:

(i) shifting the 'origin' O to O′, but with O′E′ ∥ OE, O′F′ ∥ OF and s(O′E′/OE) = 1, s(O′F′/OF) = 1,

(ii) keeping the origin fixed, but replacing OE, OF by any pair of distinct lines OE′, OF′, and

(iii) keeping O, E, F fixed but selecting a new ideal line.

We shall find that transformations of all three types form groups, and combinations generate groups (the system 'generated' by a set of operations is that consisting of the resultants of all possible sub-sets of the set).

(i) $O'E' \parallel OE$, $O'F' \parallel OF$, $s(O'E'/OE) = 1$, $s(O'F'/OF) = 1$. This transformation of co-ordinates is the counterpart of 'displacement', in the sense that the pair of co-ordinate-pairs (x_0, y_0), (x_0+h, y_0+k) may be considered as having been derived the one pair from the other by two quite distinct operations: first we may suppose that the co-ordinate-system has been kept fixed, and the second pair corresponds to a point obtained from the first by the displacement $\mathscr{D}_{OK}$, where $K = (h, k)$; secondly, we may suppose that the same point is represented in different co-ordinate-systems by the two pairs of co-ordinates.

2.4 *T*8 New axes $O'E' \parallel OE$, $O'F' \parallel OF$ (with 'fixed scales').

G $\mathfrak{C}(OE, OF)$
  $\wp|O' : O' = (r, s)|\mathfrak{C}(OE, OF)$
C $E' : E' = \mathscr{D}_{OE} O'$
S $F' : F' = \mathscr{D}_{OF} O'$
S $\wp|P : P = (x_0, y_0)|\mathfrak{C}(OE, OF)$
  $P = (x_0', y_0')|\mathfrak{C}(O'E', O'F')$
T $x_0' = x_0 - r$, $y_0' = y_0 - s$
P *c* $E'' : E'' = O'F' \cap OE$
  $L : L \in OE$, $PL \parallel OF$
  $L' : L' = PL \cap O'E'$
  $N : N \in OE$, $L'N \parallel OO'$
 *t*1 $s(OE''/OE) = r$
 *t*2 $s(NL/OE) = s(OE''/OE) = r$
T $x_0' = s(O'L'/O'E') = s(ON/OE)$
  $= s(OL/OE) - s(NL/OE)$
  $= x_0 - r$
 similarly $y_0' = y_0 - s$.

This transformation may be expressed in the form

$$\boldsymbol{\xi}' = \mathbf{D}_{O'O}\, \boldsymbol{\xi},$$

where

$$\mathbf{D}_{O'O} = \begin{bmatrix} 1 & 0 & -r \\ 0 & 1 & -s \\ 0 & 0 & 1 \end{bmatrix}.$$

2.4 *D*1 (i) *Displacement group of transformations*, **D**.

The group is generated by matrices $\mathbf{D}_{O'O}$ and consists of matrices of the same pattern as $\mathbf{D}_{O'O}$. $\mathfrak{C}(O'E', O'F')$ is obtained from $\mathfrak{C}(OE, OF)$ by an operation of the displacement group if and only if $\mathscr{D}_{OE} O' = E'$, $\mathscr{D}_{OF} O' = F'$.

2.4 *X*3 If the co-ordinate-system is changed so that the only conditions satisfied are O′E′‖OE, O′F′‖OF, show that the matrix of the transformation is of the form

$$\begin{bmatrix} p & 0 & p' \\ 0 & q & q' \\ 0 & 0 & 1 \end{bmatrix}.$$

Show that these matrices also form a group.

2.4 *D*1 (ii) *Change of axes without change of origin.*

In this case there is no simple geometric operation (like displacement) which corresponds to the transformation induced by the change of co-ordinate-system.

2.4 *T*9 Relation between $\mathfrak{C}$(OE, OF) and $\mathfrak{C}$(OE′, OF′).

*G*1 $\mathfrak{C}$(OE, OF), $\mathfrak{C}$(OE′, OF′)

*C*1 E″ : E″ ∈ OE′, EE″ ‖ OF′

F″ : F″ ∈ OF′, FF″ ‖ OE′

*G*2 Constants of the transformation

$$s(OE''/OE') = e$$
$$s(OF''/OF') = f$$

equation OF′|$\mathfrak{C}$(OE, OF) : $ax + by = 0$

equation OE′|$\mathfrak{C}$(OE, OF) : $cx + dy = 0$

$$(a \neq 0, \quad d \neq 0, \quad ad - bc \neq 0)$$

S P : P = (x_0, y_0)|$\mathfrak{C}$(OE, OF)

P = (x_0', y_0')|$\mathfrak{C}$(OE′, OF′)

T $$x_0' = e\left(x_0 + \frac{b}{a}y_0\right)$$

$$y_0' = f\left(\frac{c}{d}x_0 + y_0\right)$$

P *c* (constructions from P)

*c*1 L′ : L′ ∈ OE′, PL′ ‖ OF′

*c*2 N : N = PL′ ∩ OE

*c*3 N′ : N′ ∈ OF′, PN′ ‖ OE

*t*1 $\mathscr{D}_{ON}$N′ = P (*c*2, *c*3)

*t*2 $$s(ON/OE) = x_0 + \frac{b}{a}y_0$$

p equation PN : $a(x - x_0) + b(y - y_0) = 0$

N = PN ∩ OE

*t*3 s(ON/OE) = s(OL′/OE″) (L′N ‖ EE″)

= s(OL′/OE′) s(OE′/OE″)

= x_0'/e

$$T \quad e\left(x_0+\frac{b}{a}y_0\right) = x_0' \quad (t2, t3)$$

$$\text{similarly} \quad f\left(\frac{c}{d}x_0+y_0\right) = y_0'.$$

We may write this transformation as

$$\boldsymbol{\xi}' = \mathbf{M}\boldsymbol{\xi},$$

where

$$\mathbf{M} = \begin{bmatrix} e & e' & 0 \\ f' & f & 0 \\ 0 & 0 & 1 \end{bmatrix} \quad \left(e' = \frac{eb}{a}, \quad f' = \frac{fc}{d}\right).$$

Again, the product of two matrices of the same pattern as **M** is another matrix of the same pattern, and there is a unique inverse to each **M**, namely,

$$\frac{1}{ef-e'f'} \begin{bmatrix} f & -e' & 0 \\ -f' & e & 0 \\ 0 & 0 & 1 \end{bmatrix},$$

so that, as is clear also from the geometric definition, the matrices form a group under multiplication. But it is clear that the group is not *commutative* (as the 'displacement' group is) since, in general for two of the matrices, $\mathbf{MM}' \neq \mathbf{M'M}$.

If the operations of types (i) and (ii) are combined we obtain the most general type of transformation of co-ordinates (i.e. one in which neither origins, axes nor scales in the two systems are specially related) which leaves the ideal line fixed. The products of the two corresponding matrices are:

$$\mathbf{T} = \mathbf{DM} = \begin{bmatrix} e & e' & -r \\ f' & f & -s \\ 0 & 0 & 1 \end{bmatrix}$$

and

$$\mathbf{MD} = \begin{bmatrix} e & e' & -re-se' \\ f' & f & -rf'-sf \\ 0 & 0 & 1 \end{bmatrix}$$

—two different transformations of the same type. It is easily verified that the matrices T form a group under multiplication.

2.4 ***X*4** Given $\mathbf{T} = \mathbf{DM}$ as above find $\mathbf{M}'$ and $\mathbf{D}'$ such that

$$\mathbf{T} = \mathbf{M'D} = \mathbf{MD}'.$$

2.4 *X*5 Thence prove (without using any explicit forms) that any member of the group of matrices **T** can be expressed in a unique way as **DM**.

2.4 *D*2 *Affine group* of transformations: the group of the linear transformations of co-ordinates which leave the ideal line unchanged, generated by and consisting of the non-singular matrices

$$\mathbf{T} = \begin{bmatrix} a & b & c \\ a' & b' & c' \\ 0 & 0 & 1 \end{bmatrix}.$$

2.4 *D*1 (iii) *Equations corresponding to change only of ideal line.*

2.4 *T*10

*G*1 $\mathfrak{C}(OE_U, OF_V)$, $\mathfrak{C}(OE_{U'}, OF_{V'})$
*G*2 equation $U'V'|\mathfrak{C}(OE_U, OF_V): lx+my+1 = 0$†
S $P: P = (x_0, y_0)|\mathfrak{C}(OE_U, OF_V)$
$P = (x_0', y_0')|\mathfrak{C}(OE_{U'}, OF_{V'})$
T $x_0' = (l+1)x_0/(lx_0+my_0+1)$
$y_0' = (m+1)y_0/(lx_0+my_0+1)$
P *c*1 $L: L \in OE,\ PL\|_{V'}OF$
*c*2 $L': L' = V'P \cap OE$ (i.e. $PL'\|_{V'}OF$)
*t*1 $s_U(OU'/OE) = -1/l$ (*G*2)
*t*2 equation $V'P|\mathfrak{C}(OE_U, OF_V)$:
$x(my_0+1) - mx_0y - x_0 = 0$
*t*3 $s_U(OL'/OE) = x_0/(my_0+1)$ (*t*2)
T $x_0' = s_{U'}(OL'/OE)$

$$= \frac{x_0}{my_0+1}\left(-\frac{1}{l}-1\right)\bigg/\left(-\frac{x_0}{my_0+1}-\frac{1}{l}\right) \quad (2.2\,T24)$$

$$= \frac{(l+1)x_0}{lx_0+my_0+1}$$

similarly $$y_0' = \frac{(m+1)y_0}{lx_0+my_0+1}.$$

The expressions obtained here cannot be written immediately in the matrix form used in types (i) and (ii) because of the linear function in the denominators. But in the earlier matrix forms the third row is nugatory, since it states simply $1 = 1$. In this case we may obtain a useful equation from the third row and construct a matrix form for the transformations of type (iii) by introducing a scalar multiplier ρ and writing

$$\rho\boldsymbol{\xi}' = \mathbf{U}\boldsymbol{\xi},$$

† This form is always possible (with perhaps either $l = 0$ or $m = 0$) since O is essentially proper and the new ideal line may not be selected to pass through O.

where

$$U = \begin{bmatrix} l+1 & 0 & 0 \\ 0 & m+1 & 0 \\ l & m & 1 \end{bmatrix}.$$

The elimination of the multiplier ρ from the three linear relations which constitute the matrix equation leads to the forms for x'_0 and y'_0 stated in the theorem.

2.4*X*6 Prove that both $\mathbf{U}^{-1}$ and the product of two matrices of the pattern of **U** are matrices of the pattern of **U**.

2.4*X*7 Can a matrix **D′** or **M′** be found such that, for given **U**, **D** and **M**

$$\mathbf{UD} = \mathbf{D'U}, \quad \mathbf{UM} = \mathbf{M'U}?$$

2.4*X*8 Do matrices of the same pattern as (i) **UD**, (ii) **UM** form a group?

The product of matrices of all three types is a non-singular three-by-three matrix with elements selected without restriction from $\mathfrak{F}_R$ or $\mathfrak{F}_p$ as the case may be. The set of such matrices clearly forms a group. In its geometric interpretation it is the 'Projective Group of the plane' (over the relevant field).

2.4*D*3 *Projective group* of transformations: the group of all transformations between co-ordinate systems

$$\mathfrak{C}_{UV}(OE, OF) \quad \text{and} \quad \mathfrak{C}_{U'V'}(O'E', O'F'),$$

represented by the equations obtained from the matrix equation

$$\rho\boldsymbol{\xi}' = \mathbf{K}\boldsymbol{\xi}$$

on the elimination of ρ, where **K** is any non-singular three-by-three matrix.

We next find a function, $\Gamma_{123}, = \Gamma(\boldsymbol{\xi}_1, \boldsymbol{\xi}_2, \boldsymbol{\xi}_3)$, of the co-ordinates of three points which is such that the ratio of its values in two different co-ordinate-systems connected by an affine transformation (i.e. defined in relation to the same ideal line) is independent of the choice of the three points. Γ is called a 'relative invariant' of the three points under affine transformations; from it we are to derive immediately an 'absolute invariant' of a set of four points, i.e. a number which is the same for the given four points in all choices of co-ordinate-system (with fixed ideal line). This invariant is the counterpart of the span-ratio of three points on a line.

2.4*T*11 Proof of the 'invariant' property of Γ.

*G*1 $\wp|P_1P_2P_3$, $P_i = \xi_i|\mathfrak{C}(OE, OF)$
D $\Gamma(\xi_1, \xi_2, \xi_3) = \det[\xi_1, \xi_2, \xi_3]$†
*G*2 $\mathfrak{C}(O'E', O'F'): \xi' = T\xi$ (*D*2)
T $\Gamma(\xi_1', \xi_2', \xi_3') = (\det T)\Gamma(\xi_1, \xi_2, \xi_3)$
P $[\xi_1', \xi_2', \xi_3'] = T[\xi_1, \xi_2, \xi_3]$
$\Rightarrow \det[\xi_1', \xi_2', \xi_3'] = \det T \det[\xi_1, \xi_2, \xi_3]$.

It follows that any four points, arranged in any way as a pair of ordered sets of three, determine a number which is invariant under affine transformations. The numbers formed thus provide the analogue of the number $s_U(AB/AC)$ determined by three collinear points independently of the choice of the zero and unit points O and E.

2.4*D*4 *G* $\mathscr{L}_3\wp|P_1P_2P_3P_4$, $P_i = \xi_i|\mathfrak{C}(OE, OF)$
D $\Gamma_{123}/\Gamma_{124}$ is an *invariant* under affine transformations.

The four quantities Γ_{ijk} are connected by the relation

$$\Gamma_{123} - \Gamma_{124} + \Gamma_{134} - \Gamma_{234} = \begin{vmatrix} 1 & 1 & 1 & 1 \\ x_1 & x_2 & x_3 & x_4 \\ y_1 & y_2 & y_3 & y_4 \\ 1 & 1 & 1 & 1 \end{vmatrix} = 0,$$

so that any four proper points, no three collinear, define two functionally independent numbers under affine transformations. Since $\Gamma_{234} \neq 0$ these may be taken to be $\Gamma_{123}/\Gamma_{234}$, $\Gamma_{124}/\Gamma_{234}$. All other ratios of pairs of Γ_{ijk} can be expressed in terms of these.

2.4*X*9 Taking $P_1 = 0$, $P_2 = E$, $P_3 = F$ and $P_4 = (a, b)|\mathfrak{C}(OE, OF)$, find the value of each Γ_{ijk}, and show that, effectively, the two independent invariants may be taken to be a and b.

In the rest of this Chapter we investigate for the geometry of the *rational net* $\mathfrak{N}_R$ the interpretation in co-ordinates of the ideas of 'regions' and 'sense'. In Chapter 1.7 a half-plane was defined as follows:

G $\wp|l$; $\wp|A$, $A \notin l$
D(i) $\ulcorner l(A)$:
$\{X \in \ulcorner l(A)\} \Leftrightarrow \{XA \cap l \;\; \mathscr{B} \;\; AX, \text{ or } X = A\}$
(ii) $\ulcorner l(\tilde{A})$:
$\{Y \in \ulcorner l(\tilde{A})\} \Leftrightarrow \{YA \cap l \;\; \mathscr{B} \;\; AY\}$.

† $[\xi_1, \xi_2, \xi_3]$ is the matrix of which the columns are ξ_1, ξ_2, ξ_3 and

$$\det[\xi_1, \xi_2, \xi_3] = \begin{vmatrix} x_1 & x_2 & x_3 \\ y_1 & y_2 & y_3 \\ 1 & 1 & 1 \end{vmatrix}.$$

The equivalent conditions in co-ordinates are given by

2.4 *T*12 Half-planes determined by a line.

G $\mathfrak{C}(\mathrm{OE}, \mathrm{OF})$
$1 : \alpha^{\mathrm{T}}\xi = 0$
S $\mathrm{P}_1 = \xi_1, \quad \mathrm{P}_2 = \xi_2, \quad \{\mathrm{P}_1, \mathrm{P}_2\} \notin 1$
T $\mathrm{P}_2 \in \ulcorner 1(\mathrm{P}_1) \Leftrightarrow \alpha^{\mathrm{T}}\xi_2/\alpha^{\mathrm{T}}\xi_1 > 0$
P *c* $\mathrm{P} : \mathrm{P} = \mathrm{P}_1\mathrm{P}_2 \cap 1$
assume $\mathrm{s}(\mathrm{P}_1\mathrm{P}/\mathrm{PP}_2) = \lambda$, i.e.
*t*1 $\mathrm{P} = (\xi_1 + \lambda\xi_2)/(1+\lambda)$ (*T* 5)
*t*2 $\mathrm{P} \;\mathscr{B}\; \mathrm{P}_1\mathrm{P}_2 \Leftrightarrow \lambda < 0$ (2.2 *T* 31 (i))
*t*3 $\mathrm{P} \in 1 \Leftrightarrow \lambda = -\alpha^{\mathrm{T}}\xi_1/\alpha^{\mathrm{T}}\xi_2$
T $\mathrm{P}_2 \in \ulcorner 1(\mathrm{P}_1) \Leftrightarrow \alpha^{\mathrm{T}}\xi_2/\alpha^{\mathrm{T}}\xi_1 > 0.$

In particular, when l does not pass through the zero-point 0, points P_1 in the same half-plane as 0 are such that, if $\alpha = [\mathrm{u}, \mathrm{v}, 1]$ $\mathrm{ux} + \mathrm{vy} + 1 > 0$.

In Chapter 1.8 two ordered sets of three non-collinear points are said to determine the same sense in the plane if, on any line in relation to which all six points lie in the same half-plane, the joins of the pairs of the two sets cut out on the line two 'cyclically conformably ordered' sets of three points (p. 83). Formally:

1.8 *D*3 *G* $\mathscr{L}\wp|\mathrm{P}_1\mathrm{P}_2\mathrm{P}_3, \quad \mathscr{L}\wp|\mathrm{P}_1'\mathrm{P}_2'\mathrm{P}_3'$
S $1 : \ulcorner 1(\mathrm{P}_1) \supset \{\mathrm{P}_1, \mathrm{P}_i'\}$
C $\mathrm{L}_i = \mathrm{P}_j\mathrm{P}_k \cap 1, \quad \mathrm{L}_i' = \mathrm{P}_j'\mathrm{P}_k' \cap 1$
$\{\mathrm{i}, \mathrm{j}, \mathrm{k}\} = \{1, 2, 3\}$
D $\partial(\mathrm{P}_1\mathrm{P}_2\mathrm{P}_3/\mathrm{P}_1'\mathrm{P}_2'\mathrm{P}_3') = \mathrm{I} \Leftrightarrow$
$\partial(\mathrm{L}_1\mathrm{L}_2/\mathrm{L}_1'\mathrm{L}_2')\,\partial(\mathrm{L}_1\mathrm{L}_3/\mathrm{L}_1'\mathrm{L}_3')\,\partial(\mathrm{L}_2\mathrm{L}_3/\mathrm{L}_2'\mathrm{L}_3') = \mathrm{I}.$

2.4 *T*13 Sense and the relative invariants.

*G*1 $\mathscr{L}\wp|\mathrm{P}_1\mathrm{P}_2\mathrm{P}_3, \quad \mathscr{L}\wp|\mathrm{P}_1'\mathrm{P}_2'\mathrm{P}_3'$
*G*2 $\{\mathrm{P}_i = \xi_i, \mathrm{P}_i' = \xi_i'\}|\mathfrak{C}(\mathrm{OE}, \mathrm{OF})$
T $\partial(\mathrm{P}_1\mathrm{P}_2\mathrm{P}_3/\mathrm{P}_1'\mathrm{P}_2'\mathrm{P}_3') = \mathrm{I}$
$\Leftrightarrow \Gamma(\xi_1, \xi_2, \xi_3)/\Gamma(\xi_1', \xi_2', \xi_3') > 0$
P *t*1 Since $\Gamma(\xi_1, \xi_2, \xi_3)/\Gamma(\xi_1', \xi_2', \xi_3')$ is independent of the choice of co-ordinates (*T* 11) we need to prove the result only for some selected co-ordinate system
*s*1 $\mathfrak{C}(\mathrm{OE}, \mathrm{OF}) : \{\mathrm{y}_i, \mathrm{y}_i'\} > 0$ $(\mathrm{i} = 1, 2, 3)$, (i.e. select OE so that all six points are in the same half-plane in relation to OE)
*c*1 $\mathrm{L}_i = \mathrm{OE} \cap \mathrm{P}_j\mathrm{P}_k$
$\mathrm{L}_i' = \mathrm{OE} \cap \mathrm{P}_j'\mathrm{P}_k', \quad \{\mathrm{i}, \mathrm{j}, \mathrm{k}\} = \{1, 2, 3\}$

*t*2 $\partial(P_1P_2P_3/P_1'P_2'P_3')$
$= \partial(L_1L_2/L_1'L_2')\,\partial(L_1L_3/L_1'L_3')\,\partial(L_1L_2/L_1'L_2')$

*t*3 $L_1 = \left(\dfrac{y_k x_j - y_j x_k}{y_k - y_j}, 0\right)$, L_1' similar

*t*4 $\partial(L_1L_2/L_1'L_2') = I$

$$\Leftrightarrow \frac{\dfrac{y_1x_3-y_3x_1}{y_1-y_3}-\dfrac{y_2x_3-y_3x_2}{y_2-y_3}}{\dfrac{y_1'x_3'-y_3'x_1'}{y_1'-y_3'}-\dfrac{y_2'x_3'-y_3'x_2'}{y_2'-y_3'}} > 0 \quad (2.2T31\,(\text{iii}))$$

$$\Leftrightarrow \frac{y_3\Gamma(\xi_1,\xi_2,\xi_3)}{(y_1-y_3)(y_2-y_3)}\,\frac{(y_1'-y_3')(y_2'-y_3')}{y_3'\Gamma(\xi_1',\xi_2',\xi_3')} > 0$$

*t*5 $\partial(P_1P_2P_3/P_1'P_2'P_3') = I$

$$\Leftrightarrow \frac{y_1y_2y_3\{\Gamma(\xi_1,\xi_2,\xi_3)\}^3(y_1'-y_3')^2(y_2'-y_3')^2(y_1'-y_2')^2}{y_1'y_2'y_3'\{\Gamma(\xi_1',\xi_2',\xi_3')\}^3(y_1-y_3)^2(y_2-y_3)^2(y_1-y_2)^2} > 0$$

$$\Leftrightarrow \Gamma(\xi_1,\xi_2,\xi_3)/\Gamma(\xi_1',\xi_2',\xi_3') > 0 \quad (c1).$$

2.4*X*10 Show that $P_4 \in {}^{\ulcorner}P_1P_2(P_3) \Leftrightarrow \Gamma_{124}/\Gamma_{123} > 0$.

2.4*X*11 A symmetrical (relative) invariant form for four points is

$$\Gamma_{1234} = \Gamma_{123}\Gamma_{124}\Gamma_{134}\Gamma_{234}.$$

What geometrical property distinguishes sets of points for which $\Gamma_{1234} > 0$ from those for which $\Gamma_{1234} < 0$?

CHAPTER 2.5

INVOLUTIONS: THE GEOMETRIC TRANSFORMER, $\mathscr{J}$

The subject of this section may be supposed to originate, as many of the developments of mathematics do, as a mathematician's jeu d'esprit, in which he has no thought of its having any connection with the physical world. But this little mathematical game, when it is completed, is found to provide precisely what is required for the establishment of a satisfactory representation of physical 'length'. So far we have been able to devise a system of mathematical relations which can represent ratios of 'lengths' only for segments on the same or parallel lines. The vital link establishing a connection between two segments on non-parallel lines is missing. This link is eventually to be provided by 'reflections'; for reflections we need 'perpendiculars'; for perpendiculars we need 'involutions'.

In the sense in which the term is to be used here initially, an involution is a collinear family of pairs of points, the points of a pair being such that the product of their span-ratios relative to a given zero, unit and ideal points is constant. This first definition, which can be written in such a form as to require only a fixed ideal point, can be replaced by a second which uses only cross-ratios; this in turn is replaced by a definition which uses a purely incidence construction, and which it may therefore be possible to extend beyond primary nets.

There is also a second development of considerable importance foreshadowed in the structure of an involution, namely, the necessity for the consideration of geometries in which there are points not contained in one of the primary nets. In this Chapter the question is considered almost wholly from the geometric point of view, the algebraically equivalent process of 'extending the field' being reserved for Book 3.

2.5 ***D*1** Involution of pairs of points on a line: definition by span-ratios.

G	$\mathscr{L}$|KAA′U & ideal|U (fixed) & A′ ∈ $\mathfrak{N}$(K, A/U) & # |KA′U
S	$\wp$|P, P ∈ $\mathfrak{N}$(K, A)
C	P′ : s(KP/KA) = s(KA′/KP′)
D	P, P′ are a pair in an involution.

2.5*N*1 (i) K: the *centre* of the involution

(ii) P, P′ are *conjugates* in the involution, or P is conjugate to P′

(iii) $\mathscr{I}_{K;A,A'}$ (or $\mathscr{I}$) is the *operator* or *transformer* which transforms a point P into its conjugate P′; i.e. $\mathscr{I}_{K;A,A'}P = P'$.

For the present we ignore special cases and consider only involutions satisfying the full conditions *G* of *D*1.

2.5*C*1 Construction of pairs of an involution given the centre and one pair.

G $\mathscr{L}\#\wp|KAA'$, $A' \in \mathfrak{N}(K, A)$

*S*1 $\wp|B$, $B \in \mathfrak{N}(K, A)$ & $\#|BK$

*S*2 $\wp|P$, $P \notin KA$

*C*1 $Q : Q \in KP$ & $QA \parallel BP$

*C*2 $B' : B' \in KA$ & $B'Q \parallel A'P$

t $s(KB/KA) = s(KP/KQ)$ (2.3*T*2)

$= s(KA'/KB')$

T $B' = \mathscr{I}_{K;A,A'}B$.

2.5*T*1 The relation among A, A′, B, B′ is a relation between the pair of unordered pairs {A, A′} and {B, B′}.

G $\{A', B\} \subset \mathfrak{N}(K, A)$ & $B' = \mathscr{I}_{K;A,A'}B$

T (i) $\mathscr{I}_{K;A,A'} = \mathscr{I}_{K;A',A}$

(ii) $\mathscr{I}_{K;B,B'}A = A'$

P These relations come directly from manipulation of the span-ratios, thus

$$s(KB/KA) = s(KA'/KB')$$

$$\Rightarrow s(KB/KA)\,s(KA/KA') = s(KA/KA')\,s(KA'/KB')$$

T (i) $\Rightarrow s(KB/KA') = s(KA/KB')$

T (ii) $\Rightarrow s(KA/KB) = s(KB'/KA')$.

Since we may write the relation

$$s(KA/KB) = s(KB'/KA')$$

as

$$s(KB/KA)\,s(KB'/KA) = s(KA'/KA),$$

the construction 2.5*C*1 must be effectively the same as the 'product' construction 2.2*C*4 (p. 133). In fact, the corresponding points in the two constructions are:

(2.5*C*1)	K	A	B	B′	A′	.	S	Q	R	U	P
(2.2*C*4)	O	E	P	Q	R	H	X	M	Z	U	V,

where in 2.5*C*1 URS is the ideal line, and in 2.2*C*4 UV is the ideal line.

2.5*T*2 Construction of centre of involution given two pairs.

G $\mathscr{L}\#\wp|$AA′BB′, {B, B′} ⊂ $\mathfrak{N}$(A, A′) & s(AB/B′A′) ≠ 1
S $\wp|$P, P ∉ AA′
*C*1 Q : QA ∥ PB & QB′ ∥ PA′
*C*2 K : K = PQ ∩ AA′
T s(KA/KB) = s(KB′/KA′).

That is, K is the centre of the involution of which A, A′; B, B′ are two pairs. If s (AB/B′A′) = 1, then ideal | K, and we obtain the 'mid-point involution' (2.5*N*8, p. 166) to which 2.5*C*1 does not apply. If this type of involution is included, then it follows that the involution is determined by any two pairs of points of a primary net, neither the points of a pair nor the pairs themselves being ordered.

2.5*N*2 $\mathscr{I}_{A, A'; B, B'}$: the *transformer* determined by the two pairs A, A′; B, B′, which transforms a point into its conjugate in the involution determined by the pairs {A, A′}, {B, B′}.

With U as ideal point we may write the relation

$$s_U(KA/KB) = s_U(KB'/KA') \text{ of } D1, \text{ as}$$

$$Ɽ(KU/AB) = Ɽ(KU/B'A').$$

We are to show, in *T*3, that this implies

$$Ɽ(BB'/AK) = Ɽ(BB'/UA'),$$

i.e. that if {A, A′}, {B, B′} are pairs in the involution with centre K relative to ideal|U, then {K, U}, {A, A′} are pairs in the involution with centre B relative to ideal|B′. Thus {K, U} is a pair in the involution no different from the others when neither K nor U is taken to be ideal. The two symbols, $\mathscr{I}_{K,U;A,A'}$ & ideal|U, and $\mathscr{I}_{K;A,A'}$, are equivalent. We have to prove, then, that

2.5*T*3 $s_U(KA/KB) = s_U(KB'/KA') \Rightarrow s_{B'}(BA/BK) = s_{B'}(BU/BA')$.

*G*1 $\mathscr{L}\#$|KEU, ideal |U
{A, A′, B, B′} ⊂ $\mathfrak{N}$(K, E/U)
*G*2 Ɽ(KU/AB) = Ɽ(KU/B′A′)
T Ɽ(BB′/AK) = Ɽ(BB′/UA′)
P assume $s_U(KA/KE) = a$, $s_U(KB/KE) = b$, etc.
t aa′ = bb′ (*s*, *G*2)

T $$\text{R}(BB'/AK) = \frac{(b-a)\,b'}{b(b'-a)} \quad (2.2X4)$$
$$= \frac{aa'-ab'}{aa'-ab} \quad (t)$$
$$= \frac{a'-b'}{a'-b}$$
$$= s_U(A'B'/A'B)$$
$$= \text{R}(UA'/BB').$$

Renaming U, K as C, C′ we may therefore define the involution in the following terms:

2.5*D*2 Involution defined by cross-ratios

G $\mathscr{L}\#|AA'BB'$, $\{B, B'\} \subset \mathfrak{N}(A, A')$
D $\mathscr{I}_{A,A';B,B'}$ is the transformer transforming C into
$C' = \mathscr{I}_{A,A';B,B'}C$, such that
$$\text{R}(AA'/BC) = \text{R}(A'A/B'C').$$

The definition of an involution by means of cross-ratios may be linked with the original definition in a more geometrical fashion by rewriting 2.5*C*1 without any specific ideal line. Before we proceed to this we define a configuration to which we shall need to refer frequently:

2.5*N*3 Complete quadrangle: *Q*{P, Q, R, S/X, Y, Z}.

G $\mathscr{L}_3|PQRS$
C $X : X = QR \cap PS$
$Y : Y = RP \cap QS$
$Z : Z = PQ \cap RS$
Quadrangle: {P, Q, R, S}
Diagonal triangle of complete quadrangle: {X, Y, Z}.

Using the complete quadrangle, we obtain a purely incidence definition and construction of an involution:

2.5*D*3 The 'quadrangular' involution:

*G*1 $\mathscr{L}_3|PQRS$†
*G*2 $A : A \in \mathfrak{N}(Q, R)$
$A' : A' \in \mathfrak{N}(P, S)$

† We may assume that the planar net is defined by P, Q, R, S as in 2.3*T*9.

C $\{^{\mathcal{C}}\mathrm{B, B', C, C'}\} = \mathrm{AA'} \cap \{^{\mathcal{C}}\mathrm{PR, QS, PQ, RS}\}$
D A, A′; B, B′; C, C′ are pairs in an involution.

As we see immediately in 2.5*C*2 (which is equivalent to 2.5*C*1),

$$\mathrm{C'} = \mathscr{I}_{\mathrm{A,A';B,B'}}\mathrm{C}.$$

2.5*C*2 'Quadrangular' construction for an involution.

G $\mathscr{L}\#|\mathrm{AA'BB'C}, \quad \{\mathrm{B, B', C}\} \subset \mathfrak{N}(\mathrm{A, A'})$
*S*1 $\mathrm{P : P} \notin \mathrm{AA'}$
*S*2 $\mathrm{R : R} \in \mathfrak{N}(\mathrm{B, P}) \ \& \ \#|\mathrm{RBP}$
*C*1 $\mathrm{Q : Q = CP} \cap \mathrm{AR}$
*C*2 $\mathrm{S : S = A'P} \cap \mathrm{B'Q}$
$\mathrm{C' : C' = RS} \cap \mathrm{AA'}$
T $\mathrm{C'} = \mathscr{I}_{\mathrm{A,A';B,B'}}\mathrm{C}.$

The construction above is exactly that of 2.5*T*2 in which given A, A′; B, B′, the conjugate K of the ideal point U is determined, the points U, R, S being ideal in 2.5*T*2.

The cross-ratio theorem on which 2.5*D*2 is based is an immediate consequence of 2.5*C*2:

2.5*T*4 Cross-ratios as consequence of quadrangular construction.

G all of 2.5*C*2 & X, Y, Z : $Q\{\mathrm{P, Q, R, S/X, Y, Z}\}$
T $\mathrm{R\!\!\!\!\;/}(\mathrm{AA'/BC}) = \mathrm{R\!\!\!\!\;/}(\mathrm{A'A/B'C'})$
P *t*1 $\mathrm{AA'BC} \overset{\mathrm{P}}{\barwedge} \mathrm{AXRQ} \overset{\mathrm{S}}{\barwedge} \mathrm{AA'C'B'}$
T $\mathrm{R\!\!\!\!\;/}(\mathrm{AA'/BC}) = \mathrm{R\!\!\!\!\;/}(\mathrm{AA'/C'B'}) = \mathrm{R\!\!\!\!\;/}(\mathrm{A'A/B'C'}).$
Similarly, $\mathrm{R\!\!\!\!\;/}(\mathrm{BB'/AC}) = \mathrm{R\!\!\!\!\;/}(\mathrm{BB'/C'A'}).$

This theorem provides an alternative proof of 2.5*T*3

with	K	U	A	A′	B	B′
replaced by	A	A′	C	C′	B	B′.

2.5*T*5 If three pairs of points in involution on one line are in perspective with three pairs of points on another line, then this second set of three pairs are also in involution.

*G*1 $\mathscr{L}|\mathrm{AA'BB'CC'} \quad \mathrm{C'} = \mathscr{I}_{\mathrm{A,A';B,B'}}\mathrm{C}$
*G*2 $\mathscr{L}|\mathrm{LL'MM'NN'}; \{^{\mathcal{C}}\mathrm{L, L', M, M', N, N'}\} \barwedge \{^{\mathcal{C}}\mathrm{A, A', B, B', C, C'}\}$
T $\mathrm{N'} = \mathscr{I}_{\mathrm{L,L';M,M'}}\mathrm{N}$
P $\mathrm{R\!\!\!\!\;/}(\mathrm{AA'/BC}) = \mathrm{R\!\!\!\!\;/}(\mathrm{AA'/C'B'}) \Rightarrow \mathrm{R\!\!\!\!\;/}(\mathrm{LL'/MN}) = \mathrm{R\!\!\!\!\;/}(\mathrm{LL'/N'M'}).$

We may therefore define

2.5 *D*4 Set of three pairs of lines, all concurrent, in involution.

*G*1 $\mathscr{P}|a \cap a' \cap b \cap b' \cap c \cap c' = V$
*G*2 $l : l \not\supset V$
$\{^{\mathscr{O}}A, A'; B, B'; C, C'\} = l \cap \{^{\mathscr{O}}a, a'; b, b'; c, c')$
*G*3 $\mathscr{J}_{A,A';B,B'}C = C'$
D a, a′; b, b′; c, c′ are pairs of lines in involution.

2.5 *N*4 $\mathscr{J}_{a,a';b,b'}c = c'$.

From the construction 2.5 *C* 2 we may devise various special involutions by taking lines in special relations to the vertices of the quadrangle $Q\{P, Q, R, S/X, Y, Z\}$. Thus, taking $l = AA'$, we find:

(i) $l \supset S : S = A' = B' = C'$. S is one component of every pair and the second point of the pair is arbitrary. This is a *singular* involution.

(ii) $l = PS$: The involution is completely evanescent. X, = A, may be any point of the line, but every point of the line satisfies the geometrical condition of being conjugate to any A.

(iii) $l \supset Z$: pairs are A, A′; B, B′ and $Z = C = C'$.

(iv) $l = YZ$: pairs are A, A′; $Y = B = B'$; $Z = C = C'$.

In case (iii) Y is its own conjugate, and in case (iv) Y and Z are both their own conjugates; points having this property in relation to an involution will be called *Double-points*.

2.5 *N*5 *Double-point* in an involution $\mathscr{J}$: a point D which is self-conjugate, $\mathscr{J}D = D$.

The construction above suggests that at least some involutions have double-points, and we prove next:

2.5 *T*6 The pairs of points harmonically conjugate with regard to two fixed points form an involution of which the fixed points are the double-points.

*G*1 $\#|DD'$
*G*2 A, A′; B, B′; C, C′ : $AA' \mathscr{H} DD'$ & $BB' \mathscr{H} DD'$
& $CC' \mathscr{H} DD'$
T $\mathscr{J}_{A,A';B,B'}C = C'$
P *s* ideal|D′
X The theorem is easily proved arithmetically, but there is a straightforward geometric proof using the harmonic construction 1.4 *C* 1, as follows:
P *s* $N : N \notin DD'$
$M : M \in DN$ & $\# |MDN$
*c*1 $\{^{\mathscr{O}}P, Q, R, X\} = D'N \cap \{^{\mathscr{O}}AM, BM, CM, A'M\}$

$c2$ $\{^{\mathscr{C}}\mathrm{P}', \mathrm{Q}', \mathrm{R}', \mathrm{X}'\} = \mathrm{D'M} \cap \{^{\mathscr{C}}\mathrm{DP}, \mathrm{DQ}, \mathrm{DR}, \mathrm{DX}\}$

$t1$ $\mathrm{DD}' \cap \{^{\mathscr{C}}\mathrm{NP}', \mathrm{NQ}', \mathrm{NR}', \mathrm{NX}'\}$

$= \{^{\mathscr{C}}\mathrm{A}', \mathrm{B}', \mathrm{C}', \mathrm{A}\}$ (1.4*C*1, 1.4*X*3)

$t2$ $\mathrm{ABCA}' \barwedge^{\mathrm{M}} \mathrm{PQRX} \barwedge^{\mathrm{D}} \mathrm{P'Q'R'X'} \barwedge^{\mathrm{N}} \mathrm{A'B'C'A}$

T $\mathrm{R}(\mathrm{A'A/BC}) = \mathrm{R}(\mathrm{AA'/B'C'})$.

In an involution which has two double-points, D and D′, take D to be the zero point in the net, and D′ as the ideal point; if {A, A′} is any pair of the involution, then AA′ $\mathscr{H}$ DD′, and D = mid-⌜AA′⌝.

2.5*N*6 *Mid-point involution*: the set of pairs of points {X, X′} in a collinear set, for which, in relation to fixed points D, D′,

$$s_{D'}(XD/DX') = 1.$$

Assume next that there is a third double-point D″, then, with {D, D′} as zero and ideal points,

$$s(DD''/D''D) = 1,$$

and therefore either

(i) the primary net to which the points belong is $\mathfrak{N}_2$ and the three points on the line constitute the 'pairs' each being a double-point, or

(ii) the assumption that the involution is a mid-point involution is untenable. The involution evanesces and every point of the line is a double-point.

Thus

2.5*T*7 A non-evanescent involution in a primary net other than $\mathfrak{N}_2$ has at most two double-points.

There remains to be proved the theorem that if an involution in a primary net has one double-point then it has exactly one other double-point unless either (i) it is singular, or (ii) it evanesces. Dealing first with exceptional cases, we have:

(i) A singular involution may be defined by a centre K and a pair A, A′ of which one member coincides with K, say K = A′. Then for any other pair P, P′

$$s(KP/KA) = s(KA'/KP') = 0$$
$$\Rightarrow P = K:$$

K is the conjugate of every point including itself, and therefore is the only 'double-point' (cf. case (i) after *C*2).

(ii) An evanescent involution is such that $\mathscr{J}$P = P for all P, so that every point is a double-point.

The main theorem we prove directly from the quadrangular construction of an involution by specifying that one of the given pairs is a double-point.

2.5*T*8 If a non-singular involution has one double-point, then it has a second double-point, and consists of pairs of points harmonically separating the double-points.

G $\mathscr{L}\#|\text{XBB'C}$, $\{\text{X, C}\} \subset \mathfrak{N}(\text{B, B'})$
C (as in 2.5*C*2 with {A, A'} replaced by X)
*S*1 $\text{P}:\text{P}\notin\text{XB}$
*S*2 $\text{R}:\text{R}\in\mathfrak{N}(\text{B, P})\ \&\ \#|\text{RBP}$
*C*1 $\text{Q}:\text{Q} = \text{CP}\cap\text{XR}$
*C*2 $\text{S}:\text{S} = \text{XP}\cap\text{B'Q}$
*C*3 $\text{Y, Z}:\mathcal{Q}\{\text{P, Q, R, S/X, Y, Z}\}$
*C*4 $\text{X'}:\text{X'} = \text{YZ}\cap\text{XB}$
*t*1 $\text{BB'}\ \mathscr{H}\ \text{XX'}$
*C*5 $\text{C'}:\text{C'} = \text{RZ}\cap\text{XB}$
*t*2 $\text{CC'}\ \mathscr{H}\ \text{XX'}$
*t*3 $\mathscr{I}_{\text{X,X'};\text{B,B'}}$ consists of pairs harmonically conjugate with regard to X and X'
T X' is a second double-point, and
$\#|\text{BB'X} \Rightarrow \#|\text{XX'}$ (except in $\mathfrak{N}_2$).

We have now proved that if an involution has two double-points, then they are harmonically conjugate with regard to every pair, and, conversely, that the pairs of points of a primary net which are harmonic conjugates with regard to two given points form an involution. The question now arises: does every involution have double-points?

Let us investigate first the consequences of assuming that an involution does have double-points.

2.5*T*9 A geometrical consequence of the existence of a double-point.

*G*1 $\mathscr{L}\#|\text{KAA'DU}$, $\{\text{A', D}\} \subset \mathfrak{N}(\text{K, A/U})$
S $\text{L}:\text{L}\notin\text{KA}$
*C*1 $\text{M}:\mathscr{D}_{\text{KA}}\text{L} = \text{M}$
$\text{N}:\mathscr{D}_{\text{KD}}\text{L} = \text{N}$
*C*2 $\text{X}:\text{X} = \text{DM}\cap\text{A'N}$
T The necessary and sufficient condition for s(KA/KD) = s(KD/KA') is $\text{X}\in\text{KL}$

Sufficient
G (suff.) $\text{X}\in\text{KL}$
T (suff.) s(KA/KD) = s(LM/LN) = s(KD/KA')

Necessary
G (nec.) s(KA/KD) = s(KD/KA')
C $\text{X'}:\text{X'} = \text{DM}\cap\text{KL}$
$\text{A''}:\text{A''} = \text{X'N}\cap\text{KA}$

P *t*1 s(KA/KD) = s(KD/KA″) (*T* suff.)
*t*2 A′ = A″ ⇒ A′N = A″N ⇒ X′ = X
T (nec.) DM ∩ A′N ∈ KL.

Thus, given A ∈ $\mathfrak{N}$(K, E), the problem of constructing double-points for the involution $\mathscr{J}_{K;E,A}$ can be stated in the following form

2.5*C*3 *Problem.*

G $\mathscr{L}$ # ℘|KEA, A ∈ $\mathfrak{N}$(K, E)
L : L ∉ KE
*C*1 M : $\mathscr{D}_{LM} = \mathscr{D}_{KE}$
Problem *C* X : X ∈ KL
D : D = XM ∩ KE
N : $\mathscr{D}_{LN} = \mathscr{D}_{KD}$ & $\mathscr{L}$|XNA.

Involutions in a finite net $\mathfrak{N}_p$

At this stage we have to treat the finite nets and the rational net separately; by far the simpler are the finite nets.

We have seen that if a non-singular involution has one double-point, then (except in $\mathfrak{N}_2$) it has exactly two double-points. We assume p > 2. If D and D′ are the double-points of $\mathscr{J}_{K;E,A}$ then DD′ $\mathscr{H}$ KU and DD′ $\mathscr{H}$ EA. Thus if we regard K, E, U as fixed, then for each other point A, ∈ $\mathfrak{N}$(K, E/U), either $\mathscr{J}_{K;E,A}$ has two double-points or it has none. There is thus a one-to-one correspondence between the points of the set {A : A such that $\mathscr{J}_{K;E,A}$ has double-points} and the set of pairs of points harmonically conjugate with regard to K and U. The number of such harmonic pairs, including {E, E′}, where EE′ $\mathscr{H}$ KU, is $\frac{1}{2}$(p − 1), so that there are $\frac{1}{2}$(p − 1) points A (including E), for which an involution, with centre K and {E, A} as one pair, has double-points. There are therefore also $\frac{1}{2}$(p − 1) positions of A for which $\mathscr{J}_{K;A,E}$ does not have double-points. Thus

2.5*T*10 Of the involutions in $\mathfrak{N}_p$ which contain one fixed pair of points, {K, U}, $\frac{1}{2}$(p − 1) have double-points and $\frac{1}{2}$(p − 1) do not have double-points.

We distinguish these two types of involutions by the names appropriate to corresponding types in the rational net:

2.5*N*7 *Involutions in* $\mathfrak{N}_p$.

Elliptic Involution: $\mathscr{J}$ consists of $\frac{1}{2}$(p + 1) distinct pairs of distinct points.

Hyperbolic Involution: $\mathscr{J}$ consists of $\frac{1}{2}(p-1)$ distinct pairs of distinct points and two double-points.

2.5*X*1 $\Gamma^{(5)}$ is defined by the difference table

0	1	11	19	26	28
30	0	10	18	25	27
20	21	0	8	15	17
12	13	23	0	7	9
5	6	16	24	0	2
3	4	14	22	29	0

(The subscripts of points of a collinear set are determined by the rows.) Take $K = P_0$, $E = P_1$, $L = P_2$, ideal$|U = P_{28}$, ideal$|V = P_{24}$, and test the possibility of the construction 2.5*C*3 for the three remaining points in $\mathfrak{R}(K, E/U)$. Find the pairs in the two elliptic and the two hyperbolic involutions which have K as centre.

The rational net $\mathfrak{N}_R$

We find first, in terms of the separation relation, a necessary condition for the existence of double-points in an involution in $\mathfrak{N}_R$ (or in any net which satisfies the four separation axioms).

2.5*T*11 If the involution $\mathscr{J}_{K;A,A'}$ in $\mathfrak{N}_R$ has double-points, then K $\not\mathscr{S}$ AA′.

*G*1 Construction as in 2·5*C*3, except that E is re-named A′
*G*2 U ∈ KA, V ∈ KL & ideal|UV
T KU $\mathscr{S}$ AA′
P *t*1 KADU $\overset{X}{\barwedge}$ LNMU $\overset{V}{\barwedge}$ KDA′U
*t*2 Assume in turn the three possible relations of separation among {K, A, D, U}.
(i) KA $\mathscr{S}$ DU ⇒ KD $\mathscr{S}$ A′U (*t*1)
⇒ KA $\mathscr{S}$ UA′ (*A*$\mathscr{S}$2, p. 46)
(ii) KD $\mathscr{S}$ AU ⇒ KA′ $\mathscr{S}$ AU
(iii) KU $\mathscr{S}$ DA ⇒ KU $\mathscr{S}$ AA′
T KU $\mathscr{S}$ AA′.

For $\mathfrak{N}_R$ therefore we have two distinct problems: (i) when K $\not\mathscr{S}$ AA′, i.e. when A′ ∈ ⌜K(A), there may be double-points in $\mathfrak{N}_R$, but usually there are not, and (ii) when K ∈ ⌜AA′⌝ there cannot be double-points in any geometry which preserves the separation relation.

In algebraic terms, if s(KA′/KA) = a and s(KD/KA) = d, then D is a double-point of $\mathscr{I}_{K;A,A'}$, if and only if $d^2 = a$. The distinction made above is between rationals $a > 0$ and rationals $a < 0$. Whether or not the double-points exist in $\mathfrak{N}_R$ when $a > 0$, the sign of a determines a universal separation relation between any two pairs of the involution.

2.5*T*12 Separation relation for pairs of an involution in $\mathfrak{N}_R$.

G $\mathscr{I}_{K;X_i,X_i'}$ $i \in \{1, 2, \ldots\}$
$s(KX_i/KE) = x_i$, $s(KX_i'/KE) = x_i'$ & $x_i x_i' = a$

T $a > 0 \Rightarrow X_iX_i' \not\mathscr{S} X_jX_j'$
$a < 0 \Rightarrow X_iX_i' \mathscr{S} X_jX_j'$

P sign of $\text{R}(X_iX_i'/X_jX_j')$
$= \text{sign of } [(x_i - x_j)(x_i' - x_j')(x_i - x_j')(x_i' - x_j)]$
$= \text{sign of } [a\{(x_i + x_i') - (x_j + x_j')\}^2]$
$= \text{sign of } a$
thence by 2.2*T*31 (iv).

In physical terms, so long as $a > 0$, we can find an approximation d' to d, such that $d'^2 < a < (d' + \epsilon)^2$ for an arbitrarily selected positive number ϵ, and could therefore place objects in such a way as to satisfy the required conditions to the order of accuracy of our possible observations. On these grounds we would be justified in adjoining to the net $\mathfrak{N}_R$ a point corresponding to the number d, so that, by some iterative process, we could specify the solution to the problem as the common limit point of a certain two sequences of points. We prescribe such a process in 2.5*C*4.

On the other hand, if $a < 0$, we cannot do so without abandoning the ideas of betweenness and order.

The iterative process that we proceed to describe produces two sequences of points $\mathscr{B}|A_1A_2\ldots A_r\ldots$ and $\mathscr{B}|A_1'A_2'\ldots A_s'$ such that

(i) $\mathscr{O}|A_1A_rA_s'A_1'$ for all pairs of indices r, s,

(ii) The required point D satisfies the conditions $\mathscr{B}|A_rDA_s'$ for all pairs r, s,

(iii) by taking a large enough value of r, we can ensure that $s(A_nA_n'/AA') < \epsilon$ when $n > r$ for any prescribed value $\epsilon > 0$.

We then extend the system $\mathfrak{N}_R$ by adjoining D, and shall show that all constructions in $\mathfrak{N}_R$ can be carried out in the extended system.

2.5*C*4 Iterative construction for delimiting the double-points of an involution $\mathscr{I}_{K;A,A'}$ in $\mathfrak{N}_R$ when $A' \in \ulcorner K(A)$.

*G*1 $\mathscr{L} \# \wp|KAA'$, $A' \in \mathfrak{N}_R(K, A)$
*G*2 $s(KA/KA') = a > 1$†

† This is only a matter of naming the pair A and A′ so that $A' \mathscr{B} KA$.

$G3$ $\mathscr{J} = \mathscr{J}_{K;A,A'}$

$C1$ $A_1 : A_1 = \text{mid} - \ulcorner AA' \urcorner$

$A'_1 : A'_1 = \mathscr{J} A_1$

$C2$ $\mathscr{O}|A_1 A_2 \dots A_r \dots : A_r = \text{mid} - \ulcorner A_{r-1} A'_{r-1} \urcorner$

$\mathscr{O}|A'_1 A'_2 \dots A'_s : A'_s = \mathscr{J} A_s$

$C3$ $s(KA_r/KA') = a_r$, $s(KA'_r/KA') = a'_r$

T(i) $1 < a'_1 < a'_2 < \dots < a'_s < \dots < a_r < \dots < a_2 < a_1 < a$

(ii) given $\epsilon > 0$, r can be found such that

$$s(A'_n A_n/A'A) < \epsilon \quad \text{for} \quad n > r$$

(iii) $D \subset \ulcorner AA' \urcorner \Rightarrow D \in \ulcorner A_r A'_r \urcorner$

where D is the point postulated as satisfying the conditions of $2.5C3$

P $t1$ $a_r a'_r = a$ $(C_2, C_3, 2.5C1)$

$t2$ $a_r = \frac{1}{2}(a_{r-1} + a'_{r-1})$ $(C2)$

$$= \frac{1}{2}\left(a_{r-1} + \frac{a}{a_{r-1}}\right)$$

$t3$ $1 < a'_1 < a_1 < a$

p $a_1 = \frac{1}{2}(1+a)\begin{cases} < \frac{1}{2}(a+a) \\ > \frac{1}{2}(1+1) \end{cases}$ $(C1, G2)$

$$a'_1 = \frac{2a}{1+a} = 1 + \frac{a-1}{a+1} > 1$$

$$a_1 - a'_1 = \frac{1}{2}(1+a) - \frac{2a}{1+a} = \frac{(a-1)^2}{2(a+1)} > 0$$

$t4$ $a_r > 0, a'_r > 0$ (i.e. $\{A_r, A'_r\} \subset \ulcorner K(A))$

p (induction) assume $a_{r-1} > 0$

$a_{r-1} > 0 \Rightarrow a'_{r-1} > 0 \Rightarrow a_r > 0$

$a_1 > 0$ $(t3)$

$t5$ $a_r^2 > a$

$$p \quad a_r^2 = \frac{1}{4}\left(a_{r-1} + \frac{a}{a_{r-1}}\right)^2$$

$$= a + \frac{1}{4}\left(a_{r-1} - \frac{a}{a_{r-1}}\right)^2$$

$t6$ $a_s'^2 < a$

p $a_s'^2 a_s^2 = a^2 \ \& \ a_s^2 > a$ $(t1, t5)$

T(i) $1 < a'_1 < \dots < a'_s \dots < a_r < \dots < a_1 < a$

$$p \quad a_r - a_{r+1} = \frac{1}{2a_r}(a_r^2 - a) > 0$$

$$a'_{s+1} - a'_s = a\left(\frac{1}{a_{s+1}} - \frac{1}{a_s}\right) > 0$$

and $t5$ and $t6$

*t*7 given $\epsilon > 0$, we can find r such that $\dfrac{a_r - a_r'}{a-1} < \epsilon$

$$p \quad a_r - a_r' = \tfrac{1}{2}(a_{r-1} + a_{r-1}') - \frac{2a}{a_{r-1} + a_{r-1}'}$$

$$= \frac{(a_{r-1} - a_{r-1}')^2}{2(a_{r-1} + a_{r-1}')}$$

$$< \tfrac{1}{2}(a_{r-1} - a_{r-1}')$$

$$\Rightarrow (a_r - a_r') < \frac{1}{2^{r-1}}(a_1 - a_1') = \frac{1}{2^r}(a-1)$$

T(ii) $s(A_r'A_r/A'A) < \epsilon$ for $2^r > 1/\epsilon$ i.e. $r > \log_2(1/\epsilon)$.

*t*9 *c* $D' : \mathscr{D}_{D'K} = \mathscr{D}_{KD}$

t $DD' \mathscr{H} A_r A_r'$ 2.5*T*8

*t*10 $D \mathscr{B} A_r A_r'$

p $D \mathscr{B} AA' \Rightarrow D \in \ulcorner K(A) \Rightarrow D' \in \ulcorner K(\tilde{A})$

$\Rightarrow D' \mathscr{B} A_r A_r' \Rightarrow D \mathscr{B} A_r A_r'$

T(iii) $D \in \ulcorner A_r A_r' \urcorner$.

It is to be noticed that even if the double-points belong to $\mathfrak{N}_R$ (i.e. already belong to the geometry) they cannot be finitely constructed when we are given the centre and a pair of the involution. For example, suppose we had been given in the above construction that

$$s(KA/KA') = 4,$$

then $\qquad a_2 = \frac{5}{2}, \quad a_2' = \frac{8}{5}, \quad a_3 = \frac{41}{20}, \quad a_3' = \frac{80}{41}, \ldots;$

for all r, s, we have $a_s' < 2 < a_r$. That is, the construction 2.5*C*4 can never be completed within the framework of $\mathfrak{N}_R$.

As these points later play an essential part in the definition of congruence, we state an axiom which extends $\mathfrak{N}_R$ by adjoining the double-points of every involution $\mathscr{J}_{A,A';B,B'}$ for which the defining points satisfy $AA' \mathscr{S} BB'$. Since the double-points, D, D′, harmonically separate both the pairs we may express their relation to $\{A, A', B, B'\}$ in the form

$$DD' \mathscr{H} AA' \ \& \ DD' \mathscr{H} BB',$$

or, if we wish to isolate one of them:

$$D : \mathscr{H}_{AA'}D = \mathscr{H}_{BB'}D \ \& \ AA' \mathscr{S} DB'.$$

The preliminary form of the axiom is:

$A\,\mathfrak{R}$ (D) POSTULATION OF DOUBLE-POINTS OF INVOLUTIONS

G $\mathfrak{R}_R(K, E/U)$

S $\# | AA'BB' \in \mathfrak{R}$ & $AA' \mathscr{S} BB'$

A A point D exists such that

$$\mathscr{H}_{AA'}D = \mathscr{H}_{BB'}D \;\&\; AA' \mathscr{S} DB'.$$

This is equivalent to the definition by span-ratios

$$s_{B'}(BA/BD) = s_{B'}(BD/BA') \;\&\; D \mathscr{B}_{B'} AA'.$$

and by involutions as

$$D = \mathscr{J}_{A,A';B,B'}D$$

with a separation condition.

For any of these points D we can determine its betweenness relation to any two points of $\mathfrak{R}_R$ and therefore to any two points whether they belong to $\mathfrak{R}_R$ or are introduced through $A\mathfrak{R}(D)$.

Take $\delta = s(KD/KE)$, where $\delta^2 = d > 0$ and $d \in \mathfrak{F}_R$. If we apply the displacement operation to D and the points of $\mathfrak{R}_R$ we shall obtain an augmented net $\mathfrak{R}_R(K, E, D)$ consisting of the points with span-ratios $r + s\delta$ where $(r, s) \in \mathfrak{F}_R$. Among any three of these there is a determinate betweenness relation. Adjoin to $\mathfrak{R}_R(K, E, D)$ a point D′ given by $\delta'^2 = d' > 0$ and $d' \in \mathfrak{F}_R$; again by displacement operations we shall obtain a further augmentation to a net $\mathfrak{R}_R(K, E, D, D')$ containing points with span ratios $r + s\delta + s'\delta'$, where $(r, s, s') \in \mathfrak{F}_R$.

Now consider the points D, D′ in relation to the 'product' construction 2.2*C*4. First we need to establish a relation among the orders of the points concerned

2.5*T*13*X* Betweenness relations for 'products'.

*G*1 $\mathscr{L}\# | KEU$ & ideal$|U$

*G*2 $\{i, j\} \subset \{1, 2, \ldots\}$ & $i < j$

*G*3 $\{P_i\} \subset {}^{\ulcorner}K(E)$ & $\mathscr{L}\mathscr{B}|KP_iP_j$

$\{Q_i\} \subset {}^{\ulcorner}K(E)$ & $\mathscr{L}\mathscr{B}|KQ_iQ_j$

S $\wp|H, H \notin KE$

*C*1 $X_i : X_i = \mathscr{D}_{KP_i}H$

*C*2 $M_i : M_i = KH \cap EX_i$

*C*3 $Z_i : Z_i = HX_i \cap M_iQ_i$

*C*4 $R_i : R_i \in KE$ & $R_iZ_i \| KH$

T $\{R_i\} \subset {}^{\ulcorner}K(E)$ & $\mathscr{B}|KR_iR_j$

P (use separation relations and $A\mathscr{S}3$)

2.5 *T*14 Betweenness relations for the point constructed from D and D′ by the 'product' construction.

*G*1 $\mathfrak{R}_R(K, E/U)$

*G*2 $\{i, j\} \subset \{1, 2, \ldots\} \;\&\; i < j$

*G*3 $(A_i, A_i', B_i, B_i') \in \mathfrak{R}_R(K, E)$
$\&\; A_i \in \ulcorner K(E) \;\&\; \mathscr{B}|KA_iA_jA_i'$
$\&\; B_i \in \ulcorner K(E) \;\&\; \mathscr{B}|KB_iB_jB_i'$

*C*1 $A_j', B_j' : A_j' = \mathscr{I}_{K;A_i,A_i'}A_j,\; B_j' = \mathscr{I}_{K;B_i,B_i'}B_j$

*C*2 $D : D = \mathscr{I}_{K;A_i,A_i'}D \;\&\; D \in \ulcorner K(E)$
$D' : D' = \mathscr{I}_{K;B_i,B_i'}D' \;\&\; D' \in \ulcorner K(E)$

*C*3 (2.2*C*4) $C_i : C_i = \mathscr{I}_{K;A_i,B_i}E$
$C_i' = \mathscr{I}_{K;A_i',B_i'}E$
$D^* : \mathscr{I}_{K;D,D'}E = D^*$

T $\mathscr{B}|KC_iD^*C_i'$ (*T* 13)

Thus the 'product' construction 2.2*C*4 applies also to any two points adjoined to the rational net under the definition $A\mathfrak{R}(D)$, and the point constructed from them by 2·2*C*4 may also be constructed by the 'double-point' construction 2.5*C*4.

Now that we have defined double-points for all involutions $\mathscr{I}_{A,A';B,B'}$ for which AA′ $\mathscr{S}$ BB′ and have already seen that the postulation of double-points when AA′ $\mathscr{S}$ BB′ would contradict the separation axioms, we can classify involutions in the non-finite net in the same way as we classified them in $\mathfrak{R}_p$, namely:

2.5*N*8 *Elliptic involution*: $\mathscr{I}_{A,A';B,B'}$ where AA′ $\mathscr{S}$ BB′ and there are no double-points.

Hyperbolic involution: $\mathscr{I}_{A,A';B,B'}$ where AA′ $\mathscr{S}$ BB′ and double-points can be defined consistently with the separation relations.

Effectively we have now added to the points of a collinear set which was initially $\mathfrak{R}_R$ all points D such that $s(KD/KE) = a \pm \sqrt{b}$ where a is rational and b is a positive rational. But this system is not closed under the possible geometric operations, because, clearly, we can apply operations $\mathscr{D}$ on the points of the set, to produce

$$a + \sqrt{b} + \sqrt{c} + \sqrt{d} \ldots.$$

All such points can be arranged in a betweenness order, and we can apply the multiplication operation 2.2*C*4, which however does not produce new points. This system then is closed under the two geometric operations.

But we may further apply the postulation process of $A\mathfrak{R}(D)$ to produce $a + \sqrt{(b + \sqrt{c})}$ and then repeat operations of displacement, product-forming and adjunction of double-points. This iteratively defined system (which does not cover the whole of the algebraic numbers) contains all the points that we shall need (with one exception discussed towards the end of Book 4) in the development of the geometry which is to be a mathematical model of 'objects' and 'line-of-sight' observations. In Book 3 we investigate other types of geometries, but for the present we define

A $\mathfrak{N}_s$ THE SURDIC NET

G $\mathfrak{N}_R(K, E/U)$

A The surdic net is formed from $\mathfrak{N}_R$ by adjoining the set of points $\{Z\}$ defined recursively by the two operations

(i) displacement

(ii) postulating in the system the point Z such that

$$\mathscr{H}_{XX'}Z = \mathscr{H}_{YY'}Z \ \& \ XX' \mathscr{S} ZY'$$

where $\{X, X', Y, Y'\}$ is a set of points already constructed and is such that $XX' \mathscr{S} YY'$

A The points of $\mathfrak{N}_S$ satisfy determinable betweenness relations.

CHAPTER 2.6

PERPENDICULARS

In the course of his division of the paddock up into smaller ⁑areas by fences, the F-G has noticed a difference between runs of fence which make ⁑acute angles with each other and those which make ⁑obtuse angles with each other, while, as a boundary between these two cases, are runs of fence which are 'perpendicular'. Moreover at the back of his mind, as a possible way of solving his outstanding problem of relating ⁑distances on non-parallel lines, has been the idea of using ⁑reflections, an essential ingredient in which seems to be this relation of ⁑perpendicularity. He sets out therefore to construct much in the same way as he did for the relation of parallelism, a mathematical model of the relation of perpendicularity.

From consideration of the fences, he sees that the mathematical model should have these properties:

(i) The property is mutual, i.e. if a is perpendicular to b then b is perpendicular to a.

(ii) Through any given proper point a single line can be drawn perpendicular to a given proper line,

(iii) Two lines both perpendicular to a given line are parallel to one other,

(iv) No line is perpendicular to itself.

Considering, then, lines through a point, we see that they can be arranged as perpendicular pairs, each pair determined by either of its members, and moreover if we consider lines parallel respectively to two given perpendicular lines, we see that they form perpendicular families of lines. That is, perpendicularity is essentially a property of ideal points and we may replace the statements (i), (ii), (iii) above by the statement that ideal points can be arranged in pairs such that two lines are perpendicular if and only if their ideal points form one of the pairs. We are required then to devise a system of pairs of points on a line which would form a basis for a definition of a geometrical relation of perpendicularity that leads to a satisfactory model of the physical relation.

In Chapter 2.5 we investigated the system of pairs of points in involution, and this, we shall see, does in fact provide a satisfactory basis for the model.†

† The involution is by no means the only possible system of pairs. Consider the examples given at foot of facing page.

2.6 *D* 1 *Postulate for perpendiculars.*

(i) In a net $\mathfrak{N}_p$ or $\mathfrak{N}_R$ or $\mathfrak{N}_S$ a line is selected to be the ideal line and on it an *elliptic involution* $\mathcal{J}$ of pairs of points is prescribed to be the *orthogonality involution.*

(ii) Two lines are perpendicular if and only if their ideal points are conjugates in $\mathcal{J}$.

(iii) The relation of perpendicularity does not apply to the ideal line itself; the ideal line is neither perpendicular nor not perpendicular to any line.

2.6 *N* 1 *Perpendiculars.*

G ideal|u, $\mathcal{J} \subset$ u

N $a \perp b \Leftrightarrow a \cap u = \mathcal{J}(b \cap u)$.

2.6 *T* 1 $a \perp b \;\&\; b \perp c \Rightarrow a \parallel c$

P c $\{{}^{\mathcal{O}}A, B, C\} = u \cap \{{}^{\mathcal{O}}a, b, c\}$

T $\mathcal{J}A = B \;\&\; \mathcal{J}B = C \Rightarrow \mathcal{J}^2 A = C \Rightarrow A = C.$

2.6 *T* 2 There is a single perpendicular through a point to a line.

G $\wp$|a, P

T $b \supset P \;\&\; b \perp a \Leftrightarrow b = \langle P, \mathcal{J}(a \cap u)\rangle$.

2.6 *T* 3 In $\mathfrak{N}_R$ and $\mathfrak{N}_S$ two concurrent pairs of perpendicular lines separate each other.

G $\mathscr{P}$|$a \cap a' \cap b \cap b' \;\&\; a \perp b \;\&\; a' \perp b' \;\&\; \#$|aa′

T ab $\mathscr{S}$ a′b′

P The orthogonality involution is elliptic & 2.5 *T* 12.

(i) In $\mathfrak{N}_p$ we could name $\frac{1}{2}(p+1)$ pairs of points in a collinear set quite randomly.

(ii) In $\mathfrak{N}_p$ for a value of p for which -1 is a square the pairs could be

$$\{a^2, 1/a^2\} \quad \text{for squares}$$

and

$$\{b, -1/b\} \quad \text{for non-squares.}$$

(iii) In $\mathfrak{N}_R$, if u, v are integers without a common factor, the pairs could be

$$\left\{\frac{u}{v}, -\frac{u+v}{v}\right\} \quad \text{when v is prime}$$

and

$$\left\{\frac{u}{v}, \frac{u-v}{v}\right\} \quad \text{when v is composite.}$$

(iv) If the system is extended to the reals, pairs {x, y} could be defined by relations such as

$$x^3 + y^3 = 1.$$

If the properties deducible from the definition of perpendiculars by pairs of an involution did not correspond to those which our observations of the physical system had led us to expect, then we should need to investigate some other methods of prescribing pairs, such as those proposed above.

2.6*T*4 A special form of the separation theorem in $\mathfrak{N}_R$ and $\mathfrak{N}_S$.

*G*1 $\mathscr{P}|a \cap a' \cap b \cap b' \cap c \cap c' = V$
$\& \; a \perp a' \; \& \; b \perp b' \; \& \; c \perp c' \; \& \; \# |ab \; \& \; \# |ab'$
*G*2 $h : h \parallel c' \; \& \; \# |hc'$
*C*1 $\{^{o}A, A', B, B', C\} = h \cap \{^{o}a, a', b, b', c\}$
T (i) $s(CA/CB) = s(CB'/CA')$
(ii) $\mathscr{B}|ACA' \; \& \; \mathscr{B}|BCB'$
P (i) A, A′; B, B′ are pairs of an involution of which C is the centre
(ii) The involution is elliptic.

2.6*N*2 Acute and obtuse angular regions in $\mathfrak{N}_R$ or $\mathfrak{N}_S$.

G $\mathscr{L}\wp|VAB$
C $A' : A' \in VA \; \& \; A'B \perp VA$
N (i) $\angle V(AB)$ is *acute* $\Leftrightarrow V \; \not{\mathscr{B}} \; AA'$
(ii) $\angle V(AB)$ is *obtuse* $\Leftrightarrow V \; \mathscr{B} \; AA'$
(iii) $\angle V(AB)$ is a *right angle* (angular region) $\Leftrightarrow VA \perp VB$.

This Chapter is to be concluded by proving four theorems for perpendiculars, one of which leads to a 'practical' construction of pairs of perpendicular lines when two pairs of perpendicular lines have been assigned, and the other three are theorems essential in the development of the properties of Congruence in Book 4.

2.6*T*5 The altitudes of a triangle are concurrent.

G $\mathscr{L}\wp|ABC \; \& \; \text{ideal}|u \; \& \; \mathscr{J} \subset u$
C $H : HA \perp BC \; \& \; HB \perp CA$
T $HC \perp AB$
P *c*1 $u \cap \{^{o}BC, CA, AB, AH, BH, CH\} = \{^{o}U, V, W, A', B', C'\}$
*c*2 $F : F = AB \cap CH$
t $A'B'C'W \overline{\overline{\wedge}}\, {}^{H}ABFW \overline{\overline{\wedge}}\, {}^{C}VUC'W$
$\Rightarrow R(A'B'/C'W) = R(UV/WC')$
$\Rightarrow \mathscr{J}_{A',U;B',V}C' = W \Rightarrow \mathscr{J}C' = W$
T $HC \perp AB$.

2.6*T*6 The construction, when two pairs of perpendiculars are assigned, of the perpendicular to a given line through a given point.

*G*1 $\mathscr{L}\#\wp|KLMN \; \& \; KL \parallel MN \; \& \; KM \parallel LN$
(as in 1.5*T*1)
*G*2 $a : a \supset K \; \& \; a \perp KL \; \& \; \# |a, KM$
$b : b \supset K \; \& \; b \perp KM$

$S \quad \wp|\mathrm{H, p,} \quad \mathrm{p} \nparallel \mathrm{a}, \quad \mathrm{p} \nparallel \mathrm{b}$

$C1 \quad \mathrm{A, B : A = p \cap a, \quad B = p \cap b}$

$C2 \quad \mathrm{C : CA \parallel KM \;\&\; CB \parallel KL}$ (as in 1.5T1, 2)

$\quad t \quad \mathrm{CK \perp AB}$ (2.6T5)

$C3 \quad \mathrm{h : h \supset H \;\&\; h \parallel CK}$

$T \quad \mathrm{h \perp p \;\&\; h \supset H}.$

2.6*T*7 Equivalent eventually to 'the reflection of a right angle is a right angle'.

$G \quad \mathscr{L}\wp|\mathrm{ABP \;\&\; AP \perp PB}$

$C1 \quad \mathrm{N : N \in AB \;\&\; PN \perp AB}$

$C2$ (sufficient) $\mathrm{Q} = \mathscr{D}_{\mathrm{PN}}\mathrm{N}$

T (sufficient) $\mathrm{AQ \perp QB}$

P (sufficient) $c \quad \{^{c}\mathrm{A', A'', B', B'', U, V}\}$

$\qquad = \mathrm{u} \cap \{^{c}\mathrm{AP, AQ, BP, BQ, AB, PQ}\}$

$\quad t1 \quad \mathrm{VN} \;\mathscr{H}\; \mathrm{PQ}$ (C2 (sufficient))

$\quad t2 \quad \mathrm{UV} \;\mathscr{H}\; \mathrm{A'A''} \;\&\; \mathrm{UV} \;\mathscr{H}\; \mathrm{B'B''}$

$\qquad p \quad \mathrm{UVA'A''} \overline{\wedge}\, {}^{\mathrm{A}}\mathrm{NVPQ} \overline{\wedge}\, {}^{\mathrm{B}}\mathrm{UVB'B''}$

$\quad t3 \quad \mathrm{R}(\mathrm{UV/A'A''}) = \mathrm{R}(\mathrm{VU/B'B''}) = -1$

$\quad t4 \quad \mathscr{J}_{\mathrm{U,V;A',B'}}\mathrm{A''} = \mathrm{B''}$

T (sufficient) $\mathrm{A'', B''}$ are conjugates in $\mathscr{J}$

T (necessary) $\mathrm{AQ \perp QB \;\&\; Q \in PN \;\&\; \#|PQ \Rightarrow Q} = \mathscr{D}_{\mathrm{PN}}\mathrm{N}$

P (necessary) X.

2.6*T*8 The perpendiculars to the side-segments of a triangle through their mid-points are concurrent.

$G \quad \mathscr{L}\wp|\mathrm{ABC}$

$C1 \quad \mathrm{E, F : E = mid\text{-}\ulcorner AC \urcorner \;\&\; F = mid\text{-}\ulcorner AB \urcorner}$

$C2 \quad \mathrm{H : HE \perp AC \;\&\; HF \perp AB}$

$C3$ (sufficient) $\mathrm{D : D = mid\text{-}\ulcorner BC \urcorner}$

T (sufficient) $\mathrm{HD \perp BC}$

P (sufficient) $c1 \quad \{^{c}\mathrm{A', B', C', F'}\} = \mathrm{u} \cap \{^{c}\mathrm{BC, CA, AB, FH}\}$

$\quad c2 \quad \mathrm{F'' : F'' = DE \cap FH}$

$\quad t1 \quad \mathrm{A' \in EF, \quad B' \in FD, \quad C' \in DE}$

$\qquad p \quad \mathrm{s(AE/EC) = s(AF/FB)}$, etc.

$\quad t2 \quad \mathrm{DEF''C'} \overline{\wedge}\, {}^{\mathrm{F}}\mathrm{B'A'F'C'}$

$\quad t3 \quad \mathrm{R}(\mathrm{HD, HE/HF', HC'}) = \mathrm{R}(\mathrm{HA', HB'/HC', HF'})$

T (sufficient) $\mathscr{J}_{\mathrm{HD,HA';HE,HB'}}\mathrm{HF} = \mathrm{HC'}$

T (necessary) $\mathrm{D \in BC \;\&\; HD \perp BC \Rightarrow D = mid\text{-}\ulcorner BC \urcorner}$

P (necessary) X.

In the next theorem we have to make explicit use of the double-points of an involution $\mathscr{J}_{\mathrm{O;H,K}}$ defined by its centre and a pair. In

$\mathfrak{N}_p$ and $\mathfrak{N}_R$ the theorem is therefore restricted to specific cases in which the double-points exist. In $\mathfrak{N}_S$ we can show that a consequence of prescribing that the orthogonality involution is elliptic is that $\mathscr{J}_{O;H,K}$ is hyperbolic, so that, in $\mathfrak{N}_S$, the construction is always possible.

2.6*T*9 The theorem leading in $\mathfrak{N}_S$ to the construction of the 'reflection' of a given segment on to a given line.

*G*1 All elements belong to $\mathfrak{N}_S$ and the orthogonality involution $\mathscr{J}$ is prescribed
*G*2 $\mathscr{L}\wp|$OAB & OA $\not\perp$ OB.
*C*1 H : H $\in$ OB & OH $\perp$ AH
K : K $\in$ OB & OA $\perp$ AK
*t*1 H $\mathscr{B}$ OK (2.6*T*4(ii)) i.e. $\mathscr{J}_{O;H,K}$ is hyperbolic
*C*2 D : D $\mathscr{B}$ HK & s(OD/OH) = s(OK/OD) ($A\mathfrak{N}_S$)
D′ : D′ = $\mathscr{D}_{DO}$O
*C*3 A′ : A′ = $\mathscr{D}_{AO}$O
T AD $\perp$ A′D & AD′ $\perp$ A′D′
P *c* {$^{\mathscr{O}}$A*, B*, H*, K*, U, V}
= u $\cap$ {$^{\mathscr{O}}$OA, OB, AH, AK, AD, AD′}
*t*1 HK $\mathscr{H}$ DD′ (2.5*T*8)
*t*2 H*K*$\mathscr{H}$UV
p HKDD′ $\overline{\wedge}$ ^{A}H*K*UV
*t*3 AD $\|$ A′D′ & A′D $\|$ AD′
p 1 = s(AO/OA′) = s(A′O/OA) = s(DO/OD′)
*t*4 A*B* $\mathscr{H}$ UV
p OA* $\mathscr{H}$ AA′ & OA*AA′ $\overline{\wedge}$ ^{D}B*A*UV
*t*5 $\mathscr{J}_{A^*,K^*;B^*H^*}$U = V
p *t*2, *t*4 & 2.6*T*7 (*t*3, *t*4)
*t*6 AU $\perp$ AV, i.e. AD $\perp$ AD′
T AD $\perp$ AD′ & A′D $\perp$ A′D′.

2.6*T*10*X*

G and *C* as in *T*9
*C*4 L, M : L = mid-⌜AD⌝, M = mid-⌜AD′⌝
T OL $\perp$ AD & OM $\perp$ A′D.

2.6*X*1 *T*10 solves the following problems: given two proper points A, A′ and a line p, p $\supset$ mid-⌜AA′⌝, to construct X, X $\in$ p, such that AX $\perp$ A′X. D, D′ are two positions of X. Prove that these are the only positions of X.

EXCURSUS ON BOOK 2

EXCURSUS 2.1

CO-ORDINATE GEOMETRY IN PRIMARY NETS

Exc. 2.1.1 An algebraic basis for geometry

In the main text, on the basis of axioms $\mathscr{A}\mathscr{I}$ 1, 2, 3* and with a designated ideal line, we have been able to devise a system of co-ordinates in which the proper points of the planar net $\mathfrak{N}_p$ or $\mathfrak{N}_R$ are in one-to-one correspondence with the ordered pairs of elements of a primary field $\mathfrak{F}_p$ or $\mathfrak{F}_R$. In this Excursus we reverse the process and show that we may define points as ordered pairs of elements of a field and define a collinear set by an algebraic relation between the two elements, and then deduce $\mathscr{A}\mathscr{I}$ 1, 2, 3 and Pappus.

A point of a planar net $\mathfrak{N}$ is defined as a vector $\boldsymbol{\xi}$ where

$$\boldsymbol{\xi} = \begin{bmatrix} \mathrm{x} \\ \mathrm{y} \\ 1 \end{bmatrix} \quad \text{or} \quad \begin{bmatrix} \mathrm{x} \\ 1 \\ 0 \end{bmatrix} \quad \text{or} \quad \begin{bmatrix} 1 \\ 0 \\ 0 \end{bmatrix}$$

and x and y are elements of a primary field $\mathfrak{F}$. A collinear set is the set satisfying the condition

$$\boldsymbol{\alpha}^T\boldsymbol{\xi} = 0,$$

where $\boldsymbol{\alpha}^T = [\mathrm{u}, \mathrm{v}, 1]$ or $[\mathrm{u}, 1, 0]$ or $[1, 0, 0]$ and u and v are elements of $\mathfrak{F}$.

For convenience we shall write in the text (x, y, 1) for $\boldsymbol{\xi}$, instead of the column-vector. So long as the elements belong to a field (not necessarily a primary or even commutative field), for any pair of elements $\{\lambda, \mu\}$ for which $\lambda\mu \neq 0$, we have

$$\boldsymbol{\alpha}^T\boldsymbol{\xi} = 0 \Leftrightarrow (\lambda\boldsymbol{\alpha}^T)(\boldsymbol{\xi}\mu) = 0.$$

We may therefore, with an obvious change of notation, replace the linear equations

$$\mathrm{ux}+\mathrm{vy}+1 = 0, \quad \mathrm{ux}+\mathrm{y} = 0, \quad \mathrm{x} = 0$$

and the ideal line [0, 0, 1], which is not given by any of these equations, by

$$\lambda(\mathrm{ux}+\mathrm{vy}+\mathrm{wz})\mu = 0,$$

where $\lambda w = 1$ if $w \neq 0$, $z\mu = 1$ if $z \neq 0$,

$$\lambda v = 1 \text{ if } w = 0 \ \& \ v \neq 0, \quad y\mu = 1 \text{ if } z = 0 \ \& \ y \neq 0,$$

$$\lambda u = 1 \text{ if } v = w = 0, \quad x\mu = 1 \text{ if } y = z = 0.$$

(In Excursus 3.3 we investigate a system in which the replacement of $\boldsymbol{\alpha}^T\boldsymbol{\xi}$ by $(\lambda\boldsymbol{\alpha}^T)(\boldsymbol{\xi}\mu)$ is not permissible.)

One minor advantage that the standard form of co-ordinates has over the homogeneous forms (x, y, z), [u, v, w] is that, in the finite nets, points are much more easily identified. Thus, in $\mathfrak{N}_{11}$, (3, −4, 1) can appear in ten different homogeneous disguises, (−5, 3, 2), (−2, −1, 3), ...

We are to verify now that the subsets of points $\boldsymbol{\xi}$ defined by relations

$$\boldsymbol{\alpha}^T\boldsymbol{\xi} = 0$$

satisfy the conditions $A\mathscr{I}$ 1, 2. First, given two distinct points $\boldsymbol{\xi}_1$, $\boldsymbol{\xi}_2$, either there is a relation of the form

$$\boldsymbol{\alpha}^T\boldsymbol{\xi} = \lambda(ux+vy+wz)\mu = 0$$

satisfied by them both, namely that given by

$$u = y_1 - y_2, \quad v = x_2 - x_1, \quad w = x_1y_2 - x_2y_1,$$

or, if both points are ideal, say $(x_1, 1, 0)$, $(x_2, 1, 0)$, there is no such relation but both points lie on the ideal line, $u = v = 0$, $\lambda w = 1$.

Suppose next that there is a second such relation so that, writing them in homogeneous form, we have the equations

$$ux_1+vy_1+wz_1 = 0, \quad ux_2+vy_2+wz_2 = 0,$$

$$u'x_1+v'y_1+w'z_1 = 0, \quad u'x_2+v'y_2+w'z_2 = 0.$$

From the relations

$$w'(ux_i+vy_i+wz_i)-w(u'x_i+v'y_i+w'z_i) = 0 \quad (i \in \{1, 2\}),$$

we deduce that

$$(w'u-wu')x_1+(w'v-wv')y_1 = 0,$$

$$(w'u-wu')x_2+(w'v-wv')y_2 = 0,$$

and therefore that either

$$w'u-wu' = 0 \ \& \ w'v-wv' = 0$$

or

$$x_1y_2-x_2y_1 = 0.$$

There are two similar sets of relations, and from them we deduce that there exists either a non-zero multiplier θ such that

$$(\theta u, \theta v, \theta w) = (u', v', w')$$

or a non-zero multiplier ϕ such that

$$(x_1\phi, y_1\phi, z_1\phi) = (x_2, y_2, z_2).$$

We can see immediately that the point $(vw'-w'v, wu'-w'u, uv'-vu')\mu$ satisfies both relations $\boldsymbol{\alpha}^T\boldsymbol{\xi} = 0$, $\boldsymbol{\alpha}'^T\boldsymbol{\xi} = 0$, so that we have in fact proved that the relations stated in both $A\mathscr{I}1$ and $A\mathscr{I}2$ are consequences of the definition, namely that: (1) if the two points $\boldsymbol{\xi}_1, \boldsymbol{\xi}_2$ are distinct there is one and only one collinear set containing them both, (2) if the two collinear sets $\boldsymbol{\alpha}, \boldsymbol{\alpha}'$ are distinct, then they have one and only one common member.

If $\boldsymbol{\xi}_3$ is any point of the collinear set $\langle\boldsymbol{\xi}_1, \boldsymbol{\xi}_2\rangle$, then

$$x_3(y_1z_2-y_2z_1)+y_3(z_1x_2-z_2x_1)+z_3(x_1y_2-x_2y_1) \equiv \det[\boldsymbol{\xi}_1, \boldsymbol{\xi}_2, \boldsymbol{\xi}_3] = 0,$$

where $[\boldsymbol{\xi}_1, \boldsymbol{\xi}_2, \boldsymbol{\xi}_3]$ is the matrix whose columns are the vectors $\boldsymbol{\xi}_i$. We may rewrite this relation as

$$(x_2y_3-x_3y_2)z_1+(x_3y_1-x_1y_3)z_2+(x_1y_2-x_2y_1)z_3 = 0.$$

Also there are the two identities

$$(x_2y_3-x_3y_2)x_1+(x_3y_1-x_1y_3)x_2+(x_1y_2-x_2y_1)x_3 \equiv 0,$$

$$(x_2y_3-x_3y_2)y_1+(x_3y_1-x_1y_3)y_2+(x_1y_2-x_2y_1)y_3 \equiv 0,$$

so that, if $\mathscr{L}|\boldsymbol{\xi}_1, \boldsymbol{\xi}_2, \boldsymbol{\xi}_3$, then there is a (non-zero) vector $\boldsymbol{\varkappa}$, one form of which is $(x_2y_3-x_3y_2, x_3y_1-x_1y_3, x_1y_2-x_2y_1)$, such that

$$[\boldsymbol{\xi}_1, \boldsymbol{\xi}_2, \boldsymbol{\xi}_3]\boldsymbol{\varkappa} = \mathbf{o}.$$

Conversely, if such a vector exists, then $\det[\boldsymbol{\xi}_1, \boldsymbol{\xi}_2, \boldsymbol{\xi}_3] = 0$, and there exists a vector $\boldsymbol{\alpha}$ such that $\boldsymbol{\alpha}^T\boldsymbol{\xi}_i = 0$.

In this excursus and in Excursus 2.3 we shall usually use the homogeneous forms for the point and collinear set, and write

$$\boldsymbol{\alpha}^T\boldsymbol{\xi} = [u, v, w]\begin{bmatrix} x \\ y \\ z \end{bmatrix}.$$

The choice of ideal line is irrelevant. The basis of the co-ordinate system is the set of four points

$$\boldsymbol{\epsilon}_1 = \begin{bmatrix} 1 \\ 0 \\ 0 \end{bmatrix}, \quad \boldsymbol{\epsilon}_2 = \begin{bmatrix} 0 \\ 1 \\ 0 \end{bmatrix}, \quad \boldsymbol{\epsilon}_0 = \begin{bmatrix} 0 \\ 0 \\ 1 \end{bmatrix}, \quad \boldsymbol{\upsilon} = \begin{bmatrix} 1 \\ 1 \\ 1 \end{bmatrix}.$$

Any four points, no three collinear, may be taken as basis: if we wish to transfer from a given basis with co-ordinates $\boldsymbol{\xi}$ to a basis with co-ordinates $\boldsymbol{\xi}$ in which the base points are $\boldsymbol{\pi}_1, \boldsymbol{\pi}_2, \boldsymbol{\pi}_0, \boldsymbol{\pi}$, where $\boldsymbol{\pi}_i = (p_i, q_i, r_i)$, in the $\boldsymbol{\xi}$-system, we use the transformation

$$\rho\boldsymbol{\xi} = [\lambda_1\boldsymbol{\pi}_1, \lambda_2\boldsymbol{\pi}_2, \lambda_0\boldsymbol{\pi}_0]\boldsymbol{\xi}$$

$$= \begin{bmatrix} p_1 & p_2 & p_0 \\ q_1 & q_2 & q_0 \\ r_1 & r_2 & r_0 \end{bmatrix} \begin{bmatrix} \lambda_1 & 0 & 0 \\ 0 & \lambda_2 & 0 \\ 0 & 0 & \lambda_0 \end{bmatrix} \boldsymbol{\xi} = \boldsymbol{\Pi\Lambda\xi},$$

where the elements λ_i are selected so that

$$\rho\boldsymbol{\upsilon} = \boldsymbol{\Pi\Lambda\pi},$$

that is

$$\rho\boldsymbol{\Pi}^{-1}\boldsymbol{\upsilon} = \boldsymbol{\Lambda\pi}.$$

This equation determines (the ratios of) the elements λ_i.

Besides considering changes of basis (transformations of co-ordinates) we shall be considering transformations of the plane into itself in which every point $\boldsymbol{\xi}$ is paired off with a point $\boldsymbol{\eta}$ by means of a relation

$$\rho\boldsymbol{\eta} = \mathbf{M}\boldsymbol{\xi}$$

in which $\mathbf{M}$ is a non-singular matrix. Such a transformation is called a *projectivity* and corresponds to the general change of co-ordinates discussed in 2.4*D*3 in the same way as the displacement transformer $\mathscr{D}$ corresponds to a change of co-ordinates $\mathfrak{C}$(OE, OF) to $\mathfrak{C}$(O′E′, O′F′) in which the ideal line is unchanged & $\mathscr{D}_{\mathrm{OE}} = \mathscr{D}_{\mathrm{O'E'}}$ & $\mathscr{D}_{\mathrm{OF}} = \mathscr{D}_{\mathrm{O'F'}}$ (2.4*T*7, 8).

Consider next the effect of a change of co-ordinate system on a projectivity. Take the change of co-ordinate system to be given by

$$\rho\bar{\boldsymbol{\xi}} = \mathbf{H}\boldsymbol{\xi}$$

and the projectivity by

$$\sigma\boldsymbol{\eta} = \mathbf{M}\boldsymbol{\xi}.$$

Then

$$\sigma\rho\bar{\boldsymbol{\eta}} = \sigma\mathbf{H}\boldsymbol{\eta} = \mathbf{HM}\boldsymbol{\xi} = \rho\mathbf{HMH}^{-1}\bar{\boldsymbol{\xi}}.$$

That is, the projectivity in the $\bar{\boldsymbol{\xi}}$ co-ordinate system is given by

$$\sigma\bar{\boldsymbol{\eta}} = \bar{\mathbf{M}}\bar{\boldsymbol{\xi}},$$

where

$$\bar{\mathbf{M}} = \mathbf{HMH}^{-1}.$$

Matrices $\bar{\mathbf{M}}$ and $\mathbf{M}$ connected by a relation of this type are designated 'similar'.

Exercise 1 Prove that similarity is an equivalent relation among matrices.

Exc. 2.1.2 The Desargues figure

In the homogeneous notation for the co-ordinates take the points K, A, B, C of the standard Desargues figure ($A\mathscr{I}3$) to be $K = \mathbf{\upsilon}$, $A = \boldsymbol{\epsilon}_1$, $B = \boldsymbol{\epsilon}_2$, $C = \boldsymbol{\epsilon}_0$.

Since $A' \in KA$, A′ may be taken to be

$$A' = \mathbf{\upsilon} + (\alpha - 1)\boldsymbol{\epsilon}_1 = (\alpha, 1, 1),$$

and similarly $B' = (1, \beta, 1)$ and $C' = (1, 1, \gamma)$.

We assume $(\alpha, \beta, \gamma) \notin \{0, 1\}$. Since BC is $x = 0$, L is the point

$$(x, y, z) = \rho(1, \beta, 1) + \sigma(1, 1, \gamma)$$

for which $x = 0$; L is therefore $(0, \beta - 1, 1 - \gamma)$.

Similarly M and N are $(1 - \alpha, 0, \gamma - 1)$ and $(\alpha - 1, 1 - \beta, 0)$. Since

$$\begin{bmatrix} 0 & 1-\alpha & \alpha-1 \\ \beta-1 & 0 & 1-\beta \\ 1-\gamma & \gamma-1 & 0 \end{bmatrix} \begin{bmatrix} 1 \\ 1 \\ 1 \end{bmatrix} = \mathbf{o}$$

we have $\mathscr{L}|$LMN; the line LM is

$$(1-\beta)(1-\gamma)x + (1-\gamma)(1-\alpha)y + (1-\alpha)(1-\beta)z = 0.$$

The special forms of the Desargues figure are obtained by suitable selection of the elements α, β, γ. Thus in the simply-special case in which $\mathscr{L}|$BCLA′ we have $A' = (0, 1, 1)$. The argument above is unaffected if $\alpha = 0$, likewise it is unaffected if also $\beta = 0$ or $\beta = 0$ & $\gamma = 0$, corresponding to the doubly- and triply-special figures.

Exercise 2 Discuss the case $\mathscr{L}|$KLMN.

In the case of the triply-special figure, the line LM is $x + y + z = 0$, and we shall obtain the quadruply-special figure, with $\mathscr{L}|$KLMN, only when $(1, 1, 1)$ lies on this line. It follows that the only primary net in which there can be a quadruply-special Desargues figure is $\mathfrak{N}_3$, and in $\mathfrak{N}_3$ all non-degenerate Desargues figures are quadruply-special.

Exercise 3 Prove that in $\mathfrak{N}_5$ Desargues figures can be found in which no four of the ten points are collinear.

Exc. 2.1.3 The Pappus figure

At one place in Book 3 the Pappus figure plays a prominent part, namely, in distinguishing between geometric systems in which multiplication of the elements of the underlying field is commutative and those in which it is not. Since we are at present concerned only with primary fields, there is no question of non-commutativity, but it is instructive to write all the products with the factors in their proper order. We base the whole proof on the theorem that a necessary and sufficient condition that three points $\boldsymbol{\pi}_1$, $\boldsymbol{\pi}_2$, $\boldsymbol{\pi}_3$ should belong to a collinear set is that there should exist (post-) multipliers λ_1, λ_2, λ_3 not all zero such that

$$\boldsymbol{\pi}_1\lambda_1+\boldsymbol{\pi}_2\lambda_2+\boldsymbol{\pi}_3\lambda_3 = \mathbf{o}.$$

We use post-multipliers for the reason that in geometry over a non-commutative field, the conditions

$$\alpha^{\mathrm{T}}\boldsymbol{\xi} = 0 \quad \text{and} \quad (\lambda\boldsymbol{\alpha}^{\mathrm{T}})(\boldsymbol{\xi}\mu) = 0$$

are identical, but $\boldsymbol{\alpha}^{\mathrm{T}}(\mu\boldsymbol{\xi}) = 0$ may well be different.

The Pappus construction is (2.3*T* 10).

G $\mathscr{L}\#|$ABC & $\mathscr{L}\#|$A′B′C′ & $\#|$AB, A′B′
C A″ = BC′ ∩ B′C, B″ = CA′ ∩ C′A, C″ = AB′ ∩ A′B
T $\mathscr{L}|$A″B″C″.

To emphasize the order of multiplication, we shall write each point as a column vector. Take A, B′, C, B″ as basis, that is

$$\mathrm{A} = \begin{bmatrix}1\\0\\0\end{bmatrix}, \quad \mathrm{B}' = \begin{bmatrix}0\\1\\0\end{bmatrix}, \quad \mathrm{C} = \begin{bmatrix}0\\0\\1\end{bmatrix}, \quad \mathrm{B}'' = \begin{bmatrix}1\\1\\1\end{bmatrix}.$$

B ∈ AC and C′ ∈ AB″:

Take

$$\mathrm{B} = \begin{bmatrix}1\\0\\\beta\end{bmatrix}, \quad \mathrm{C}' = \begin{bmatrix}\gamma\\1\\1\end{bmatrix}.$$

$$\mathrm{A}' = \mathrm{B}''\mathrm{C} \cap \mathrm{B}'\mathrm{C}' = \begin{bmatrix}1\\1\\1\end{bmatrix}\gamma + \begin{bmatrix}0\\0\\1\end{bmatrix}(1-\gamma) = \begin{bmatrix}\gamma\\1\\1\end{bmatrix} + \begin{bmatrix}0\\1\\0\end{bmatrix}(\gamma-1) = \begin{bmatrix}\gamma\\\gamma\\1\end{bmatrix},$$

$$\mathrm{C}'' = \mathrm{A}'\mathrm{B} \cap \mathrm{A}\mathrm{B}' = \begin{bmatrix}\gamma\\\gamma\\1\end{bmatrix}\beta - \begin{bmatrix}1\\0\\\beta\end{bmatrix} = \begin{bmatrix}1\\0\\0\end{bmatrix}(\gamma\beta-1) + \begin{bmatrix}0\\1\\0\end{bmatrix}\gamma\beta = \begin{bmatrix}\gamma\beta-1\\\gamma\beta\\0\end{bmatrix},$$

$$\mathrm{A}'' = \mathrm{B}\mathrm{C}' \cap \mathrm{B}'\mathrm{C} = \begin{bmatrix}1\\0\\\beta\end{bmatrix}\gamma - \begin{bmatrix}\gamma\\1\\1\end{bmatrix} = \begin{bmatrix}0\\0\\1\end{bmatrix}(\beta\gamma-1) - \begin{bmatrix}0\\1\\0\end{bmatrix} = \begin{bmatrix}0\\-1\\\beta\gamma-1\end{bmatrix}.$$

One point of A″C″ is

$$\begin{bmatrix}\gamma\beta-1\\ \gamma\beta\\ 0\end{bmatrix}+\begin{bmatrix}0\\ -1\\ \beta\gamma-1\end{bmatrix}=\begin{bmatrix}\gamma\beta-1\\ \gamma\beta-1\\ \beta\gamma-1\end{bmatrix}.$$

That is: $\mathscr{L}|A''B''C''$ provided $\beta\gamma = \gamma\beta$, a condition which is satisfied in the primary nets at least.

Exc. 2.1.4 The Menelaus theorem

$G1$ $\mathscr{L}\wp|P_1P_2P_3$

$G2$ $\{Q_1, Q_2, Q_3\}: Q_1 \in P_2P_3$ & $Q_2 \in P_3P_1$ & $Q_3 \in P_1P_2$ & $s(P_2Q_1/Q_1P_3) = k_1$ & $s(P_3Q_2/Q_2P_1) = k_2$ & $s(P_1Q_3/Q_3P_2) = k_3$ & $(k_1, k_2, k_3) \notin \{0, -1\}$

T $\mathscr{L}|Q_1Q_2Q_3 \Leftrightarrow k_1k_2k_3 = -1.$

This is a theorem about span-ratios; we therefore use the co-ordinates in their standard form in relation to a fixed ideal line.

Take $P_i = \pi_i$; then $\mathscr{L}|P_1P_2P_3 \Rightarrow$ 'there is no set of multipliers ρ_i such that $\Sigma\rho_i\pi_i = \mathbf{o}$'.

$$Q_1 = (\pi_2+k_1\pi_3)(1+k_1)^{-1} \quad \text{etc.}$$

$\mathscr{L}|Q_1Q_2Q_3 \Leftrightarrow$ there exist multiplier λ_i, not all zero, such that

$$\lambda_1(\pi_2+k_1\pi_3)+\lambda_2(\pi_3+k_2\pi_1)+\lambda_3(\pi_1+k_3\pi_2) = \mathbf{o}.$$

Moreover, under the prescribed conditions, namely, that no point Q_i coincides with a point P_j, none of the multipliers λ_i can be zero. The conditions are therefore

$$(\lambda_3+\lambda_2k_2)\pi_1+(\lambda_1+\lambda_3k_3)\pi_2+(\lambda_2+\lambda_1k_1)\pi_3 = \mathbf{o},$$

and $$\lambda_1\lambda_2\lambda_3 \neq 0.$$

Since there is no relation $\Sigma\rho_i\pi_i = \mathbf{o}$ with non-zero coefficients, it follows that there is a set of multipliers λ_1, λ_2, λ_3, none zero, such that

$$\lambda_1\lambda_2\lambda_3 = (-\lambda_3k_3)(-\lambda_1k_1)(-\lambda_2k_2) = -\lambda_1\lambda_2\lambda_3k_1k_2k_3.$$

That is, $$\mathscr{L}|Q_1Q_2Q_3 \Leftrightarrow k_1k_2k_3 = -1.$$

EXCURSUS 2.2

A NON-DESARGUESIAN CO-ORDINATE GEOMETRY

A counter-example is usually somewhat artificial, and the geometry to be described here, based on that devised by Hilbert and modified by Moulton, is no exception. At the first introduction of the Desargues construction in Chapter 1.3 some hint was given of the way in which the geometry of physical space might be modified to provide a geometry in which the final collinearity relation among the ten points of the figure was not a consequence of the prescribed nine collinearities. It was suggested that if we could devise a geometry in which, in some given Desargues figure, one of the ten collinear sets could be transformed by a process corresponding to physical refraction, while the others were unchanged, we would have a system which satisfied the conditions of $A\mathcal{I}$ 1 and 2 but not those of $A\mathcal{I}$ 3.

We are to devise therefore a geometry $\mathfrak{M}$, which is based on the co-ordinate geometry of the planar net $\mathfrak{N}_R$ established in Chapter 2.4. Divide the planar net into three zones by two parallel lines, namely, into the zones

$$H_1 : y \leqslant 0$$
$$H_0 : 0 \leqslant y \leqslant 2$$
$$H_2 : y \geqslant 2.$$

Each boundary line is included in the two zones whose points it otherwise lies between. We use for the equations of the collinear sets the forms

$$y = mx + c \quad \text{(including } y = c \text{ and } y = 0\text{)},$$
$$x = h \quad \text{and} \quad x = 0.$$

The ideal line has to be considered separately. Divide the collinear sets in $\mathfrak{N}_R$ into two subsets:

$$\text{l-lines:} \quad y = mx + c \quad \text{with } m \leqslant 0 \ \& \ c \text{ unrestricted,}$$
$$\text{r-lines:} \quad y = mx + c \quad \text{with } m > 0 \ \& \ c \text{ unrestricted.}$$

Every ordered pair of values (m, c) identifies a collinear set in $\mathfrak{N}_R$. The lines $x = h$ (including $x = 0$) and the ideal line are included in the subset of l-lines and need separate consideration.

In the geometry $\mathfrak{M}$ we use the same co-ordinates as in $\mathfrak{N}_R$ and divide the net into the same zones and the lines into two corresponding subsets, λ-lines and ρ-lines. Each λ-line is identical with an l-line, namely:

λ-line: $y = mx+c$ with $m \leqslant 0$ & unrestricted c together with $x = h$ (all h) and the ideal line.

Each ρ-line consists of parts of three r-lines, r_1, r_0, r_2, one part in each zone.

ρ-line: in zone H_1, the half-line

$$(r_1) \quad y = mx+c \quad \text{with } m > 0 \ \& \ y \leqslant 0,$$

in zone H_0, the segment

$$(r_0) \quad y = 2(mx+c) \quad \text{with } m > 0 \ \& \ 0 \leqslant y \leqslant 2,$$

in zone H_2, the half-line

$$(r_2) \quad y = mx+c+1 \quad \text{with } m > 0 \ \& \ y \geqslant 2.$$

The three parts of the ρ-line link at the boundaries of the zones, namely, at the points $(-c/m, 0)$ and $((1-c)/m, 2)$, and because $r_1 \parallel r_2$, a ρ-line has a single point common with the ideal line.

We have to show that, in geometry $\mathfrak{M}$, there is through any two distinct points a single line (either a ρ-line or a λ-line), and that any two lines of whatever subsets have a single common point. If we prove that there is a single line through any two points we shall need to prove only that any two lines have at least one common point.

Let K_0, K_1 be any two points (x_0, y_0) and (x_1, y_1) named in such a way that $y_0 \leqslant y_1$.

(i) If $y_0 = y_1$ or if $x_0 = x_1$ the line is a λ-line.

(ii) If $y_0 < y_1$ and $x_0 > x_1$ then there is one and only one λ-line containing K_0 and K_1, namely, that determined by the values of m and c satisfying

$$y_i = mx_i + c \quad (i \in \{0, 1\}),$$

because
$$y_1 - y_0 = m(x_1 - x_0) \Rightarrow m < 0.$$

If K_0, K_1 are in the same zone, then there is clearly no ρ-line containing them. If they are in different zones then there is a ρ-line containing them only if the appropriate pair of the following set of equations leads to a value of $m > 0$:

$$y_0 \leqslant 0 \ \& \ y_0 = mx_0 + c,$$

$$0 \leqslant y_0 \leqslant 2 \ \& \ y_0 = 2(mx_0+c) \quad \text{or} \quad 0 \leqslant y_1 \leqslant 2 \ \& \ y_1 = 2(mx_1+c),$$

$$y_1 \geqslant 2 \ \& \ y_1 = mx_1+c+1.$$

On eliminating c, we find as the three possible relations:

$$y_0-\tfrac{1}{2}y_1 = m(x_0-x_1) \;\&\; y_0 \leqslant 0 \;\&\; 0 \leqslant y_1 \leqslant 2,$$
$$y_0-y_1+1 = m(x_0-x_1) \;\&\; y_0 \leqslant 0 \;\&\; y_1 \geqslant 2,$$
$$\tfrac{1}{2}y_0-y_1+1 = m(x_0-x_1) \;\&\; 0 \leqslant y_0 \leqslant 2 \;\&\; y_1 \geqslant 2.$$

In all three cases $m < 0$, and there is therefore no ρ-line containing K_0 and K_1.

(iii) If $y_0 < y_1$ and $x_0 < x_1$ there is clearly no λ-line containing K_0 and K_1, while the relations established in case (ii) show that there is always one and only one ρ-line containing them.

We now have to show that any two lines have a common point. Let the lines p, p′ be given by the ordered pairs (m, c), (m′, c′). If $m = m'$ they have a common ideal point. Assume $m > m'$. The point on p at which $y = \eta > 2$ is given by

$$x = (\eta-k)/m,$$

where $k = c$ or $c+1$. For these co-ordinates

$$Y \equiv y-m'x-c' = \eta-\frac{m'}{m}(\eta-k)-c'.$$

Similarly at the point on p where $y = \eta' < 0$, $x = (\eta'-c)/m$ and for these co-ordinates

$$Y' \equiv y-m'x-c' = \eta'-\frac{m'}{m}(\eta'-c)-c'.$$

For large enough η and η' the signs of Y and Y′ are determined by $\eta(1-m'/m)$ and $\eta'(1-m'/m)$, that is they are one positive and one negative. Thus there are points of p in each of the regions determined by p′, and therefore at least one point of p lies on p′.

We have now only to construct a figure in the geometry $\mathfrak{M}$ in which all but one of the relevant segments coincide with the segments of lines in $\mathfrak{R}_R$, whilst the last segment lies part on each of two lines in $\mathfrak{R}_R$. A simple numerical case is sufficient. Take

$$A = (-1, 0), \qquad B = (1, 0), \qquad C = (0, 1),$$
$$A' = (-1, -3), \quad B' = (1, -3), \quad C' = (0, -2),$$

$AA' \parallel BB' \parallel CC'$ and all are λ-lines,

$AB \parallel A'B'$ and $BC \parallel B'C'$ and all are λ-lines.

The Desargues condition is that $AC \parallel A'C'$ and, if it is satisfied, the figure is simply-special. AC and A′C′ are both ρ-lines; their equations are

	AC	A′C′
$H_1 : y \leqslant 0$	$y = \tfrac{1}{2}(x+1)$	$y = x-2$
$H_0 : 0 \leqslant y \leqslant 2$	$y = x+1$	$y = 2(x-2)$
$H_2 : y \geqslant 2$	$y = \tfrac{1}{2}(x+3)$	$y = x-1.$

These are clearly not parallel: they have common the point (5, 4) in zone H_2.

Exercise Select in each of the following cases a set of ten points (i) which has nine of the collinearity properties of the non-special Desargues figure, (ii) which has all but one of the collinearity properties of the triply-special Desargues figure, (iii) which spans the three zones, contains both λ-lines and ρ-lines and satisfies all the conditions of the Desargues configuration.

EXCURSUS 2.3

PROJECTIVITIES AND THE CYCLIC GENERATION OF SOME FINITE PRIMARY PLANAR NETS

Exc. 2.3.1 The sequence of powers of a matrix

We assume throughout this excursus that we are given a fixed basis $\mathfrak{N}_p(O, E, F/u)$ and we take the corresponding co-ordinates in the homogeneous form, $\boldsymbol{\xi} = (x, y, z)$ for points and $\boldsymbol{\alpha}^T = [u, v, w]$ for collinear sets. The point $\boldsymbol{\xi}$ lies in the collinear set $\boldsymbol{\alpha}$ if and only if $\boldsymbol{\alpha}^T\boldsymbol{\xi} = 0$. We use

$$\boldsymbol{\epsilon}_1 = \begin{bmatrix}1\\0\\0\end{bmatrix}, \quad \boldsymbol{\epsilon}_2 = \begin{bmatrix}0\\1\\0\end{bmatrix}, \quad \boldsymbol{\epsilon}_0 = \begin{bmatrix}0\\0\\1\end{bmatrix}, \quad \boldsymbol{\upsilon} = \begin{bmatrix}1\\1\\1\end{bmatrix}$$

as the basis of the co-ordinate system.

We are to consider the sequence of projectivities determined by the powers of some given three-by-three non-singular matrix $\mathbf{M}$. From a given point $P_0 = \boldsymbol{\pi}$ we derive a sequence of points P_0, $P_1 = \mathbf{M}\boldsymbol{\pi}$, $P_2 = \mathbf{M}^2\boldsymbol{\pi}, \ldots$. If, in $\mathfrak{N}_p$, $\# | P_m P_n$ for every pair $\{m, n\}$ such that $m < n \leqslant p^2+p$, then we shall obtain in turn all the points of the system. Moreover we shall have named the points in such a way that

$$\mathscr{L}|P_r P_s P_t \Rightarrow \mathscr{L}|P_{r+k} P_{s+k} P_{t+k},$$

where $k \in \{0, \ldots, p^2+p\}$ and $r+k$, $s+k$, $t+k$ are reduced modulo p^2+p+1. Thus, suppose $P_r P_s$ is the collinear set

$$\boldsymbol{\alpha}^T\boldsymbol{\xi} = 0,$$

then
$$\boldsymbol{\alpha}^T \mathbf{M}^r \boldsymbol{\pi} = \boldsymbol{\alpha}^T \mathbf{M}^s \boldsymbol{\pi} = \boldsymbol{\alpha}^T \mathbf{M}^t \boldsymbol{\pi} = 0$$

so that P_{r+k}, P_{s+k}, P_{t+k} belong to the collinear set

$$(\boldsymbol{\alpha}^T \mathbf{M}^{-k})\boldsymbol{\xi} = 0.$$

Thus part of a cyclic table for $\mathfrak{N}_p$ is

$$\begin{matrix} \cdots & P_{r-1} & P_r & P_{r+1} & \cdots \\ \cdots & P_{s-1} & P_s & P_{s+1} & \cdots \\ \cdots & P_{t-1} & P_t & P_{t+1} & \cdots \\ \cdots & \cdots & \cdots & \cdots & \cdots \end{matrix}$$

In successive parts of this excursus we exhibit some matrices **M** which have the required property, and before considering matrices for p = 7 we discuss some general principles governing the construction of such matrices and the conditions that they have to satisfy.

Exc. 2.3.2 A cyclic generation of $\Gamma^{(2)}$

Take in $\mathfrak{F}_2$

$$\mathbf{M} = \begin{bmatrix} 0 & 1 & 0 \\ 1 & 0 & 1 \\ 0 & 1 & 1 \end{bmatrix}.$$

The powers of **M** are

$\mathbf{M}^0$	$\mathbf{M}^1$	$\mathbf{M}^2$	$\mathbf{M}^3$	$\mathbf{M}^4$	$\mathbf{M}^5$	$\mathbf{M}^6$
1 0 0	0 1 0	1 0 1	0 0 1	0 1 1	1 1 0	1 1 1
0 1 0	1 0 1	0 0 1	0 1 1	1 1 0	1 1 1	1 0 0
0 0 1	0 1 1	1 1 0	1 1 1	1 0 0	0 1 0	1 0 1

and $\mathbf{M}^7 = \mathbf{1}$.

The first column in $\mathbf{M}^0(=\mathbf{1})$ is $\boldsymbol{\epsilon}_1$, so that the first columns of the succeeding matrices are $\mathbf{M}\boldsymbol{\epsilon}_1$, $\mathbf{M}^2\boldsymbol{\epsilon}_1, \ldots$. That is, if we take $P_0 = \boldsymbol{\epsilon}_1$, they are the points $P_1, P_2, \ldots$. Also, the second column of $\mathbf{M}^0$ is $\boldsymbol{\epsilon}_2 = \mathbf{M}\boldsymbol{\epsilon}_1$, so that the second column of $\mathbf{M}^r$ is $\mathbf{M}^{r+1}\boldsymbol{\epsilon}_1$, and, similarly, the third column of $\mathbf{M}^r$ is $\mathbf{M}^{r+3}\boldsymbol{\epsilon}_1$. It follows that if the first columns of three of the matrices are such that

$$\mathbf{M}^r\boldsymbol{\epsilon}_1 + \mathbf{M}^s\boldsymbol{\epsilon}_1 + \mathbf{M}^t\boldsymbol{\epsilon}_1 = \mathbf{0},$$

then
$$\mathbf{M}^r + \mathbf{M}^s + \mathbf{M}^t = \mathbf{0}.$$

We can therefore discover the collinear sets by finding sets of three first columns whose sum is the zero vector. We find:

$$\mathbf{1} + \mathbf{M} + \mathbf{M}^5 = \mathbf{0}$$

$$\mathbf{1} + \mathbf{M}^2 + \mathbf{M}^3 = \mathbf{0}$$

$$\mathbf{1} + \mathbf{M}^4 + \mathbf{M}^6 = \mathbf{0},$$

giving the collinear sets

$\{P_0, P_1, P_5\}$, namely, $\boldsymbol{\alpha}^T = [0, 0, 1]$,

$\{P_0, P_2, P_3\}$, $[0, 1, 0]$,

$\{P_0, P_4, P_6\}$, $[0, 1, 1]$.

These may be displayed in the difference table

0	1	5
6	0	4
2	3	0

If we had used

$$\mathbf{N} = \mathbf{M}^{-1} = \mathbf{M}^6 = \begin{bmatrix} 1 & 1 & 1 \\ 1 & 0 & 0 \\ 1 & 0 & 1 \end{bmatrix}$$

as the initial matrix, we should have found

$$\mathbf{1}+\mathbf{N}+\mathbf{N}^3 = \mathbf{1}+\mathbf{M}^6+\mathbf{M}^4 = \mathbf{0}$$

leading to the table

$$\begin{matrix} 0 & 1 & 3 \\ 6 & 0 & 2 \\ 4 & 5 & 0 \end{matrix}$$

in the form in which we used it in Excursus 1.1.3.

Exc. 2.3.3 A cyclic generation of $\Gamma^{(3)}$

Take in $\mathfrak{F}_3$

$$\mathbf{K} = \begin{bmatrix} 0 & 1 & 0 \\ 1 & 0 & 1 \\ 0 & 1 & -1 \end{bmatrix}.$$

The sequence of powers is

$$\begin{matrix} \mathbf{K}^0 & \mathbf{K}^1 & \mathbf{K}^2 & \mathbf{K}^3 & \mathbf{K}^4 & \mathbf{K}^5 & \mathbf{K}^6 \\ \begin{matrix} 1 & 0 & 0 \\ 0 & 1 & 0 \\ 0 & 0 & 1 \end{matrix} & \begin{matrix} 0 & 1 & 0 \\ 1 & 0 & 1 \\ 0 & 1 & -1 \end{matrix} & \begin{matrix} 1 & 0 & 1 \\ 0 & -1 & -1 \\ 1 & -1 & -1 \end{matrix} & \begin{matrix} 0 & -1 & -1 \\ -1 & -1 & 0 \\ -1 & 0 & 0 \end{matrix} & \begin{matrix} -1 & -1 & 0 \\ -1 & -1 & -1 \\ 0 & -1 & 0 \end{matrix} & \begin{matrix} -1 & -1 & -1 \\ -1 & 1 & 0 \\ -1 & 0 & -1 \end{matrix} & \begin{matrix} -1 & 1 & 0 \\ 1 & -1 & 1 \\ 0 & 1 & 1 \end{matrix} \end{matrix}$$

$$\begin{matrix} \mathbf{K}^7 & \mathbf{K}^8 & \mathbf{K}^9 & \mathbf{K}^{10} & \mathbf{K}^{11} & \mathbf{K}^{12} & \\ \begin{matrix} 1 & -1 & 1 \\ -1 & -1 & 1 \\ 1 & 1 & 0 \end{matrix} & \begin{matrix} -1 & -1 & 1 \\ -1 & 0 & 1 \\ 1 & 1 & 1 \end{matrix} & \begin{matrix} -1 & 0 & 1 \\ 0 & 0 & -1 \\ 1 & -1 & 0 \end{matrix} & \begin{matrix} 0 & 0 & -1 \\ 0 & -1 & 1 \\ -1 & 1 & -1 \end{matrix} & \begin{matrix} 0 & -1 & 1 \\ -1 & 1 & 1 \\ 1 & 1 & -1 \end{matrix} & \begin{matrix} -1 & 1 & 1 \\ 1 & 0 & 0 \\ 1 & 0 & -1 \end{matrix} & \mathbf{K}^{13} = \mathbf{1}. \end{matrix}$$

The second and third columns of $\mathbf{K}^r$ are the first columns of $\mathbf{K}^{r+1}$ and $-\mathbf{K}^{r-3}$ respectively. The matrices satisfy:

$$\mathbf{1}-\mathbf{K}+\mathbf{K}^6 = \mathbf{0}$$

and

$$\mathbf{1}+\mathbf{K}+\mathbf{K}^4 = \mathbf{0},$$

so that the key collinear set is $\{P_0, P_1, P_4, P_6\}$ and the difference table is that used in Excursus 1.1.3, namely,

0	1	4	6.	Collinear set	[0, 0, 1]
12	0	3	5		[0, −1, 1]
9	10	0	2		[0, 1, 0]
7	8	11	0		[0, 1, 1].

Exercise 1 Show that the matrix

$$\mathbf{H} = \begin{bmatrix} 0 & 1 & 0 \\ 1 & 0 & 1 \\ 0 & 1 & 1 \end{bmatrix}$$

also leads to a cyclic generation of $\Gamma^{(3)}$ and the same difference table. The points P_r above are re-arranged in the order 0, 1, 2, 11, 6, 7, 4, 5, 12, 9, 10, 3, 8. The point P_1 and the collinear set $\{P_9, P_{10}, P_0, P_2\}$ occur in the same places in both arrangements, and the remaining points are interchanged by pairs; {4, 6}, {5, 7}, {3, 11}, {8, 12} and are such for example that $\mathscr{L}|P_1P_0P_4P_6$. Prove that

$$-\mathbf{H} = \mathbf{DKD}^{-1},$$

where D is a certain diagonal matrix.

Exercise 2 Show that the projectivity matrix

$$\mathbf{S} = \begin{bmatrix} 0 & 1 & 0 \\ 0 & 0 & 1 \\ 1 & 0 & 0 \end{bmatrix}$$

leaves one point V fixed, and permutes the others in four cycles of three. Verify that one of these sets is collinear with V and the other three are in perspective with each other from V.

Exercise 3 Show that the projectivity matrix

$$\mathbf{T} = \begin{bmatrix} 0 & 1 & 0 \\ 1 & 0 & 0 \\ 1 & 1 & 1 \end{bmatrix}$$

leaves two points V, W fixed and permutes the others in cycles of two, three and six points. Verify that the collinear set VW contains the cycle of two points, and that the points of the cycle of three are collinear with V (or W).

Exc. 2.3.4 A cyclic generation of $\Gamma^{(5)}$

Take again, but this time in $\mathfrak{F}_5$, the matrix

$$\mathbf{K} = \begin{bmatrix} 0 & 1 & 0 \\ 1 & 0 & 1 \\ 0 & 1 & -1 \end{bmatrix}.$$

We find for this matrix the sequence of points $P_r = \mathbf{K}^r \boldsymbol{\epsilon}_1$ to be

r				r				r			
0	1	0	0								
1	0	1	0	11	−1	1	0	21	2	2	1
2	1	0	1	12	1	−1	1	22	2	−2	1
3	0	2	−1	13	−1	2	−2	23	−2	−2	2
4	2	−1	−2	14	2	2	−1	24	−2	0	1
5	−1	0	1	15	2	1	−2	25	0	−1	−1
6	0	0	−1	16	1	0	−2	26	−1	−1	0
7	0	−1	1	17	0	−1	2	27	−1	−1	−1
8	−1	1	−2	18	−1	2	2	28	−1	−2	0
9	1	2	−2	19	2	1	0	29	−2	−1	−2
10	2	−1	−1	20	1	2	1	30	−1	1	1

$\mathbf{K}^{31} = \mathbf{1}.$

The second and third columns of K^r are $K^{r+1}\epsilon_1$ and $-K^{r+6}\epsilon_1$ respectively.

From the table of points P_r we see that

$$\begin{aligned} 1+K+K^{26} &= 0,\\ 1+2K+K^{28} &= 0,\\ 1-2K+2K^{19} &= 0,\\ 1-K+K^{11} &= 0. \end{aligned}$$

Thus the difference table giving the collinear sets through $P_0 = (1, 0, 0)$ is

							Collinear set
0	1	11	19	26	28.		[0, 0, 1]
30	0	10	18	25	27		[0, −1, 1]
20	21	0	8	15	17		[0, 2, 1]
12	13	23	0	7	9		[0, 1, 1]
5	6	16	24	0	2		[0, 1, 0]
3	4	14	22	29	0		[0, −2, 1].

Exercise 4 Take

$$\mathbf{G} = \begin{bmatrix} 0 & 1 & 0 \\ 1 & 0 & 1 \\ 0 & 1 & 2 \end{bmatrix}.$$

Show that $\mathbf{G}^{12} = \mathbf{1}$, that the point (2, 1, 1) and the line [2, 1, 1] are invariant, and that there are two cycles of 12 points, and, on the invariant line, two cycles of three points.

Excursus 2.3.5 Some general remarks and a cyclic generation of $\Gamma^{(7)}$

We find easily that the matrices

$$\mathbf{M}_+ = \begin{bmatrix} 0 & 1 & 0 \\ 1 & 0 & 1 \\ 0 & 1 & 1 \end{bmatrix} \quad \text{and} \quad \mathbf{M}_- = \begin{bmatrix} 0 & 1 & 0 \\ 1 & 0 & 1 \\ 0 & 1 & -1 \end{bmatrix}$$

have the properties that $\mathbf{M}_+^7 = -\mathbf{21}$, $\mathbf{M}_-^7 = \mathbf{21}$, so that some other matrix has to be sought to generate the cycle of 57 points in $\Gamma^{(7)}$. It can further be checked that $\mathbf{M}_+$ and $\mathbf{M}_-$ leave invariant the points $(1, -2, 3)$ and $(1, 2, 3)$ respectively.

If a matrix $\mathbf{M}$ leaves some point $\boldsymbol{\pi}$ invariant then $\mathbf{M}\boldsymbol{\pi} = \rho\boldsymbol{\pi}$ for some element ρ of $\mathfrak{F}_7$, that is, the equations

$$(\rho\mathbf{1} - \mathbf{M})\boldsymbol{\xi} = \mathbf{o}$$

have a solution $\boldsymbol{\xi} = \boldsymbol{\pi}$, so that

$$\det(\rho\mathbf{1} - \mathbf{M}) = 0.$$

This equation, cubic in ρ, is the *characteristic equation* of $\mathbf{M}$. We have therefore, in the present context, to exclude any matrix which is such that the characteristic equation $\det(\rho\mathbf{1}-\mathbf{M}) = 0$ has a root in $\mathfrak{F}_p$ or, in particular, in $\mathfrak{F}_7$.

The form of matrix we have been using, generalised to

$$\mathbf{M} = \begin{bmatrix} 0 & 1 & 0 \\ 1 & 0 & 1 \\ 0 & 1 & a \end{bmatrix}, \quad \text{where } a \neq 0,$$

has several advantages for hand computation, among them, that it is symmetrical and therefore so are all its powers, a property which provides a useful check. Another is that successive points in a cycle are easy to compute: if $\boldsymbol{\pi} = (p, q, r)$, then $\mathbf{M}\boldsymbol{\pi} = (q, p+r, q+ar)$.

For this form of $\mathbf{M}$ the equation

$$\det(\rho\mathbf{1}-\mathbf{M}) = 0$$

becomes
$$\rho(\rho^2-2)-a(\rho^2-1) = 0.$$

To test if this equation has a root we merely have to substitute the set of values $\{\pm 1, \pm 2, \ldots, \pm\tfrac{1}{2}(p-1)\}$ for ρ and solve the resulting linear equations for a. Clearly $\rho = \pm 1$ may always be omitted. In the case $p = 7$ we are left with

$$\rho = \pm 2: \quad \pm 4 - 3a = 0,$$

$$\rho = \pm 3: \quad 0 - a = 0.$$

Thus the only cases to be excluded on these grounds are

$$\rho = 2 \;\&\; a = 1, \quad \rho = -2 \;\&\; a = -1,$$

namely, precisely the two values of a that we have already found to be unsuitable. Any of the other four, so far as we can see at present, is suitable. Take $a = 2$,

$$\mathbf{L} = \begin{bmatrix} 0 & 1 & 0 \\ 1 & 0 & 1 \\ 0 & 1 & 2 \end{bmatrix}.$$

For the purposes of practical computation the following procedure may be adopted. First calculate

$$\mathbf{L}^2, \mathbf{L}^4, \mathbf{L}^8, \mathbf{L}^{16}, \mathbf{L}^{32}, \mathbf{L}^{48} = \mathbf{L}^{16}\mathbf{L}^{32}, \mathbf{L}^{56} = \mathbf{L}^{48}\mathbf{L}^8.$$

Since $\det\mathbf{L} = -2$, $\det(\mathbf{L}^{57}) = (\det\mathbf{L})^{57} = -1$, and $\mathbf{L}^{57} = -2\mathbf{1}$, so that as a check on the calculation of these matrices we may verify that

$$\mathbf{L}^{56} = -2\mathbf{L}^{-1} = -\begin{bmatrix} 1 & 2 & -1 \\ 2 & 0 & 0 \\ -1 & 0 & 1 \end{bmatrix}.$$

From the structure of $\mathbf{L}$ we know that $\mathbf{L}^r\epsilon_2 = \mathbf{L}^{r+1}\epsilon_1$. Moreover we find that

$$\mathbf{L}^{16} = \begin{bmatrix} 0 & 1 & 2 \\ 1 & 2 & -2 \\ 2 & -2 & 3 \end{bmatrix}$$

a matrix whose first column is the same as the third column of $\mathbf{L}$, so that $\mathbf{L}^r\epsilon_0 = \mathbf{L}^{r+15}\epsilon_1$. From $P_0 = \epsilon_1$ we have therefore calculated the co-ordinates of the points P_r for

r = 1	2	4	8	16	32	48	56
	3	5	9	17	33	49	
	17	19	23	31	47	6	14

Among these sets of co-ordinates there are enough immediately observable linear relations to determine the whole difference table.

It should be observed that the matrix

$$\mathbf{L}^3 = 2\begin{bmatrix} 0 & 1 & 1 \\ 1 & 1 & 3 \\ 1 & 3 & -1 \end{bmatrix} = 2\mathbf{L}^*$$

say, is such that $(\mathbf{L}^*)^{19} = \sigma\mathbf{1}$. That is, in general terms, when p^2+p+1 is not prime, and suppose that k is a factor, we may select a matrix which passes the $\det(\rho\mathbf{1}-\mathbf{M})$ test, but still is not suitable, because it picks out only every k*th* term in the full sequence. If, for example, we had been led to select $\mathbf{L}^*$, we should have obtained a cycle of 19 points. Given $\mathbf{L}^*$ the task of finding a 'cube root' $\mathbf{L}$ (such that $\mathbf{L}^3 = \mathbf{L}^*$) is by no means easy!

Again,

$$\mathbf{L}^{19} = \begin{bmatrix} -1 & 0 & 2 \\ 0 & 1 & -3 \\ 2 & -3 & 0 \end{bmatrix} = \overline{\mathbf{L}},$$

say, so that $\overline{\mathbf{L}}^3 = -\mathbf{1}$, while $\det(\rho\mathbf{1}-\overline{\mathbf{L}}) = \rho^3+2$ and $\rho^3+2 = 0$ has no solutions in $\mathfrak{F}_7$. (The only cubes in $\mathfrak{F}_7$ are ± 1.)

Returning to the main problem of finding the difference table; instead of juggling with a particular set of powers of the matrices, we may use the columns of these matrices as checks on the terms in the sequence $P_0, P_1, P_2, \ldots$ calculated directly. We find that the points on the line $z = 0$ are

P_0	1 0 0			P_{34}	−3	1	0
P_1	0 1 0			P_{37}	3	1	0
P_5	2 1 0			P_{43}	−2	1	0
P_{27}	1 1 0			P_{45}	−1	1	0

The difference table is therefore:

0	1	5	27	34	37	43	45.	Collinear set	[0, 0, 1]
56	0	4	26	33	36	42	44		[0, −3, 1]
52	53	0	22	29	32	38	40		[0, 2, 1]
30	31	35	0	7	10	16	18		[0, −2, 1]
23	24	28	50	0	3	9	11		[0, −1, 1]
20	21	25	47	54	0	6	8		[0, 3, 1]
14	15	19	41	48	51	0	2		[0, 1, 0]
12	13	17	39	46	49	55	0		[0, 1, 1]

Exercise 5 Show that the matrix over $\mathfrak{F}_7$

$$\mathbf{S} = \begin{bmatrix} 0 & 1 & 0 \\ 0 & 0 & 1 \\ 1 & 0 & 0 \end{bmatrix}$$

leaves three non-collinear points invariant and the lines joining pairs of these points over-all invariant. The remaining 54 points form 18 cycles of three. Two of these cycles lie on each of the invariant lines.

Exercise 6 Show that, in $\Gamma^{(p)}$, the matrix

$$\mathbf{S} = \begin{bmatrix} 0 & 1 & 0 \\ 0 & 0 & 1 \\ 1 & 0 & 0 \end{bmatrix}$$

leaves three points invariant if and only if $p-3$ is a square.

Exercise 7 Show that for $\mathfrak{F}_{11}$ the standard equation

$$\det(\rho\mathbf{1}-\mathbf{M}) \equiv \rho(\rho^2-2)-a(\rho^2-1) = 0$$

has no roots only when $a = \pm 1$.

Use the matrix

$$\begin{bmatrix} 0 & 1 & 0 \\ 1 & 0 & 1 \\ 0 & 1 & 1 \end{bmatrix}$$

over $\mathfrak{F}_{11}$ to find a difference table for $\Gamma^{(11)}$.

Exercise 8 For $\mathfrak{F}_p$ show that there is always at least one pair of values of a for which $\rho(\rho^2-2)-a(\rho^2-1) = 0$ has no root.

Exercise 9 Prove that the 40-point geometry of three dimensions (Excursus 1.5.2) is generated by the matrix

$$\begin{bmatrix} 0 & 1 & 0 & 0 \\ -1 & 0 & 1 & 0 \\ 0 & -1 & 0 & 1 \\ 0 & 0 & -1 & 1 \end{bmatrix},$$

and find a matrix which generates the Fano 15-point geometry (Excursus 1.5.1).

EXCURSUS 2.4

POLAR SYSTEMS IN THE PLANAR NET

The direct generalisation of the system of pairs of points in involution on a line is the 'polarity' system in the plane. In physical Euclidean geometry this system is the system of ⁑poles and ⁑polars with regard to a ⁑conic, the conic playing the part of the double-points in the involution. But just as we have had to distinguish involutions which have double-points from those which have not, so we have to recognise the possibility of defining a polarity system in which there is no conic. In physical geometry we distinguish among ⁑ellipses, ⁑hyperbolas and ⁑parabolas—a distinction which depends only on the choice of ideal line—and ignore the more fundamental distinction between quadratic forms in real rectangular Cartesian co-ordinates which lead to the equations of ⁑conics such as

$$x^2/a^2 \pm y^2/b^2 - 1 = 0$$

and quadratic forms such as $x^2/a^2 + y^2/b^2 + 1$ which are always positive.

But in the development of a plane geometry which provides a model for Euclidean geometry we see in Chapters 2.5 and 4.1 that an involution which has no double-points plays an essential part in defining first perpendiculars and then congruence. If we were setting up a model for geometry in three-dimensional Euclidean space, we should find that we require in a planar net a polar system which, again, has no double-points. In Excursus 4.1 we discuss an alternative definition of perpendiculars which is based on such a polar system.

We cannot therefore set out to construct conics directly, rather we have to construct the plane polar system in which each point is associated with its polar line, and from this define the conic, when it exists, as the locus of the point which lies on its polar line.

By way of introduction (Excursus 2.4.1) we set up first an algebraic treatment of involutions, and then (Excursus 2.4.2) develop a geometric theory of the polar system, basing it on the definition of an involution by the products of span-ratios. In Excursus 2.4.3 we translate this into algebraic terms, and in Excursus 2.4.4 develop briefly a classification of polar systems and conics by the double-points of the defining involutions.

Exc. 2.4.1 Note on the algebraic treatment of involutions

Let us assume a basis K (zero point), E, U on a line and take the first definition of an involution, namely, as a set of pairs of points $\{P_i, P'_i\}$ on a line such that

$$s(KP_i/KP_j) = s(KP'_j/KP'_i) \quad (\{i, j\} \subset \{1, 2, ...\}).$$

Write $$s(KP_i/KE) = x_i, \quad s(KP'_i/KE) = x'_i,$$

then the involution is determined by pairs $\{x_i, x'_i\}$ of elements of the field $\mathfrak{F}$ such that

$$x_i x'_i = k, \quad \text{where k is constant.}$$

With a change of K and E, that is, with the replacement of x by $\alpha x + \beta$, this relation is replaced by

$$(\alpha x_i + \beta)(\alpha x'_i + \beta) = k,$$

or $$a x_i x'_i + h(x_i + x'_i) + b = 0. \tag{1}$$

This may be written as

$$[x_i, 1]\begin{bmatrix} a & h \\ h & b \end{bmatrix}\begin{bmatrix} x'_i \\ 1 \end{bmatrix} = 0;$$

we take (1) to be the basic algebraic definition of the involution. The involution has centre $x_i = -h/a$ (the corresponding value x'_i is undefined), and has double-points only if there is a solution in $\mathfrak{F}$ for the equation

$$ax^2 + 2hx + b = 0.$$

The involution is singular if the matrix

$$\begin{bmatrix} a & h \\ h & b \end{bmatrix}$$

is singular, because if $ab - h^2 = 0$ (and $h \neq 0$), then

$$h\{a x_i x'_i + h(x_i + x'_i) + b\} \equiv (a x_i + h)(h x'_i + b),$$

and the point conjugate to any x_i is the point $x'_i = -b/h$. The algebraic equivalent of the statement: 'if an involution is prescribed to have a double-point, then either it has a second double-point or is singular' is: 'if a quadratic function is prescribed to have one zero in a field then either it has a second zero or it is the square of a linear function'. It is easily verified that if x_0 is a zero of $ax^2 + 2hx + b$, then (if $a \neq 0$), so is $x_1 = -x_0 - 2h/a$. $x_1 \neq x_0$ provided that $x_0 \neq -h/a$, that is, provided that $ab - h^2 \neq 0$.

If the involution is prescribed to have a double-point, and we select that point as the ideal point, then $a = 0$, and the involution becomes the mid-point involution

$$h(x_i + x_i') + b = 0,$$

of which the second double-point (the mid-point) is $-b/2h$.

Another way of expressing condition (1) is this: let $\{x_i, x_i'\}$ be the roots of the equation

$$\lambda_i x^2 + \mu_i x + \nu_i = 0.$$

Then, since $\quad x_i x_i' = \nu_i/\lambda_i \quad \text{and} \quad x_i + x_i' = -\mu_i/\lambda_i,$

we must have $\quad a\nu_i - 2h\mu_i + b\lambda_i = 0.$

That is, the involution consists of the pairs of points given by the roots of the equations

$$\lambda x^2 + \mu x + \nu = 0 \tag{2.1}$$

in which the coefficients are subject to the condition

$$a\nu - 2h\mu + b\lambda = 0. \tag{2.2}$$

Since the set $\{\lambda, \mu, \nu\}$ satisfies one linear condition there are two linearly independent sets, say $\{\lambda_0, \mu_0, \nu_0\}$ and $\{\lambda_1, \mu_1, \nu_1\}$, and every other set is a linear combination of them. That is, a third way of defining the pairs of the involution is as the pairs of points corresponding to the roots of

$$(\lambda_0 x^2 + \mu_0 x + \nu_0) + \kappa(\lambda_1 x^2 + \mu_1 x + \nu_1) = 0. \tag{3}$$

In this form the involution is directly determined by two of its pairs. The centre is given by $\kappa = -\lambda_0/\lambda_1$, since for this value of κ the expression is linear in x, and the 'second root' is undefined. The involution has double-points if

$$(\mu_0 + \kappa\mu_1)^2 - 4(\lambda_0 + \kappa\lambda_1)(\nu_0 + \kappa\nu_1) = 0,$$

as an equation in κ, has roots in $\mathfrak{F}$. A value of κ which is a root, when substituted into equation (3), gives the square of a linear function—that is, it gives the co-ordinate of the double-point corresponding to the root.

Exc. 2.4.2 Polarity as a geometric transformer

The basis of a polarity is two collinear sets with a prescribed involution on each. To simplify the description of the construction and the proofs of the theorems we use the definition of the involution in which the centre K and ideal point U are prescribed as one pair of the involution, and the involution consists of pairs of points $\{A_i, A_i'\}$ for which

$$s_U(KA_i/KA_j') = s_U(KA_j/KA_i').$$

All results obtained in this way may be readily interpreted in terms of cross-ratios and of the quadrangular construction for involutions.

While the arguments used in this Excursus apply to primary nets $\mathfrak{N}_p$ and $\mathfrak{N}_R$ in general (and in fact to the extensions of these nets), they do not apply to $\mathfrak{N}_2$ in which there are no non-singular involutions, and have to be modified for $\mathfrak{N}_3$. $\mathfrak{N}_5$ has six points in a collinear set and in general terms the arguments apply in $\mathfrak{N}_5$ without modification.

Exc. 2.4.2 ***D*1** *Construction of the line defined to be the polar of a point in a polarity.*

*G*1 $\mathscr{L}|\mathrm{KUV}$ & ideal$|\mathrm{UV}$ (fixed)
$\mathrm{a} = \mathrm{KU}, \quad \mathrm{b} = \mathrm{KV}$
*G*2 $\mathscr{I}_\alpha = \mathscr{I}_{\mathrm{K;A,A'}} \subset \mathrm{a}$
$\mathscr{I}_\beta = \mathscr{I}_{\mathrm{K;B,B'}} \subset \mathrm{b}$
S $\wp|\mathrm{P_i}, \quad \mathrm{P_i} \notin \mathrm{a}$ & $\mathrm{P_i} \notin \mathrm{b}$
*C*1 $\mathrm{L_i} : \mathrm{L_i} \in \mathrm{a}$ & $\mathrm{L_i P_i} \parallel \mathrm{b}$
$\mathrm{M_i} : \mathrm{M_i} \in \mathrm{b}$ & $\mathrm{M_i P_i} \parallel \mathrm{a}$
*C*2 $\mathrm{L'_i} : \mathrm{L'_i} = \mathscr{I}_\alpha \mathrm{L_i}$
$\mathrm{M'_i} : \mathrm{M'_i} = \mathscr{I}_\beta \mathrm{M_i}$
D $\mathrm{p_i} : \mathrm{p_i} = \mathrm{L'_i M'_i}$.

Given $\mathscr{I}_\alpha$, $\mathscr{I}_\beta$ and an arbitrary line $\mathrm{p_i}$, the steps above in the reverse order provide a construction for the point $\mathrm{P_i}$ of which $\mathrm{p_i}$ is the polar.

Exc. 2.4.2 ***N*1**

Polar: $\mathrm{p_i}$ is the polar of $\mathrm{P_i}$ in the *polarity*, $\mathscr{Q}_{\alpha,\beta} = \mathscr{Q}_{\beta,\alpha}$, determined by $\mathscr{I}_\alpha, \mathscr{I}_\beta$, $\mathrm{p_i} = \mathscr{Q}_{\alpha,\beta}\mathrm{P_i}$,
pole: $\mathrm{P_i}$ is the *pole* of $\mathrm{p_i}$ in the polarity, $\mathrm{P_i} = \mathscr{Q}_{\alpha,\beta}\mathrm{p_i}$,
centre: K is the *centre* of $\mathscr{Q}$,
diameter: any collinear set containing K.

Exc. 2.4.2 ***T*1** Special polars.

(i) The polar of a point of a is parallel to b.

$$\mathrm{P_i} \in \mathrm{a} \Rightarrow \mathrm{M_i} = \mathrm{K} \Rightarrow \mathrm{M'_i} = \mathrm{V} \Rightarrow \mathrm{p_i} \parallel \mathrm{b}.$$

(ii) The polar of K is the ideal line.

$$\mathrm{P_i} = \mathrm{K} \Rightarrow \mathrm{L'_i} = \mathrm{U} \;\&\; \mathrm{M'_i} = \mathrm{V} \Rightarrow \text{ideal}|\mathrm{p_i}.$$

Exc. 2.4.2 ***T*2** The polars of points on a diameter are parallel.

G $\mathscr{Q}_{\alpha,\beta}$ determined by $\mathscr{I}_\alpha \subset \mathrm{a}, \quad \mathscr{I}_\beta \subset \mathrm{b}, \quad \mathrm{a} \cap \mathrm{b} = \mathrm{K}$
S $\#\wp|\mathrm{P_1P_2}, \quad \mathscr{L}|\mathrm{KP_1P_2}$
C $\{\mathrm{p_1}, \mathrm{p_2}\} : \mathrm{p_i} = \mathscr{Q}_{\alpha,\beta}\mathrm{P_i}$

T $p_1 \| p_2$

P *c*1 $\{L_i\} : \{L_i\} \subset a \;\&\; P_iL_i \| b$

$\{M_i\} : \{M_i\} \subset b \;\&\; P_iM_i \| a$

t $s(KL_1'/KL_2') = s(KL_2/KL_1)$

$= s(KP_2/KP_1) = s(KM_2/KM_1)$

$= s(KM_1'/KM_2')$.

Exc. 2.4.2*N*2 Conjugate diameter.

*G*1 $\mathscr{Q}_{\alpha,\beta}P = p$

*G*2 $c : c = KP$

$d : d \supset K \;\&\; d \| p$

N d is the diameter *conjugate* to c.

In particular a and b are mutually conjugate since the polar of any point on either is parallel to the other.

Exc. 2.4.2*T*3 Conjugacy of diameters is a mutual property.

*G*1 $\mathscr{Q}_{\alpha,\beta}, \wp | P_i$

*G*2 $Q : KQ \| \mathscr{Q}_{\alpha,\beta}P_i$

T $KP_i \| \mathscr{Q}_{\alpha,\beta}Q$

*C*1 L_i, M_i, L_i', M_i' as in 2.4.2*D*1

*C*2 $Q_i : Q_i \in P_iL_i \;\&\; KQ_i \| L_i'M_i'$

*C*3 $N_i : N_i \in b \;\&\; Q_iN_i \| a$

$N_i' : N_i' = \mathscr{I}_\beta N_i$

T $L_i'N_i' \| KP_i$

P $s(KL_i/KL_i') = s(L_iQ_i/M_i'K)$

$= s(KN_i/M_i'K) = s(KM_i/N_i'K)$

$= s(L_iP_i/N_i'K)$

T $L_i'N_i' \| KP_i$.

Exc. 2.4.2*T*4 Pairs of conjugate diameters form a family of pairs in involution.

G $\mathscr{Q}_{\alpha,\beta}$ determined by $\mathscr{I}_\alpha \subset a \;\&\; \mathscr{I}_\beta \subset b$

*S*1 $\wp | h : h \| b \;\&\; \# | hb$

*S*2 $\wp | P_i : P_i \in h$

*C*1 $\{L, L'\} : L = h \cap a, \quad L' = \mathscr{I}_\alpha L$

*C*2 $M_i : M_i \in b \;\&\; P_iM_i \| a$

$M_i' : M_i' = \mathscr{I}_\beta M_i$

*C*3 $Q_i : Q_i \in h \;\&\; KQ_i \| L'M_i'$

(i.e. KQ_i is the diameter conjugate to KP_i)

T $\mathscr{I}_{L;P_1,Q_1}P_j = Q_j$

P $s(Q_i L/KM_i') = s(KL/KL') = s(Q_j L/KM_j')$
$s(LP_j/KM_i') = s(LP_i/KM_j')$
$\Rightarrow s(LQ_i/LP_j) = s(LQ_j/LP_i)$

T $\mathscr{I}_{L;P_i,Q_i} P_j = Q_j$.

That is, $\{\{KP_i, KQ_i\}, i \in \{1, 2, \ldots\}\}$ is a set of pairs in an involution in which a, b is one pair.

Exc. 2.4.2*T*5 $\mathscr{Q}_{\alpha,\beta}$ induces on any diameter an involution whose centre is K.

*G*1 $\mathscr{Q}_{\alpha,\beta}$ determined by $\mathscr{I}_\alpha \subset$ a, $\mathscr{I}_\beta \subset$ b
*G*2 $\wp$|c, c ⊃ K & # |abc
S $\wp|P_i$, $P_i \in$ c & # $|KP_i$
C L_i, M_i, L_i', M_i' as in Exc. 2.4.2*D*1
$P_i' : P_i' = (\mathscr{Q}_{\alpha,\beta} P_i) \cap KP_i$
T $\mathscr{I}_{K;P_i,P_i'} P_j = P_j'$
P $s(KP_i'/KP_j') = s(KL_i'/KL_j')$
$= s(KL_j/KL_i) = s(KP_j/KP_i)$
T $\mathscr{I}_{K;P_i,P_i'} P_j = P_j'$.

Exc. 2.4.2*T*6 If the polar of P_i contains P_j, then the polar of P_j contains P_i.

G $\mathscr{Q}_{\alpha,\beta}$ determined by $\mathscr{I}_\alpha \subset$ a, $\mathscr{I}_\beta \subset$ b
S $\wp|P_i$
$\wp|P_j$, $P_j \in \mathscr{Q}_{\alpha,\beta} P_i$
T $P_i \in \mathscr{Q}_{\alpha,\beta} P_j$
C L_i, M_i, L_i', M_i' as in Exc. 2.4.2*D*1
P $s(M_i P_i/KL_j') = s(KL_i/KL_j')$
$= s(KL_j/KL_i') = s(M_j M_i'/KM_i')$
$= 1 - s(KM_j/KM_i') = 1 - s(KM_i/KM_j')$
$= s(M_i M_j'/KM_j')$
T $\mathscr{L}|L_j' M_j' P_i$.

Exc. 2.4.2*N*3 Conjugate points and lines.

If $P_j \in \mathscr{Q}_{\alpha,\beta} P_i$, then P_i, P_j are *conjugate* points and $\mathscr{Q}_{\alpha,\beta} P_i$, $\mathscr{Q}_{\alpha,\beta} P_j$ are conjugate lines.

Each line contains the pole of the other. Conjugate diameters are a pair of conjugate lines specially related to the ideal line.

Exc. 2.4.2*N*4 Self-polar triads.

$P_i P_j P_k$ form a self-polar triad if $\mathscr{Q}_{\alpha,\beta} P_i = P_j P_k$ & P_j, P_k are conjugate.

Each point is conjugate to the other two, and each of the three collinear sets determined by pairs of the points is conjugate to the other two. The original triad, KUV, used in *D*1 to generate the polarity, is a self-polar triad, the pair of conjugate diameters and the ideal line being the collinear sets in the triad.

Exc. 2.4.2*T*7 If c, d are conjugate diameters in $\mathcal{Q}_{\alpha,\beta}$, and $\mathcal{I}_\gamma$, $\mathcal{I}_\delta$ are the involutions induced on them by $\mathcal{Q}_{\alpha,\beta}$, and $\mathcal{Q}_{\gamma,\delta}$ is the polarity determined by $\mathcal{I}_\gamma, \mathcal{I}_\delta$, then $\mathcal{Q}_{\gamma,\delta} = \mathcal{Q}_{\alpha,\beta}$.

*G*1 $\mathcal{Q}_{\alpha,\beta}$ determined by $\mathcal{I}_\alpha \subset \mathrm{a}, \mathcal{I}_\beta \subset \mathrm{b}$
*G*2 c, d conjugate diameters: $\mathcal{I}_\gamma$, $\mathcal{I}_\delta$ the involutions induced on them by $\mathcal{Q}_{\alpha,\beta}$
S $\wp|\mathrm{P}, \mathrm{P} \notin \mathrm{c}, \mathrm{P} \notin \mathrm{d}$
C $\mathrm{p} : \mathrm{p} = \mathcal{Q}_{\alpha,\beta}\mathrm{P}$
T $\mathcal{Q}_{\gamma,\delta}\mathrm{P} = \mathrm{p}$
P *c*1 $\{\mathrm{C}, \mathrm{D}\} : \mathrm{C} = \mathrm{p} \cap \mathrm{c}, \quad \mathrm{D} = \mathrm{p} \cap \mathrm{d}$
*c*2 $\mathrm{C}' : \mathrm{C}' = \mathrm{c} \cap \mathcal{Q}_{\alpha,\beta}\mathrm{C}$
$\mathrm{D}' : \mathrm{D}' = \mathrm{d} \cap \mathcal{Q}_{\alpha,\beta}\mathrm{D}$
*t*1 $\mathcal{I}_\gamma \mathrm{C}' = \mathrm{C}, \quad \mathcal{I}_\delta \mathrm{D}' = \mathrm{D}$
*t*2 $\mathcal{Q}_{\alpha,\beta}\mathrm{C} = \mathrm{C}'\mathrm{P}, \quad \mathcal{Q}_{\alpha,\beta}\mathrm{D} = \mathrm{D}'\mathrm{P}$
p $\mathrm{C}' \in \mathcal{Q}_{\alpha,\beta}\mathrm{C}$ (*c*2)
$\mathrm{C} \in \mathcal{Q}_{\alpha,\beta}\mathrm{P} \Rightarrow \mathrm{P} \in \mathcal{Q}_{\alpha,\beta}\mathrm{C}$
*t*3 $\mathcal{Q}_{\gamma,\delta}\mathrm{P} = \mathrm{CD} = \mathrm{p}$
p $\mathrm{PC}' \| \mathrm{d}$ & $\mathrm{PD}' \| \mathrm{c}$ & $\mathcal{I}_\gamma \mathrm{C}' = \mathrm{C}$ & $\mathcal{I}_\delta \mathrm{D}' = \mathrm{D}$
T $\mathcal{Q}_{\gamma,\delta} = \mathcal{Q}_{\alpha,\beta}$.

Exc. 2.4.2*T*8*X* $\mathcal{Q}_{\alpha,\beta}$ induces an involution in every collinear set in the planar net.

The theorem is proved already for diameters (*T*5) and for the ideal line, on which the involution is that cut by conjugate diameters.

G $\mathcal{Q}_{\alpha,\beta}$ determined by $\mathcal{I}_\alpha \subset \mathrm{a}, \quad \mathcal{I}_\beta \subset \mathrm{b}$
S $\mathrm{h} : \mathrm{h} \not\supset \mathrm{K}$
$\mathrm{H}_1 : \mathrm{H}_1 \in \mathrm{h}$
C $\mathrm{H}_1' : \mathrm{H}_1' = (\mathcal{Q}_{\alpha,\beta}\mathrm{H}_1) \cap \mathrm{h}$
T $\{\{\mathrm{H}_1, \mathrm{H}_1'\}, \mathrm{i} \in \{1, 2, \ldots\}\}$ are pairs in an involution
P *c*1 $\mathrm{c} : \mathrm{c} \supset \mathrm{K}$ & $\mathrm{c} \| \mathrm{h}$
$\mathrm{d} : \mathrm{d} \supset \mathrm{K}$ & d conjugate to c
*c*2 $\mathrm{H}_0 : \mathrm{H}_0 = \mathrm{d} \cap \mathrm{h}$
T $\mathcal{I}_{\mathrm{H}_0; \mathrm{H}_1, \mathrm{H}_1'}\mathrm{H}_j = \mathrm{H}_j'$.

Exc. 2.4.2*T*9*X* If r and s are any two conjugate lines, and $\mathcal{I}_\rho$, $\mathcal{I}_\sigma$ are the involutions induced on them by $\mathcal{Q}_{\alpha,\beta}$, then $\mathcal{Q}_{\rho,\sigma} = \mathcal{Q}_{\alpha,\beta}$.

Take first r and s to be d and h of T8.

In effect, then, the original triad KUV may be replaced by any self-polar triad together with the induced involutions on any two of the lines, the third line playing the part of the ideal line. Since, further, the choice of ideal line is irrelevant in the quadrangular construction of involutions, we could devise a direct incidence construction (without the intervention of span-ratios) for the polarity based on

$$\mathscr{I}_\alpha \subset \mathrm{KA} \quad \text{such that } \mathscr{I}_\alpha \mathrm{K} = \mathrm{A}$$

and
$$\mathscr{I}_\beta \subset \mathrm{KB} \quad \text{such that } \mathscr{I}_\beta \mathrm{K} = \mathrm{B}.$$

Since each of the involutions is determined by one other pair of points, we may take as the data:

$$\mathscr{L}|\mathrm{KAB} \text{ is a self-polar triad,}$$
$$\mathrm{C, d} \text{ such that d is to be the polar of C}$$

(with no unspecified incidences among these elements). The other pairs in the involutions are then:

$$\text{on} \quad \mathrm{KA} : \mathrm{KA} \cap \{\mathrm{BC, d}\},$$
$$\text{on} \quad \mathrm{KB} : \mathrm{KB} \cap \{\mathrm{CA, d}\}.$$

The following exercise provides a construction of the polarity on this basis.

Exercise 1 Construction of the polar of an arbitrary point when a self-polar triad and a point and its polar line are given.

G $\mathscr{L}|\mathrm{KAB; C; d}$
S H
$C1$ $\{\mathrm{L, L'}\} : \{\mathrm{L, L'}\} = \mathrm{KA} \cap \{\mathrm{BC, d}\}$
$\{\mathrm{M, M'}\} : \{\mathrm{M, M'}\} = \mathrm{KB} \cap \{\mathrm{AC, d}\}$
$\{\mathrm{C', N}\} : \{\mathrm{C', N}\} = \mathrm{d} \cap \{\mathrm{KC, AB}\}$
$C2$ $\mathrm{P : P = BH \cap KA}$
$\mathrm{F : F = PN \cap KC}$
$\mathrm{F' : F' = FL \cap AB}$
$\mathrm{P' : P' = F'C' \cap KA}$
$T1$ $\mathscr{I}_{\mathrm{K,A;L,L'}}\mathrm{P = P'}$
$C3$ $\mathrm{Q : Q = AH \cap KB}$
$\mathrm{G : G = QN \cap KC}$
$\mathrm{G' : G' = GM \cap AB}$
$\mathrm{Q' : Q' = G'C' \cap KB}$
$T2$ $\mathscr{I}_{\mathrm{K,B;M,M'}}\mathrm{Q = Q'}$
$T3$ $\mathscr{Q}\mathrm{H = P'Q'}$.

Exercise 2 In the net $\mathfrak{N}_R$, ABC is a proper self-polar triad in a polar system $\mathcal{Q}$ which is defined by two elliptic involutions. Prove that the centre lies in the region $\mathfrak{R}$ABC.

Exercise 3 The Desargues figure defines a unique polarity with regard to which it is its own polar:

G Ten points P_{ij}
$\mathscr{L}|P_{ij}P_{jk}P_{kl} = l_{rs}$ $\{i, j, k, r, s\} = \{1, 2, 3, 4, 5\}$

C $P'_{rs,ij} = l_{rs} \cap l_{ij}$ (fifteen points)

*T*1 On l_{rs} : $\{P_{ij}, P'_{ij,rs}\}$, $\{P_{jk}, P'_{jk,rs}\}$, $\{P_{kl}, P'_{kl,rs}\}$ are pairs of an involution

*T*2 There is a unique polarity $\mathcal{Q}$ in which each line l_{ij} is the polar of the corresponding point P_{ij}

*T*3 In $\mathcal{Q}$ a set of three points such as $\{P_{ij}, P_{rs}, P'_{ij,rs}\}$ forms a self-polar triad.

So far we have made no mention of double-points in the involutions, nor of the possibility that some of the involutions might be singular, since, in $\mathfrak{N}_R$ at least, systems can be devised in which none of the induced involutions has double-points and none is singular. Clearly constructions could be simplified if the basic involutions have double-points, since we may replace the 'involution' construction by the 'harmonic' construction.

This Excursus should have established with the reader the bona fides of a polarity as a geometric transformer which determines a one-to-one correspondence between points and lines in the plane. Further development is much simplified if we formulate the problems in algebraic terms.

Exc. 2.4.3 Algebraic representation of a polarity

We define the polarity $\mathcal{Q}_{\alpha,\beta}$ as in Excursus 2.4.2*D*1, by two involutions $\mathscr{I}_\alpha = \mathscr{I}_{K;A,A'}$, $\mathscr{I}_\beta = \mathscr{I}_{K;B,B'}$ with ideal|u fixed, and take as the co-ordinate system $\mathscr{C}$(KE, KF/u), where E ∈ KA, F ∈ KB. Let the constants of the two involutions be given by

$$s(KA/KE)\,s(KA'/KE) = \alpha,$$
$$s(KB/KF)\,s(KB'/KF) = \beta,$$

so that the equations of the involutions are

$$\mathscr{I}_\alpha \text{ on } y = 0\colon x_1x_1' = \alpha,$$
$$\mathscr{I}_\beta \text{ on } x = 0\colon y_1y_1' = \beta.$$

Let $P_1 = (x_1, y_1, 1)$ so that in the notation of Excursus 2.4.2*D*1,

$$L_1 = (x_1, 0, 1), \quad L_1' = (\alpha/x_1, 0, 1),$$
$$M_1 = (0, y_1, 1), \quad M_1' = (0, \beta/y_1, 1),$$

and the polar of P_1, $p_1 = L_1'M_1'$, has co-ordinates $[-x_1/\alpha, -y_1/\beta, 1]$. We may therefore express the polarity in the form

$$[x_1, y_1, 1]\begin{bmatrix} -1/\alpha & 0 & 0 \\ 0 & -1/\beta & 0 \\ 0 & 0 & 1 \end{bmatrix}\begin{bmatrix} x \\ y \\ 1 \end{bmatrix} = 0,$$

say, $$\boldsymbol{\xi}_1^T \boldsymbol{\Gamma} \boldsymbol{\xi} = 0,$$
where Γ is a diagonal matrix.

Now transform the co-ordinates to a system $\boldsymbol{\xi}$ given by

$$\rho\boldsymbol{\xi} = \mathbf{M}\boldsymbol{\xi},$$

where $\mathbf{M}$ is any non-singular three-by-three matrix. The equation of the polarity $\mathscr{Q}$ becomes $$\boldsymbol{\xi}_1^T \mathbf{M}^T \boldsymbol{\Gamma} \mathbf{M} \boldsymbol{\xi} = 0.$$

Write $$\bar{\boldsymbol{\Gamma}} = \mathbf{M}^T \boldsymbol{\Gamma} \mathbf{M},$$

then $$\bar{\boldsymbol{\Gamma}}^T = (\mathbf{M}^T \boldsymbol{\Gamma} \mathbf{M})^T = \mathbf{M}^T \boldsymbol{\Gamma}^T \mathbf{M} = \mathbf{M}^T \boldsymbol{\Gamma} \mathbf{M} = \bar{\boldsymbol{\Gamma}},$$

so that $\bar{\boldsymbol{\Gamma}}$ is symmetrical. Without going into detail, we may say then, that the equation of a polarity referred to a co-ordinate system to which it bears no special relation, will be expressed by an equation of the form

$$[x_1, y_1, 1]\begin{bmatrix} a & h & g \\ h & b & f \\ g & f & c \end{bmatrix}\begin{bmatrix} x \\ y \\ 1 \end{bmatrix} = 0.$$

Return now to the simple form with which we started, namely, that in which the polar of $(x_1, y_1, 1)$ is $[-x_1/\alpha, -y_1/\beta, 1]$ and in which the centres of the two defining involutions are at the origin. This relation we may also express in terms of the conditions of conjugacy. $(x_1, y_1, 1)$ is conjugate to $(x_j, y_j, 1)$ if it lies on $[-x_j/\alpha, -y_j/\beta, 1]$, so that $(x_1, y_1, 1)$, $(x_j, y_j, 1)$ are conjugate if

$$\frac{x_1 x_j}{\alpha} + \frac{y_1 y_j}{\beta} - 1 = 0$$

and similarly the lines $[u_1, v_1, 1]$, $[u_j, v_j, 1]$ are conjugate if

$$\alpha u_1 u_j + \beta v_1 v_j - 1 = 0.$$

A diameter is $[u, 1, 0]$; the polar of a point $(x_1, -ux_1, 1)$ on this diameter is the line $[-x_1/\alpha, ux_1/\beta, 1]$. The polars of points on the diameter are parallel to $[u', 1, 0]$ where $u' = -\beta/\alpha u$, namely, to the conjugate diameter. The pairs of conjugate diameters therefore form the involution given by $uu' = -\beta/\alpha$.

The conjugate of the point $(x_1, -ux_1, 1)$ in the involution induced by $\mathcal{Q}$ on the diameter [u, 1, 0] is the intersection of [u, 1, 0] and $[-x_1/\alpha, ux_1/\beta, 1]$, namely, the point determined on the line by

$$x = 1/[x_1(1/\alpha + u^2/\beta)].$$

Thus the involution on the diameter [u, 1, 0] consists of the pairs of points $(x_1, -ux_1, 1)$, $(x_1', -ux_1', 1)$ given by

$$x_1 x_1' = \alpha\beta/(\alpha u^2 + \beta).$$

Consider next the involution induced on a general line, h = [u, v, 1]. The points in which h meets KA, KB are $(-1/u, 0, 1)$, $(0, -1/v, 1)$ and a general point on h can therefore be taken as

$$[-1/u(\kappa+1),\ -\kappa/v(\kappa+1),\ 1].$$

The polar of this point is

$$[1/\alpha u(\kappa+1),\ \kappa/\beta v(\kappa+1),\ 1]$$

and meets h in the point

$$[-1/u(\kappa'+1),\ -\kappa'/v(\kappa'+1),\ 1]$$

for which κ' is given by

$$-1/\alpha u^2 - \kappa\kappa'/\beta v^2 + (\kappa+1)(\kappa'+1) = 0.$$

The involution on h therefore consists of pairs of points with parameters $\{\kappa, \kappa'\}$ related by

$$\kappa\kappa'(1-1/\beta v^2) + (\kappa+\kappa') + (1-1/\alpha u^2) = 0$$

or

$$[\kappa, 1]\begin{bmatrix} 1-1/\beta v^2 & 1 \\ 1 & 1-1/\alpha u^2 \end{bmatrix}\begin{bmatrix} \kappa' \\ 1 \end{bmatrix} = 0.$$

The involution is singular if the matrix is singular, that is, if

$$1/\beta v^2 + 1/\alpha u^2 - 1/\alpha\beta u^2 v^2 = 0$$

or

$$\alpha u^2 + \beta v^2 - 1 = 0.$$

If (x_δ, y_δ) is the double-point in this singular involution, then [u, v, 1] is its polar line. That is, [u, v, 1] and $[-x_\delta/\alpha, -y_\delta/\beta, 1]$ are the same line, and therefore $x_\delta = -\alpha u$, $y_\delta = -\beta v$. The locus of the double points in the singular involutions is therefore

$$x^2/\alpha + y^2/\beta - 1 = 0.$$

It should be remembered, however, that there may be no singular involutions in the system.

Exc. 2.4.4 Types of polarities: conics

We have to take account of the sharp difference between the finite net $\mathfrak{N}_p$ and the rational net $\mathfrak{N}_R$ in any investigation which is affected by the double-points of an involution. We take first $\mathfrak{N}_R$ and assume that it has been extended to the surdic net $\mathfrak{N}_S$ so that every hyperbolic involution has double-points.

(i) *Polarities in* $\mathfrak{N}_S$. There are three apparent types of polarity systems determined by the types of involutions, $\mathscr{I}_\alpha(x_1 x_1' = \alpha)$ and $\mathscr{I}_\beta(y_1 y_1' = \beta)$, namely:

HH: both hyperbolic, $\alpha > 0, \beta > 0$,

HE: one hyperbolic, one elliptic, $\alpha\beta < 0$,

EE: both elliptic, $\alpha < 0, \beta < 0$.

The corresponding involution of conjugate diameters

$$uu' = -\beta/\alpha,$$

is hyperbolic in type HE, and elliptic in the other two types.

The involution induced on the diameter [u, 1, 0] is given by

$$x_1 x_1' = \alpha\beta/(\alpha u^2 + \beta)$$

and is therefore always hyperbolic for type HH (there are two double-points on every diameter), always elliptic for type EE, and in type HE changes from one to the other for the values of u given by

$$\alpha u^2 + \beta = 0.$$

That is, in type HE, the two self-conjugate diameters form the boundary between the regions in which there are and in which there are not double-points on the diameters. On a self-conjugate diameter the involution is singular and the double-point is ideal.

The involution on the line [u, v, 1] has been shown to be given by parameter-pairs $\{\kappa, \kappa'\}$ satisfying the relation:

$$[\kappa, 1]\begin{bmatrix} 1 - 1/\beta v^2 & 1 \\ 1 & 1 - 1/\alpha u^2 \end{bmatrix}\begin{bmatrix} \kappa' \\ 1 \end{bmatrix} = 0$$

and is therefore elliptic or hyperbolic according to the sign of $\gamma = (\alpha u^2 + \beta v^2 - 1)/\alpha\beta$. For type EE, $\gamma < 0$ for all pairs $\{u, v\}$, so that every involution induced by the polarity is elliptic and there are no singular involutions. For types HH and HE, γ can be made to have either sign by proper choice of $\{u, v\}$.

If a point D is a double-point in an induced involution then the polar line d of D passes through D, and therefore D is a double-point in the induced involution on any line through D. Since D is conjugate to every point of d, and to no other points, d is the only line through D on which the involution is singular.

We may obtain all the double-points in $\mathscr{Q}_{\alpha,\beta}$ as the double-points in the induced involutions on the diameters. The locus of double-points is therefore represented by the equations

$$\left.\begin{aligned} ux+y &= 0, \\ x^2 &= \alpha\beta/(\alpha u^2+\beta), \end{aligned}\right\}$$

i.e. by $$x^2/\alpha+y^2/\beta-1 = 0.$$

This is the equation of the *conic* associated with the polarity (of type HH or HE); the polar line of a point of the conic (i.e. the line, through the point, on which the involution is singular) is the *tangent* to the conic. The co-ordinate system we have chosen, defined by two conjugate diameters and the ideal line, could be replaced by another based on any self-polar triad, of which one of the vertices was taken as origin and the other two as ideal points.

Exercise 3 Prove that the involutions induced on the joins of pairs of points of a self-polar triad are either all three elliptic or two hyperbolic and one elliptic.

From this exercise it follows that there are only two distinct types of polarities in $\mathfrak{R}_S$, those in which there are singular involutions and those in which there are not.

In one form or another, in working with conics, a problem which may be stated in the following terms is often posed: given two polarities $\mathscr{Q}$ and $\mathscr{Q}'$, find the points D which are double-points in both. For the problem to be meaningful both polarities must of course have associated conics. The problem is reduced to more manageable form by selecting the co-ordinate system specially in relation to one of the systems. Take (0, 0, 1) and (1, 0, 0) to be double-points in $\mathscr{Q}$ and [1, 0, 0] and [0, 0, 1] to be their polar lines. Then, assuming first a general form of equation for the polarity $\mathscr{Q}$, we find that the elements of the matrix satisfy the conditions

$$\begin{bmatrix} 1 & 0 & 0 \\ 0 & 0 & 1 \end{bmatrix}\begin{bmatrix} a & h & g \\ h & b & f \\ g & f & c \end{bmatrix} = \begin{bmatrix} 0 & 0 & 1 \\ 1 & 0 & 0 \end{bmatrix}.$$

That is, a = h = f = c = 0. By adjusting the scale (selecting the unit point) on the axes, we can reduce the equation of this polarity to

$$[x_1, y_1, 1]\begin{bmatrix} 0 & 0 & 1 \\ 0 & -2 & 0 \\ 1 & 0 & 0 \end{bmatrix}\begin{bmatrix} x \\ y \\ 1 \end{bmatrix} = 0,$$

and the equation of the associated conic to

$$y^2 - x = 0.$$

Take the other polarity $\mathcal{Q}'$ to be

$$[x_1, y_1, 1]\begin{bmatrix} a_0 & a_1 & a_2' \\ a_1 & a_2 & a_3 \\ a_2' & a_3 & a_4 \end{bmatrix}\begin{bmatrix} x \\ y \\ 1 \end{bmatrix} = 0,$$

(the matrix is a general symmetrical matrix, with the elements named in a certain pattern), then the common double-points of $\mathcal{Q}$ and $\mathcal{Q}'$ are given by

$$a_0 y^4 + 2a_1 y^3 + (a_2 + 2a_2') y^2 + 2a_3 y + a_4 = 0,$$

which is a 'general quartic equation' in y; unless the matrix is specially selected, the equation has no solutions in $\mathfrak{N}_S$. If we are given that $\mathcal{Q}$ and $\mathcal{Q}'$ have one common double-point, we could take that point to be (1, 0, 0); then $a_0 = 0$, and the equation reduces to a cubic. Again there are in general no solutions in $\mathfrak{N}_S$†.

We need therefore to extend $\mathfrak{F}_S$ to include the (real) solutions of all those cubic and quartic equations which have (real) solutions.

We can define in space of n dimensions a polarity system by means of involutions on n concurrent lines, and, except in the case in which all the involutions are elliptic, we shall obtain a set of double-points of which the locus is a 'quadric'. The common double-points of n such polarity systems will be found as the solutions of an equation of order 2^n or possibly less. In this way we may regard any polynomial equation as having been derived from a set of polarities in space of suitably high dimension. In other words, if we wish to include in the geometry the double-points common to polarity systems in spaces of unrestricted dimensions, we shall need to extend the field $\mathfrak{F}_R$ to the field, $\mathfrak{F}_A$ say, of 'algebraic numbers'.

Before we leave the question of double-points in polarities in the non-finite planar nets, we should devote a little more attention to the problem of double-points in $\mathfrak{N}_R$. We have seen that $\mathcal{Q}_{\alpha,\beta}$ has double-points if there are solutions to the equation

$$x^2/\alpha + y^2/\beta - 1 = 0.$$

Since $(\alpha, \beta, x, y) \in \mathfrak{F}_R$, we may replace them by $(c/a, c/b, \xi/\zeta, \eta/\zeta)$, where $(c, \zeta) \in$ {positive integers}, $(\xi, \eta) \in$ {non-negative integers} and $(a, b) \in$ {integers} $-$ {0}, and have then to find solutions of

$$a\xi^2 + b\eta^2 - c\zeta^2 = 0.$$

† In this case the demonstration is simpler because it is clear that $x^3 = 2$ has no solution which can be expressed finitely as an element $a + \sqrt{(b + \sqrt{(c + \dots)})}$ of $\mathfrak{N}_S$.

This may be written as

$$\mathrm{ca}\xi^2+\mathrm{cb}\eta^2-(\mathrm{c}\zeta)^2 = 0$$

and so as

$$\mathrm{p}\xi^2+\mathrm{q}\eta^2 = \omega^2.$$

In the particular case where $p = q = 1$, we recognise the Pythogoras relation, with solutions

$$\xi = \mathrm{u}^2-\mathrm{v}^2, \quad \eta = 2\mathrm{uv}, \quad \omega = \mathrm{u}^2+\mathrm{v}^2,$$

and might suppose that a corresponding form could always be found. But consider

$$3\xi^2+5\eta^2 = \omega^2.$$

In scale 10 the last digit of any square is 0, 1, 4, 9, 6 or 5 so that the last digit of $3\xi^2+5\eta^2$ is given by (0, 3, 2, 7, 8 or 5) + (0 or 5), that is, it is (0, 8, 7, 2, 3 or 5) and consequently $3\xi^2+5\eta^2$ could be a square only when ξ is a multiple of 5. But $3(5\xi')^2+5\eta^2$ could be a square only if η is a multiple of 5, and in that case ω also is a multiple of 5. We may therefore remove a factor 5^2 from the two sides of the equation, and, by repetition of this process, arrive at an equation

$$3\xi_0^2+5\eta_0^2 = \omega_0^2$$

in which ξ_0 is not a multiple of 5. That is, the polarity

$$3\mathrm{x}_1\mathrm{x}+5\mathrm{y}_1\mathrm{y}-1 = 0$$

in $\mathfrak{R}_R$ has no double-points.

Exercise 4 The polarity

$$2\mathrm{x}_1\mathrm{x}+3\mathrm{y}_1\mathrm{y} = 5$$

in $\mathfrak{R}_R$ has one obvious double-point. By considering lines through this point find parametric expressions for x_1 and y_1 which give all the double-points.

(ii) *Polarities in* $\mathfrak{R}_p$ ($p \neq 2$). Except where inequalities are used, the algebra over $\mathfrak{F}_p$ has the same appearance as that over $\mathfrak{F}_S$; in $\mathfrak{R}_p$ the distinction between hyperbolic and elliptic involutions is drawn on a basis not of inequalities but of squares and non-squares in $\mathfrak{F}_p$. The double-points of a polarity $\mathcal{Q}_{\alpha,\beta}$ in $\mathfrak{R}_p$ are determinable as the points which lie on their own polar lines, and consequently any double-point in $\mathcal{Q}_{\alpha,\beta}$ satisfies the condition

$$\mathrm{x}^2/\alpha+\mathrm{y}^2/\beta = 1.$$

We are to show that there are in fact always exactly $p+1$ such points; first we show that there is always at least one.

If α is a square, say $\alpha = \lambda^2$, then clearly $(\lambda, 0, 1)$ is a double-point, so that we need only consider the case in which both α and β are non-squares. We shall assume, as is easy to prove by assuming the contrary, that if α, β are non-squares then $\alpha\beta$ is a square. If $\alpha\beta = \gamma^2$, then we may write the relation as

$$(x/\alpha)^2 + (y/\gamma)^2 = 1/\alpha$$

or as

$$1 + (\alpha y/\gamma x)^2 = \alpha/x^2.$$

We have to show that a relation

$$1 + (\text{square}) = (\text{non-square})$$

is always satisfiable, in other words, that the statement: 'for every λ in $\mathfrak{F}_p$, λ^2+1 is a square' is false. To prove this, arrange the squares, s_i, in $\mathfrak{F}_p$ in counting order, so that, as integers,

$$0 < s_1(=1) < s_2 < \ldots < s_{\frac{1}{2}(p-1)} \leqslant p-1.$$

If the statement were true then

$$s_{\frac{1}{2}(p-1)} + 1 = 0, \quad s_2 = 2, \quad \text{implying } s_3 = 3, \ldots,$$

a clearly contradictory situation.

Thus there is always at least one solution, say $D_0 = (x_0, y_0, 1)$, to $x^2/\alpha + y^2/\beta = 1$. The polar line of this point is $xx_0/\alpha + yy_0/\beta = 1$. On every line of the remaining set of p through D_0 the induced involution is non-singular with D_0 as one double-point; this involution therefore has a second double-point, giving, so far, $p+1$ double-points in all. There can be no other double-points, because the set of $p+1$ lines through D_0 contains all the points in the plane.

Exercise 5 A point P may be related to the 'conic' in $\mathfrak{N}_p$ (that is, to the set of $p+1$ double-points) in one of three ways:

(i) P belongs to the conic, one of the lines through P is the 'tangent' and the remaining p of the lines are chords,

(ii) P is 'outside' the conic: two lines through P are tangents, $\frac{1}{2}(p-1)$ lines are chords, and $\frac{1}{2}(p-1)$ are 'non-secants' (that is, such that the induced involution on the line is elliptic),

(iii) P is 'inside' the conic: $\frac{1}{2}(p+1)$ of the lines through P are chords and $\frac{1}{2}(p+1)$ are non-secants.

Exercise 6 Find the numbers of (*a*) chords and non-secants, (*b*) inside and outside points, in relation to a conic in $\mathfrak{N}_p$.

Exercise 7 In the Desargues figure in any primary net take

$$\mathrm{K} = \upsilon, \quad \mathrm{A} = \epsilon_1, \quad \mathrm{B} = \epsilon_2, \quad \mathrm{C} = \epsilon_0,$$
$$\mathrm{A}' = (\alpha, 1, 1), \quad \mathrm{B}' = (1, \beta, 1), \quad \mathrm{C}' = (1, 1, \gamma).$$

Show that the equation of the polarity in which the figure is its own polar is

$$[\mathrm{x}_1 \quad \mathrm{y}_1 \quad 1]\begin{bmatrix} \beta\gamma - 1 & 1-\gamma & 1-\beta \\ 1-\gamma & \gamma\alpha - 1 & 1-\alpha \\ 1-\beta & 1-\alpha & \alpha\beta - 1 \end{bmatrix}\begin{bmatrix} \mathrm{x} \\ \mathrm{y} \\ 1 \end{bmatrix} = 0.$$

NOTE ON THEOREM 2.2*T*20

The proof does not make it clear why the value of r/s should be the same in the two statements '*T*'. In each of the nets $\mathfrak{N}(\mathrm{O}, \mathrm{E}/\mathrm{U})$, $\mathfrak{N}(\mathrm{U}, \mathrm{E}/\mathrm{O})$ X is assumed to be constructed in a finite sequence of steps, each step in one net having a counterpart in the other. The correspondence is made explicit in the following theorem.

*G*1	$\mathscr{L}_3$|OUVK; E, H, H′ as in 2.2*T*20	
S	X : X ∈ OU & # |OUX	
C	M : M = KU ∩ HH′	
	Y, N : {^{e}Y, N} = VM ∩ {^{e}OU, OK}	
*G*2	ideal|UV & X ∈ $\mathfrak{N}$(O, E/U)	ideal|OV & X ∈ $\mathfrak{N}$(U, E/O)
G	OK ∥ XM & KM ∥ OE & EK ∥ YM	EK ∥ YN ∥ UH′ & EU ∥ KH′
t	Y = $\mathscr{D}_{\mathrm{OE}}$X	s(UY/UX) = s(NK/UH′) = s(YE/UE) = 1 − s(UY/UE)
T	s(OY/OE) = s(OX/OE) + 1	s(UE/UY) = s(UE/UX) + 1

Consider the sequences

$$\{\mathrm{E} = \mathrm{E}_1, \mathrm{E}_2, \mathrm{E}_3, \ldots\} \in \mathfrak{N}(\mathrm{O}, \mathrm{E}/\mathrm{U})$$

and

$$\{\mathrm{E} = \mathrm{F}_1, \mathrm{F}_{\frac{1}{2}}, \mathrm{F}_{\frac{1}{3}}, \ldots\} \in \mathfrak{N}(\mathrm{U}, \mathrm{E}/\mathrm{O})$$

From the theorems above

$$\mathrm{X} = \mathrm{E}_r \Rightarrow \mathrm{Y} = \mathrm{E}_{r+1};$$

$$\mathrm{X} = \mathrm{F}_{1/r} \Rightarrow \mathrm{Y} = \mathrm{F}_{1/(r+1)} \quad \& \quad \mathrm{E} = \mathrm{E}_1 = \mathrm{F}_1 \Rightarrow \mathrm{E}_r = \mathrm{F}_{1/r}$$

The required condition

$$\mathrm{X} = \mathrm{E}_{r/s} \in \mathfrak{N}(\mathrm{O}, \mathrm{E}/\mathrm{U}) \Rightarrow \mathrm{X} = \mathrm{F}_{s/r} \in \mathfrak{N}(\mathrm{U}, \mathrm{E}/\mathrm{O})$$

now follows immediately.

BOOK 3

GEOMETRY AND ALGEBRA

CHAPTER 3.1

INTRODUCTION: THE DISPLACEMENT TRANSFORMER, $\mathscr{D}$

Up to this stage we have considered, usually, only the geometry of points in the same primary net, that is, the systems of points which satisfy the incidence axioms $A\mathscr{I}$ 1, 2 (properties of collinear sets) and 3 (Desargues), in which points of a collinear set are derived from three given base points in the set by repetitions of the single operation of displacement. These primary nets of a collinear set, defined from O, E and ideal|U are the sets:

$\mathfrak{N}_p$: $\mathscr{D}^r_{OE}E_0 = E_r : r \in \mathfrak{F}_p (= GF(p))$

(i.e. r is a remainder on division of an integer by p).

$\mathfrak{N}_R$: points $E_{r/s}$ defined by $\mathscr{D}^s_{OE_{r/s}} = \mathscr{D}^r_{OE}$

($r \in \{\text{integers}\}$, $s \in \{\text{positive integers}\}$).

To these we added in the last Chapter

$\mathfrak{N}_S$: the surdic net, derived from $\mathfrak{N}_R$ by adjoining recursively the double-points of hyperbolic involutions.

The operation of multiplication was introduced (2.2*C*4) but did not play a necessary part in defining the system, because all points could be obtained by operations $\mathscr{D}$ alone.

We are now to consider what are the consequences of postulating the existence in the geometry of points outside the primary net. When this is done the geometry loses contact with direct observation of the physical world, since, as has been explained before, the only physical measurements that can be made refer to objects represented by points in the rational net. We shall however draw figures in which the points outside the primary net of necessity are represented by marks having the appearance of belonging to it.

For these new points we shall use the terminology of span-ratios, but it must be understood that we may assume for them none of the additive or multiplicative properties of span-ratios until we have proved that there are geometrical operations completely representable by addition and multiplication.

Throughout Book 3 we shall adopt the following notation:

3.1*N*1 (i) $\mathfrak{N}(O, E)$ or $\mathfrak{N}(O, E/U)$ the basic primary net, either $\mathfrak{N}_p$ or $\mathfrak{N}_R$.

(ii) F, G, ... new basis points such that

$$\{F, G, ...\} \subset OE \quad \text{but} \quad \{F, G, ...\} \not\subset \mathfrak{N}(O, E).$$

(iii) $\mathfrak{f} = S(OF/OE)$, $\mathfrak{g} = s(OG/OE), ...$, new 'units' of which the properties have yet to be determined.

(iv) $\mathfrak{N}(O, E')$ a basic primary net on a different line through O, with $\{F', ...\} \subset OE'$ & $\{F', ...\} \not\subset \mathfrak{N}(O, E')$.

(v) $\mathfrak{F}_p$, $\mathfrak{F}_R$ the primary fields corresponding to the nets $\mathfrak{N}_p$ and $\mathfrak{N}_R$.

From the definition of the operation $\mathscr{D}$ we have, for $\mathscr{L}\wp|ABC$,

$$\mathscr{D}_{AB}C = D \Rightarrow \mathscr{D}_{AB} = \mathscr{D}_{CD},$$

so that

*G*1 $\mathscr{L}\wp|ABA'B'$ (points not necessarily in the same primary net)
*G*2 C, C′; CC′ ∥ AB, ⋕|CC′, AB
C D, D′ : $D = \mathscr{D}_{AB}C$, $D' = \mathscr{D}_{A'B'}C'$
T $\mathscr{D}_{AB} = \mathscr{D}_{CD}$, $\mathscr{D}_{A'B'} = \mathscr{D}_{C'D'}$.

We can therefore meaningfully write

$$s(AB/A'B') = s(CD/C'D')$$

even though at present no algebraic meaning has been ascribed to the symbol s(AB/A′B′). It follows that the 'first ⁑parallels-and-proportions theorem', 2.2*T*4, is still valid:

3.1*T*1

*G*1 $\mathscr{L}$⋕|OEFU & ideal|U & $F \notin \mathfrak{N}(O, E)$
*G*2 K, l : K ∉ OE & $l \parallel_U OE$ & ⋕|l, OE & K ∉ l
C {ᶜL, M, N} = l ∩ {ᶜKO, KE, KF}
T s(LN/LM) = s(OF/OE)
P Take ideal|K
In relation to ideal|UK, $\mathscr{D}_{OE} = \mathscr{D}_{LM}$, $\mathscr{D}_{OF} = \mathscr{D}_{LN}$.

To make the geometric operations resemble as closely as possible the algebraic operations we introduce a special notation, in which with each point of the line we associate two *operators*, or *transformers*, which we shall denote by $\mathscr{D}_A$ and $\mathscr{M}_A$. A 'transformer' is a geometrical construction, determined on a line by the basis O, E, U and a fixed point A and a variable point X, which leads for every X to a unique new point Y which is the 'transform' of X by the 'transformer' determined by A; we shall express these transformations by the statements $\mathscr{D}_A X = Y$ and $\mathscr{M}_A X = Y$.

3.1 *C*1 (i) The displacement transformer, $\mathscr{D}_A$ (first construction). (So far as points of a single collinear set are concerned $\mathscr{D}_A$ is identical with $\mathscr{D}_{OA}$, and the construction below is 2.1*C*2.)

G $\mathscr{L}\#|$OAU & ideal|U (E is not involved in the construction)
*S*1 $\wp|$B : B $\in$ OA
*S*2 $\wp|$L′ : L′ $\notin$ OA
*C*1 M′ : L′M′ $\parallel$ OA & AM′ $\parallel$ OL′ (M′ = $\mathscr{D}_{OA}$L′)
*C*2 P : P $\in$ OA & M′P $\parallel$ BL′ (P = $\mathscr{D}_{L'M'}$B)
D(i) P = $\mathscr{D}_A$B.

3.1 *C*1 (ii) $\mathscr{D}_A$ in relation to an unspecified ideal line.

G $\mathscr{L}\#|$OAU
*S*1 B : B $\in$ OA & $\#|$BU
*S*2 L : L $\notin$ OA
*S*3 L′ : L′ $\in$ OL & $\#|$OLL′
*C*1 M : M = UL $\cap$ BL′
*C*2 M′ : M′ = UL′ $\cap$ AL
*C*3 P : P = MM′ $\cap$ OA
T P = $\mathscr{D}_A$B with ideal|ULM.

It may be noted that, viewed as a configuration, the points above form a complete quadrangle *Q*{L, M, M′, L′/., ., U} and a collinear set O, A, B, P, U which contains the vertex U of the diagonal triangle.

3.1 *C*1 (iii) Interpretation of *C*1(ii) as $\mathscr{D}_B$A.

G $\mathscr{L}\#|$OABU, ideal|U
*S*1 $\wp|$L : L $\notin$ OA
*S*2 L′ : L′ $\in$ OL & ideal|L′
*C*1 M : M = $\mathscr{D}_{OB}$L, M′ : M′ $\in$ AL & ideal|M′
*C*2 P : P = $\mathscr{D}_{LM}$A
T P = $\mathscr{D}_B$A with ideal|UL′M′.

3.1 *T*2 $\mathscr{D}_A$B = $\mathscr{D}_B$A
G $\mathscr{L}\#|$OABU, ideal|U
C construction of 3.1*C*1(ii)
T P = $\mathscr{D}_A$B with ideal|ULM (*C*1(i))
= $\mathscr{D}_B$A with ideal|UL′M′ (*C*1(iii))
P The operation $\mathscr{D}$ on OE depends only on the ideal point U on OE and not otherwise on the ideal line.

We may define $\mathscr{D}_A$ by any one of the constructions *C*1(i), (ii) and (iii); (iii) has the advantage that the construction splits into two parts, the first depending only on the point A (and O and U) which determines the transformer $\mathscr{D}_A$, and the second only on B, the point to be transformed. We use (iii) therefore (even although it is the construction for $\mathscr{D}_{OB}$ A) in the formal definition of $\mathscr{D}_A$.

3.1*D*1 The displacement transformer $\mathscr{D}_A$.

G	$\# \wp \| OA$
*S*1	$\wp \| B, \quad B \in OA$
*S*2	$\wp \| L, L \notin OA$
*C*1	$M : ML \parallel OA \;\&\; MB \parallel OL$
*C*2	$P : P \in OA \;\&\; MP \parallel LA$
D	$P = \mathscr{D}_A B$.

(In the Euclidean diagram the construction looks like that of a ⁑triangle BMP identical with a fixed ⁑triangle OLA, with the same ⁑orientation and with its ⁑base-segment on OA.)

3.1*T*3 In 3.1*D*1, P is independent of the choice of L, and of the ideal line through U.

P 2.1*T*3 and 2.1*D*2. Only $A\mathscr{I}$1, 2 and 3* are required in the proof.

3.1*T*4 Special transformations.

T (i) $\mathscr{D}_O B = \mathscr{D}_B O = B$ for all B
(ii) $\mathscr{D}_A B = B \Rightarrow A = O$ or $B = U$.

3.1*T*5 The inverse transformation.

G	$\mathscr{L} \# \wp \| OA$
S	$L : L \notin OA$
*C*1	$N : ON \parallel AL \;\&\; LN \parallel OA$
*C*2	$\bar{A} : \bar{A} \in OA \;\&\; N\bar{A} \parallel OL$
T (i)	$\mathscr{D}_A \bar{A} = O = \mathscr{D}_{\bar{A}} A$
(ii)	$\mathscr{D}_A B = O \Rightarrow B = \bar{A}$.

3.1*T*6 The transformers $\mathscr{D}$ generate a closed system, that is, given any points $\{A, B\} \subset OE$, the point $\mathscr{D}_A B$ can be constructed.

3.1*T*7 The transformer $\mathscr{D}_A$ is symmetric in the determining point A and the point to be transformed.

T $\mathscr{D}_A B = \mathscr{D}_B A$
P This is 3.1*T*2.

3.1 *T*8 The transformers $\mathscr{D}_A$ are associative.

G $\mathscr{L}\#\wp|OABC$

S $\wp|L, \quad L\notin OA$

C $M:\mathscr{D}_{OB}L = M$

$P:\mathscr{D}_{LM}A = P$

$N:\mathscr{D}_{AO}L = N$

$Q:\mathscr{D}_{NM}C = Q$

$P':\mathscr{D}_{NL}C = P'$

T $\mathscr{D}_{OA}(\mathscr{D}_{OB}\mathscr{D}_{OC}) = (\mathscr{D}_{OA}\mathscr{D}_{OB})\mathscr{D}_{OC} = \mathscr{D}_{OQ}$

P *t*1 $\mathscr{D}_{OA}\mathscr{D}_{OB} = \mathscr{D}_{OP} = \mathscr{D}_{NM} = \mathscr{D}_{CQ}$

*t*2 $\mathscr{D}_{OB}\mathscr{D}_{OC} = \mathscr{D}_{PQ}\mathscr{D}_{AP} = \mathscr{D}_{AQ}$

*t*3 $(\mathscr{D}_{OA}\mathscr{D}_{OB})\mathscr{D}_{OC} = \mathscr{D}_{CQ}\mathscr{D}_{OC} = \mathscr{D}_{OQ}$

$= \mathscr{D}_{OA}\mathscr{D}_{AQ} = \mathscr{D}_{OA}(\mathscr{D}_{OB}\mathscr{D}_{OC})$

T $\mathscr{D}_A(\mathscr{D}_B C) = \mathscr{D}_{\mathscr{D}_A B}C.$

CHAPTER 3.2

THE MULTIPLICATION TRANSFORMER, $\mathscr{M}$

In this Chapter we introduce the second set of transformers $\{\mathscr{M}_A\}$, one transformer being determined, for a given basis O, E, U on the line, by each point A. $\mathscr{M}_A$ operates on every point of the line to produce another point. The construction for $\mathscr{M}_A$ is in fact the product construction 2.2*C*4 applied to points A, B not necessarily belonging to the net $\mathfrak{N}$(O, E/U). In 2.2*C*4 there was no necessity to prove that the point derived by the construction was independent of the choice of auxiliary points or that $\mathscr{M}_A B = \mathscr{M}_B A$, because the transformer $\mathscr{M}$ could be replaced by a sequence of constructions based on the transformers $\mathscr{D}$ and these are uniquely defined and have the property $\mathscr{D}_P Q = \mathscr{D}_Q P$. We shall prove that the construction is unique, but shall need to use the full Desargues axiom $A\mathscr{I}3$ (though we shall not prove that a proof using only $A\mathscr{I}3^*$ is not possible). We shall not prove $\mathscr{M}_A B = \mathscr{M}_B A$, and shall in fact produce a counter-example of a geometric system which conforms to $A\mathscr{I}1$, 2 and 3, but in which $\mathscr{M}_A B \neq \mathscr{M}_B A$.

In 3.1 *C*1 (ii) we noted that, when no ideal line is specified, the construction for $\mathscr{D}_A$ leads to a complete quadrilateral Q{L, M, L′, M′/ .,., U}, the collinear set on which $\mathscr{D}_A$ operates being AU. In the same way we shall find $\mathscr{M}_A$ closely linked with the complete quadrilateral. The simple Euclidean diagrams for the two constructions are very much alike. Suppose we are given $\mathscr{L}\#|$OEAB and L$\notin$OE. For $\mathscr{D}_A B = P$, we construct the ⁑triangle BMP ⁑identical with the ⁑triangle OLA with P$\in$OA & ∂(OLA/BMP) = I. For $\mathscr{M}_A B = P$ we construct the ⁑triangle BMP ⁑similar to ELA (note, ELA, not OLA), with M$\in$OL & P$\in$OA & ∂(ELA/BMP) = I.

3.2*D*1 Definition of the multiplication transformer, $\mathscr{M}_A$.

G $\mathscr{L}\#|$OEAB & ideal|u fixed

S $\wp|$L, L$\notin$OE

*C*1 M : M$\in$OL & BM$\|$EL

*C*2 P : P$\in$OE & MP$\|$AL

N $\mathscr{M}_A$: the multiplication transformer, determined in OE by an arbitrary point A on OE and the basis O, E, U, which transforms by this construction the point B into the point P

D $\mathscr{M}_A B = P$.

3.2*C*1 Definition of $\mathscr{M}_A$ with unspecified ideal line.

G $\mathscr{L}\#|$OEUAB
*S*1 L : L $\notin$ OE
*S*2 u : u $\supset$ U, $\#|$u, OE & u $\not\ni$ L
*C*1 $L_u : L_u = EL \cap u$
*C*2 $M_u : M_u = AL \cap u$
*C*3 $M : M = BL_u \cap OL$
*C*4 $P : P = MM_u \cap OE$
T P = $\mathscr{M}_A$B on basis O, E and ideal|U
P 3.2*D*1 with ideal|u.

It has first to be proved that $\mathscr{M}_A$, which is solely concerned with points of the line, is independent of the choice of L and the choice of ideal line u, through U. The proof is carried out in three stages. Let L″ be any point not on OE. Construct L′ = AL″ $\cap$ EL. First replace L by L′ on EL, and then replace L′ by L″ on AL′. If $\mathscr{M}_A$ is unaffected by each of these changes, then it is unaffected by the change from L to L″. We have finally to prove that, if L is fixed, $\mathscr{M}_A$ is independent of the choice of u through U.

3.2*T*1 First part of the uniqueness theorem for $\mathscr{M}_A$.

G figure as in 3.2*D*1
S $\wp$|L′ : L′ $\in$ EL, $\#|$ELL′
C M′ : M′ = OL′ $\cap$ BM
T PM′ $\|$ AL′
P

$$O\begin{Bmatrix}\mathrm{LM}\\ \mathrm{L'M'}\\ \mathrm{AP}\end{Bmatrix} \Rightarrow \mathscr{L}|\langle \mathrm{LL'}\cap\mathrm{MM'},\ \mathrm{LA}\cap\mathrm{MP},\ \mathrm{L'A}\cap\mathrm{M'P}\rangle$$
$$\Rightarrow \mathrm{L'A}\cap\mathrm{M'P}\in \mathrm{u}$$
$$\Rightarrow \mathrm{L'A}\,\|\,\mathrm{M'P}.$$

3.2*T*2*X* Completion of uniqueness theorem for $\mathscr{M}_A$ with fixed ideal line.

G figure as in 3.2*D*1
S L″ : L″ $\in$ AL, $\#|$ALL″
C M″ : M″ $\in$ OL″ & BM″ $\|$ EL″
T $\mathscr{L}|$MPM″.

It should be noted that in the figures used in the constructions for each of the parts of the theorem four points on the ideal line are involved, three on the pairs of parallel lines and the fourth on OE. This last point is not part of the Desargues figure used in the proof; we use in fact the full Desargues axiom $A\mathscr{I}3$ and not $A\mathscr{I}3^*$.

3.2*T*3*X* Uniqueness of $\mathscr{M}_A$ for change of ideal line.

G figure of 3.2*C*1
S $u' : u' \supset U, \ \# | u', u, OE \ \& \ u' \not\supset L$
C $L'' = u' \cap EL$
$M'' = u' \cap AL$
$\overline{M} = PM'' \cap BL''$
T $\mathscr{L} | OLM\overline{M}$.

Again the full axiom $A\mathscr{I}3$ is needed.

3.2*T*4 *Special transformations.*

T (i) $\mathscr{M}_O B = \mathscr{M}_B O = O$ for all B
(ii) $\mathscr{M}_A B = O \Rightarrow A = O$ or $B = O$
(iii) $\mathscr{M}_E B = \mathscr{M}_B E = B$ for all B
(iv) $\mathscr{M}_A B = B \Rightarrow A = E$ unless either $B = O$ or ideal|B.

3.2*T*5 The transformers $\mathscr{M}$ generate a closed system, that is, given any pair of points $\{A, B\} \subset OE$, the point $\mathscr{M}_A B, \in OE$, can be constructed.

3.2*T*6 The inverse transformer.

G $\mathscr{L}\#\wp | OEA$
S $\wp | L : L \notin OE$
C $L' : L' \in OL \ \& \ EL' \parallel AL$
$A^* : A^* \in OE \ \& \ A^*L' \parallel EL$
T (i) $\mathscr{M}_{A^*} A = E, \ \ \mathscr{M}_A A^* = E$
(ii) $\mathscr{M}_B A = E \Leftrightarrow B = A^*$
(iii) $A = O \Leftrightarrow A^* = U$, i.e. $O^* = U$.

In making the construction when $A = O$, we find: ideal|L′, and thus ideal|A*L′; that is, $A^* = U$.

3.2*T*7 Systems satisfying $A\mathscr{I}1, 2, 3$ can be devised in which $\mathscr{M}_A B \neq \mathscr{M}_B A$.

If $\{A, B\} \in \mathfrak{N}(O, E)$ then $\mathscr{M}_A B = \mathscr{M}_B A$, because the transformer $\mathscr{M}_A$ can be replaced by a construction involving only repetitions of the constructions for $\mathscr{D}_{OE}$ and $\mathscr{D}_{OA}$. A system in which $A\mathscr{I}1, 2, 3$ are valid and $\mathscr{M}_A B \neq \mathscr{M}_B A$ is described in Chapter 3.5.

3.2*T*8 The transformers $\mathscr{M}$ are associative.

G $\mathscr{L}\#\wp | OEABC$
C $P : P = \mathscr{M}_A B$
$R : R = \mathscr{M}_P C$

$Q : Q = \mathscr{M}_B C$
$R' : R' = \mathscr{M}_A Q$
T $R = R'$
P (set out the construction in detail)
s $\wp | L : L \notin OE$
*c*1 $M : M \in OL, BM \| EL$
*c*2 $P : P \in OE, PM \| AL$
*c*3 $N : N \in OM, CN \| EM$
*c*4 $R : R \in OE, RN \| PM$
*c*5 $Q : Q \in OE, QN \| BM$
t $P = \mathscr{M}_A B$ (*c*1, *c*2)
$R = \mathscr{M}_P C$ (*c*3, *c*4)
$Q = \mathscr{M}_B C$ (*c*3, *c*5)
$R = \mathscr{M}_A Q$ (*c*4, *c*5)
T $\mathscr{M}_{\mathscr{M}_A B} C = \mathscr{M}_A(\mathscr{M}_B C) = R.$

3.2*T*9 The transformers $\mathscr{M}$ are left-distributive over the transformers $\mathscr{D}$.

T $\mathscr{M}_C(\mathscr{D}_A E) = \mathscr{D}_{M_C E}(\mathscr{M}_C A) = \mathscr{D}_C(\mathscr{M}_C A)$†
G $\mathscr{L}\# | OEACU$ & ideal$|U$
S $\wp | L : L \notin OE$
*C*1 $L' : L' \in OL$ & $AL' \| EL$
*C*2 $M : M \in AL'$ & $LM \| EA$ $(\mathscr{D}_{EA} = \mathscr{D}_{LM})$
*C*3 $M' : M' \in EL$ & $L'M' \| EA$ $(\mathscr{D}_{EA} = \mathscr{D}_{M'L'})$
*C*4 $N' : N' = CL \cap L'M'$
*C*5 $N : N \in LM$ & $L'N \| LN'$ $(\mathscr{D}_{N'L'} = \mathscr{D}_{LN})$
*C*6 $S : S = OE \cap MM'$
*C*7 $T : T = OE \cap NN'$
*C*8 $A' : A' = OE \cap L'N$ $(\mathscr{D}_{CA'} = \mathscr{D}_{N'L'})$
T $\mathscr{M}_C(\mathscr{D}_A E) = \mathscr{D}_C(\mathscr{M}_C A) = T$
P *t*1 $\mathscr{P} | LL' \cap MM' \cap NN'$

p
$$\mathscr{L}\begin{Bmatrix} N'M' \cap MN \\ M'L \cap L'M \\ LN' \cap NL' \end{Bmatrix} \Rightarrow \mathscr{P} | LL' \cap MM' \cap NN'$$

c $K : K = LL' \cap MM' \cap NN'$
*t*2 $S = \mathscr{D}_A E$ (*C*1, *C*2, *C*3, *C*6)
*t*3 $\mathscr{M}_C S = T$ (from quadrangle LKM′N′ with ideal$|$M′N′)
*t*4 $\mathscr{M}_C A = A'$ (*C*1, *C*5, *C*8)
*t*5 $\mathscr{D}_C A' = T$ $(\mathscr{L}|OLL', \mathscr{D}_{CA'} = \mathscr{D}_{LN} = \mathscr{D}_{N'L'})$

† The full distribution theorem is

$$\mathscr{M}_C(\mathscr{D}_A B) = \mathscr{D}_{\mathscr{M}_C A}(\mathscr{M}_C B) = \mathscr{D}_{\mathscr{M}_C B}(\mathscr{M}_C A),$$

but we shall see that the theorem above is algebraically equivalent (p. 231).

$$T \quad \mathscr{M}_{\mathrm{C}}(\mathscr{D}_{\mathrm{A}}\mathrm{E}) = \mathscr{M}_{\mathrm{C}}\mathrm{S} = \mathrm{T}$$
$$= \mathscr{D}_{\mathrm{C}}\mathrm{A}' = \mathscr{D}_{\mathrm{C}}(\mathscr{M}_{\mathrm{C}}\mathrm{A}).$$

3.2*T*10*X* The transformers $\mathscr{M}$ are right-distributive over the transformers $\mathscr{D}$

$$T \quad \mathscr{M}_{\mathscr{D}_{\mathrm{A}}\mathrm{E}}\mathrm{C} = \mathscr{D}_{\mathrm{C}}(\mathscr{M}_{\mathrm{A}}\mathrm{C}).$$

The complete form is $\mathscr{M}_{\mathscr{D}_{\mathrm{A}}\mathrm{B}}\mathrm{C} = \mathscr{D}_{\mathscr{M}_{\mathrm{A}}\mathrm{C}}(\mathscr{M}_{\mathrm{B}}\mathrm{C})$.

CHAPTER 3.3

THE TRANSFORMERS $\mathscr{D}$ AND $\mathscr{M}$ AND ALGEBRAIC FIELDS

When O, E and U had been assigned, we were able to establish a one-to-one correspondence between the proper points of a primary net $\mathfrak{N}(O, E/U)$ and the elements either of $\mathfrak{F}_p$ or $\mathfrak{F}_R$ in the following way: if A is such that $(\mathscr{D}_{OA})^n = (\mathscr{D}_{OE})^m$ then A corresponds to the element $s(OA/OE) = m/n$ of $\mathfrak{F}_p$ or $\mathfrak{F}_R$ as the case may be.

To a point A not specified as belonging to $\mathfrak{N}(O, E/U)$ we now assign an element of an algebraic system (which, as we see shortly, is a field), using still the notation of span ratios, $s(OA/OE) = a$. The relations among these elements are prescribed to conform to the geometric operations represented by the transformers $\mathscr{D}_A$ and $\mathscr{M}_A$.

Since in particular, when we consider elements of a primary net,

$\mathscr{D}_A B = P$ corresponds to

$$s(OA/OE)+s(OB/OE) = s(OP/OE)$$

or

$$a \quad + \quad b \quad = \quad p$$

and $\mathscr{M}_A B = Q$ corresponds to

$$a \quad \times \quad b \quad = \quad q,$$

we shall use the signs + and × for the algebraic operations corresponding to $\mathscr{D}_A$ and $\mathscr{M}_A$ whether or not A belongs to the primary net.

The strict analogy between the geometric and algebraic systems could be better expressed by using the ordinary convention for representing functional relations, thus: The transformer $\mathscr{D}_A$ is the geometric function determined by A (and the basis) which associates with each point B a point $\mathscr{D}_A(B)$. $+_a$ is the algebraic function determined by an element a which associates with each element b an element $+_a(b)$. When the associated element is written in this form there is no reason for expecting *a priori* that $+_a(b)$ and $+_b(a)$ are the same element; that is, even when related to the algebra, the theorem $\mathscr{D}_A B = \mathscr{D}_B A$ is not trivial.

Once the actual roles of the symbols + and × are clear, there is no necessity to diverge from ordinary practice, and we shall therefore write: if $s(OA/OE) = a$, etc., and $\mathscr{D}_A B = P$, $\mathscr{M}_A B = Q$, then $+_a b = a+b = p$ and $\times_a b = a \times b = q$.

The properties of the algebraic operators '+' and '×' are to be precisely those of '$\mathscr{D}$' and '$\mathscr{M}$', so that we may tabulate them as follows:

	Point A : $A \in OE$ & $\wp \vert A$	Element $a = s(OA/OE)$
	(From **Chapter 3.1**)	
	Transformer $\mathscr{D}$	Operator +
T 4	$\mathscr{D}_O B = \mathscr{D}_B O = B$ for all B and	There is a unique element 0, such that $b+0 = 0+b = b$
	$\mathscr{D}_A B = B \Rightarrow A = O$	$a+b = b \Rightarrow a = 0$
	or $B = U$	no analogue
T 5	$\mathscr{D}_A \bar{A} = O = \mathscr{D}_{\bar{A}} A$ $\mathscr{D}_A B = O \Rightarrow B = \bar{A}$	There is a unique element $-a$, such that $a+(-a) = -a+a = 0$
T 6	The system generated by transformers $\mathscr{D}$ is closed	The system is closed under '+'
T 7	$\mathscr{D}_A B = \mathscr{D}_B A$	$a+b = b+a$
T 8	$\mathscr{D}_A(\mathscr{D}_B C) = \mathscr{D}_{\mathscr{D}_A B} C$	$a+(b+c) = (a+b)+c$
	(From **Chapter 3.2**)	
	Transformer $\mathscr{M}$	Operator $\times$
T 4 (i)	$\mathscr{M}_O B = \mathscr{M}_B O$ for all B	$0 \times b = b \times 0 = 0$ for all b
(ii)	$\mathscr{M}_A B = O \Rightarrow A = O$ or $B = O$	$a \times b = 0 \Rightarrow a = 0$ or $b = 0$
(iii)	$\mathscr{M}_E B = \mathscr{M}_B E = B$ for all B	$1 \times b = b \times 1 = b$ for all b
(iv)	$\mathscr{M}_A B = B \Rightarrow A = E$	$a \times b = b \Rightarrow a = 1$
	or $B = O$	or $b = 0$
	or $B = U$	no analogue
T 5	The system generated by transformers $\mathscr{M}$ is closed	The system is closed under $\times$
T 6	For each point A except O there is a unique point A* such that	For each a, $a \neq 0$, there is a unique element a^{-1} such that
	$\mathscr{M}_{A*} A = \mathscr{M}_A A^* = E$	$a^{-1} \times a = a \times a^{-1} = 1$
T 7	There exist systems in which $\mathscr{M}_A B \neq \mathscr{M}_B A$	There exist systems in which $a \times b \neq b \times a$
T 8	$\mathscr{M}_{\mathscr{M}_A B} C = \mathscr{M}_A(\mathscr{M}_B C)$	$(a \times b) \times c = a \times (b \times c)$
T 9	$\mathscr{M}_C(\mathscr{D}_A E) = \mathscr{D}_C(\mathscr{M}_C A)$	$c \times (1+a) = c + c \times a$
T 10	$\mathscr{M}_{\mathscr{D}_A E} C = \mathscr{D}_C(\mathscr{M}_A C)$	$(a+1) \times c = c + a \times c$

As a consequence of *T* 8 and *T* 9 we have, for any three elements a, b, c, and for d defined by $d = b^{-1} \times a$, $a = b \times b^{-1} \times a = b \times d$, so that

$$\begin{aligned} c \times (a+b) &= c \times (b \times d + b) = c \times [b \times (d+1)] \\ &= (c \times b) \times (d+1) = (c \times b) \times d + c \times b \\ &= c \times (b \times d) + c \times b = c \times a + c \times b. \end{aligned}$$

Thus *T* 9, and similarly *T* 10, may be replaced by the more general forms

T 9′ $\mathscr{M}_C(\mathscr{D}_A B) = D_{\mathscr{M}_C A}(\mathscr{M}_C B)$ $c \times (a+b) = c \times a + c \times b$

T 10′ $\mathscr{M}_{\mathscr{D}_A B} C = D_{\mathscr{M}_A C}(\mathscr{M}_B C)$ $(a+b) \times c = a \times c + b \times c$

As another consequence of *T* 8, 9, 10 we find

3.3 *T*1 Any element of the system commutes with any element of the primary field.

G $\mathscr{L} \# \wp | OEX$, $X \in \mathfrak{N}(O, E)$

$A : A \in OE$, $A \notin \mathfrak{N}(O, E)$

$s(OX/OE) = x = m/n$

$s(OA/OE) = a$

T $\mathscr{M}_A X = \mathscr{M}_X A$ i.e. $a \times x = x \times a$

P *g*1 $n \in \{\text{positive integers}\}$:

*t*1 $na = an$

$$\begin{aligned} na &= (1+1+ \dots +1)a \\ &= a+a+ \dots +a \\ &= a(1+1+ \dots +1) \\ &= an \end{aligned}$$

*t*2 $(-n)a = -(na) = -(an) = a(-n)$

*g*2 $m \in \{\text{integers}\}$, $n \in \{\text{positive integers}\}$

*t*3 $nam = (na)m = m(na) = m(an)$

$= man$

*t*4 $nam = man$

$\Rightarrow n^{-1}namn^{-1} = n^{-1}mann^{-1}$

$\Rightarrow amn^{-1} = n^{-1}ma$

T $ax = xa$.

The proof is written in terms of $\mathfrak{F}_R$ but can be readily adapted to $\mathfrak{F}_p$.

One difference between the geometric system of a collinear set of points belonging to a planar net $\mathfrak{N}$ constructed under $A\mathscr{I}$ 1, 2, 3, and an algebraic system satisfying the conditions listed above, must be borne in mind: there is in the collinear set an ideal point for which there is no counterpart in the algebraic system $\mathfrak{F}$. $\mathfrak{N}_p$ is the set of p collinear proper points in a collinear set which contains one more point, the ideal point, the whole set of $p+1$ points being, from the geometrical point of view, symmetrical—any three of them may be chosen to be O, E, U. On the other hand $\mathfrak{F}_p$ consists of exactly the p elements 0, 1, ..., p − 1.

In the case of $\mathfrak{R}_R$ thé distinction is not so clear. In the geometry there is an ideal point for which s(OU/OE) is undefinable, since the configuration corresponding to the operation $\mathscr{D}_{OA}$ B evanesces for all B when A = U. There is therefore no element in the set of rational numbers corresponding to the point U, but transformations of the rationals can bring U into the system at the expense of some other point. For example, the geometric transformation which interchanges A and A* (2.2*T*20 and 3.2*T*6) corresponds to an algebraic transformation which brings in U at the expense of O.

CHAPTER 3.4

THE EXTENSION OF A PRIMARY NET BY THE ADJUNCTION OF ONE POINT

In this chapter we assume throughout:

$G \quad \mathscr{L}\#|\mathrm{OEFU}$ & ideal$|$U & F $\notin \mathfrak{N}$(O, E/U).

U is fixed but the line chosen as ideal line may be varied. We shall use $\mathfrak{f}$ for the undefined algebraic element

$$\mathfrak{f} = \mathrm{s(OF/OE)}.$$

The three proper points O, E, F determine, to begin with, two nets $\mathfrak{N}$(O, E/U) and $\mathfrak{N}$(O, F/U). One of the basic properties of the geometry is:

3.4*T*1 $\mathfrak{N}$(O, E/U) and $\mathfrak{N}$(O, F/U) are isomorphic (i.e., in this context, both are $\mathfrak{N}_p$ for some p, or both $\mathfrak{N}_R$).

$G \quad \mathscr{L}\#\wp|\mathrm{OEF}$ & F $\notin \mathfrak{N}$(O, E)
$S1 \quad \wp|\mathrm{P} : \mathrm{P} \in \mathfrak{N}$(O, E)
$S2 \quad \wp|\mathrm{H} : \mathrm{H} \notin \mathrm{OE}$
$C1 \quad$ K : K $\in$ OH, PK $\|$ EH
$C2 \quad$ Q : Q $\in$ OE, QK $\|$ FH
$T \quad$ Q $\in \mathfrak{N}$(O, F) & s(OQ/OF) = s(OP/OE)
$P \quad$ s(OP/OE) = s(OK/OH) (2.3*T*2)
= s(OQ/OF).

This theorem establishes an exact one-to-one correspondence between $\mathfrak{N}$(O, E/U) and $\mathfrak{N}$(O, F/U) in which O and E correspond respectively to O and F. We assume that the net $\mathfrak{N}$(O, E) has been assigned, i.e. that s(OP/OE) is an element of $\mathfrak{F}_R$ or $\mathfrak{F}_p$ for given p; since there is an exact one-to-one correspondence between the sets {P} and {Q}, the specifications of the nets $\mathfrak{N}$(O, E) and $\mathfrak{N}$(O, F) are identical.

From P, $\in \mathfrak{N}$(O, E), and Q, $\in \mathfrak{N}$(O, F), we can construct a third net $\mathfrak{N}$(P, Q) isomorphic with these; we examine next the expression for the span-ratio of a point R, $\in \mathfrak{N}$(P, Q), in terms of the elements s(OP/OE) and the postulated 'new unit' $\mathfrak{f}$ = s(OF/OE).

3.4*T*2 The addition theorem in the compounding of primary nets.

*G*1 $\mathscr{L}\wp|\mathrm{OEF}, \mathrm{F} \notin \mathfrak{N}(\mathrm{O}, \mathrm{E})$
*G*2 $\mathrm{P}:\mathrm{P} \in \mathfrak{N}(\mathrm{O}, \mathrm{E})$
$\mathrm{Q}:\mathrm{Q} \in \mathfrak{N}(\mathrm{O}, \mathrm{F})$
$\mathrm{R}:\mathrm{R} \in \mathfrak{N}(\mathrm{P}, \mathrm{Q})$
$\mathfrak{f}:\mathfrak{f} = \mathrm{s}(\mathrm{OF}/\mathrm{OE})$
*G*3 $(\mathscr{D}_{\mathrm{OP}})^{\mathrm{l}'} = (\mathscr{D}_{\mathrm{OE}})^{\mathrm{l}}$
$(\mathscr{D}_{\mathrm{OQ}})^{\mathrm{m}'} = (\mathscr{D}_{\mathrm{OF}})^{\mathrm{m}}$
$(\mathscr{D}_{\mathrm{PR}})^{\mathrm{n}'} = (\mathscr{D}_{\mathrm{PQ}})^{\mathrm{n}}$
(for $\mathfrak{N}_{\mathrm{p}}, \mathrm{l}' = \mathrm{m}' = \mathrm{n}' = 1$ and $\mathrm{l}, \mathrm{m}, \mathrm{n} \in \{1, 2, \ldots, \mathrm{p}-1\}$)
T $\mathrm{s}(\mathrm{OR}/\mathrm{OE}) = \mathrm{r}+\mathrm{r}'\mathfrak{f}$
P

$$
\begin{aligned}
(\mathscr{D}_{\mathrm{PR}})^{\mathrm{l'm'n'}} &= (\mathscr{D}_{\mathrm{PQ}})^{\mathrm{l'm'n}} \\
\Rightarrow (\mathscr{D}_{\mathrm{PO}})^{\mathrm{l'm'n'}}(\mathscr{D}_{\mathrm{OR}})^{\mathrm{l'm'n'}} &= (\mathscr{D}_{\mathrm{PO}})^{\mathrm{l'm'n}}(\mathscr{D}_{\mathrm{OQ}})^{\mathrm{l'm'n}} \\
\Rightarrow (\mathscr{D}_{\mathrm{OR}})^{\mathrm{l'm'n'}} &= (\mathscr{D}_{\mathrm{OP}})^{\mathrm{l'm'(n'-n)}}(\mathscr{D}_{\mathrm{OQ}})^{\mathrm{l'm'n}} \\
&= (\mathscr{D}_{\mathrm{OE}})^{\mathrm{lm'(n'-n)}}(\mathscr{D}_{\mathrm{OF}})^{\mathrm{l'mn}} \\
\Rightarrow \mathrm{s}(\mathrm{OR}/\mathrm{OE}) &= \frac{1}{\mathrm{l'm'n'}}[\mathrm{lm'(n'-n)}+\mathrm{l'mn}\mathfrak{f}] \\
&= \mathrm{r}+\mathrm{r}'\mathfrak{f},
\end{aligned}
$$

where r and r′ belong to the primary field.

3.4*D*1 The first extension net, $\mathfrak{N}(\mathrm{O}, \mathrm{E}, \mathrm{F}/\mathrm{U})$.

G $\mathscr{L}\#|\mathrm{OEFU}$ & ideal|U & $\mathrm{F} \notin \mathfrak{N}(\mathrm{O}, \mathrm{E})$
D $\mathfrak{N}(\mathrm{O}, \mathrm{E}, \mathrm{F}/\mathrm{U}) = \{\mathrm{R}\}$, where R is determined by
*s*1 $\mathrm{P} \in \mathfrak{N}(\mathrm{O}, \mathrm{E}/\mathrm{U})$
*s*2 $\mathrm{Q} \in \mathfrak{N}(\mathrm{O}, \mathrm{F}/\mathrm{U})$
*s*3 $\mathrm{R} \in \mathfrak{N}(\mathrm{P}, \mathrm{Q}/\mathrm{U})$
g $\mathrm{s}(\mathrm{OF}/\mathrm{OE}) = \mathfrak{f}$
d $\mathfrak{N}(\mathrm{O}, \mathrm{E}, \mathrm{F}/\mathrm{U}) = \{\mathrm{R}:\mathrm{s}(\mathrm{OR}/\mathrm{OE}) = \mathrm{r}+\mathrm{r}'\mathfrak{f}, (\mathrm{r}, \mathrm{r}') \in \mathfrak{F}\}$.

3.4*N*1 Basis of the first extension net: the points O, E, F (and ideal|U) form a *basis* for $\mathfrak{N}(\mathrm{O}, \mathrm{E}, \mathrm{F}/\mathrm{U})$.

3.4*N*2 The algebraic field corresponding to $\mathfrak{N}(\mathrm{O}, \mathrm{E}, \mathrm{F})$ will be denoted by $\mathfrak{F}(1, \mathfrak{f})$ or $\mathfrak{F}_{\mathrm{p}}(1, \mathfrak{f})$ or $\mathfrak{F}_{\mathrm{R}}(1, \mathfrak{f})$. When there is any doubt about the field designated by $\mathfrak{F}$, the symbol $\mathfrak{F}(1)$ will be used for the primary field.

The definition, as far as *s*3, is identical in appearance with the definition of the primary planar set $\mathfrak{N}(\mathrm{O}, \mathrm{E}, \mathrm{F}/\mathrm{UV})$ in 2.3*D*3 and 2.3*N*1. The essential difference is that here $\mathscr{L}|\mathrm{OEF}$ & $\mathrm{F} \notin \mathfrak{N}(\mathrm{O}, \mathrm{E})$, while in Chapter 2.3 we have $\mathscr{L}|\mathrm{OEF}$. There are p^2 proper points in the first extension net of $\mathfrak{N}_{\mathrm{p}}$ on the line and p^2 proper points in the planar system $\mathfrak{N}_p$, but the geometric systems are symmetric over the proper and ideal points together, that is, over p^2+1 points on the line, and over p^2+p+1 points in the plane. In the linear set there is one and only one zero point, $0+0\mathfrak{f}$ or $0\times(r+r'\mathfrak{f})$, and there is one and only one ideal point and it has no algebraic counterpart. For the rest of this Chapter, the symbol $\mathfrak{N}(O, E, F)$ refers only to the first extension net on the line.

Given any element $a+a'\mathfrak{f}$ of $\mathfrak{F}(1, \mathfrak{f})$ we can determine (finitely and uniquely) the point P such that $s(OP/OE) = a+a'\mathfrak{f}$, for if A, A′ are determined by $s(OA/OE) = a$ and $s(OA'/OF) = a'$, then $P = \mathscr{D}_A A'$. We assume that when the data include a statement

$$\text{‘}G \quad P : P \in \mathfrak{N}(O, E, F)\text{’},$$

then an element $a+a'\mathfrak{f}$ of $\mathfrak{F}(1, \mathfrak{f})$ has been assigned such that

$$s(OP/OE) = a+a'\mathfrak{f}.$$

The net $\mathfrak{N}(O, E, F)$ was constructed on the basis of the points O, E, F (and the ideal point) using only the transformers $\mathscr{D}$. Under the transformers $\mathscr{M}$ the system is not necessarily closed: $\mathscr{M}_A B$ is not necessarily a point in $\mathfrak{N}(O, E, F)$. As a simple example, suppose that we begin with $\mathfrak{N}_R(O, E)$ and adjoin the point F given by $s(OF/OE) = \sqrt[3]{2}$, so that $\mathfrak{N}(O, E, F)$ consists of points with span-ratios $r+r'\sqrt[3]{2}$. Then, if $s(OA/OE) = a+a'\sqrt[3]{2}$, $s(OB/OE) = b+b'\sqrt[3]{2}$, and $P = \mathscr{M}_A B$, we find $s(OP/OE) = ab+(ab'+a'b)\sqrt[3]{2}+bb'\sqrt[3]{4}$. P is not a point of $\mathfrak{N}(O, E, F)$; we need the doubly extended net $\mathfrak{N}(O, E, F, F^{(2)})$, where $F^{(2)} = \mathscr{M}_F F$, to accommodate it.

Clearly then the geometric system is not determined completely by the specification of $\mathfrak{N}(O, E)$ as $\mathfrak{N}_p$ or $\mathfrak{N}_R$, and the assignment of F, $F\in OE$, $F\notin\mathfrak{N}(O, E)$.

We need also information about the points $F^{(2)}, F^{(3)}, \ldots, F^{(r)}, \ldots$, where $F^{(r)} = (\mathscr{M}_F)^{r-1}F$. The development can be explained more concisely in algebraic terms, that is, in terms of the new unit $\mathfrak{f}$ and its powers. We have proved that $\mathfrak{f}$ (and therefore all powers of $\mathfrak{f}$) commute with the elements of the primary field, so that we have only to develop properties of the set of elements $\{a+a'\mathfrak{f}\}$ under addition and commutative multiplication. There are three degrees of complication that we have to consider:

(i) Quadratic extensions

Let us assume that the new unit $\mathfrak{f}$ is such that there exist $(\rho, \sigma)\in\mathfrak{F}$, such that

$$\mathfrak{f}^2 = \rho\mathfrak{f}+\sigma,$$

that is, in geometric terms, that

$$\mathscr{M}_F F \in \mathfrak{N}(O, E, F).$$

By a change of reference points we may replace $\mathfrak{f}$ by $\mathfrak{f}'$, where

$$\mathfrak{f} = \mathfrak{f}' + \mathrm{m}, \quad \mathrm{m} \in \mathfrak{F}(1),$$

so that

$$\mathfrak{f}'^2 = (\rho - 2\mathrm{m})\mathfrak{f}' - \mathrm{m}^2 + \rho\mathrm{m} + \sigma.$$

By taking $2\mathrm{m} = \rho$, the relation reduces to

$$\mathfrak{f}'^2 = \tfrac{1}{4}(\rho^2 + 4\sigma).$$

We must assume therefore that $\rho^2 + 4\sigma \neq 0$. Since $\rho^2 + 4\sigma$ might be any member of $\mathfrak{F}(1)$, it will have:

in general no square root in $\mathfrak{F}_\mathrm{R}$,
a square root in the surdic field $\mathfrak{F}_\mathrm{S}$ if and only if $\rho^2 + 4\sigma > 0$,
a square root for exactly $\frac{1}{2}(\mathrm{p} - 1)$ of the non-zero members of $\mathfrak{F}_\mathrm{p}$.

If $\rho^2 + 4\sigma$ has a square root, then $\mathfrak{f}' \in \mathfrak{F}$ and $\mathfrak{f} \in \mathfrak{F}$, and the assumption that $\mathrm{F} \notin \mathfrak{N}(\mathrm{O}, \mathrm{E})$ is falsified. We are concerned therefore only with those values of ρ and σ for which $\rho^2 + 4\sigma$ is a non-square. Thus, by choice of reference points, we can reduce the significant cases of a single adjoined point to that in which that adjoined point satisfies

$$\mathscr{M}_\mathrm{F}\mathrm{F} = \mathrm{H} : \mathrm{H} \in \mathfrak{N}(\mathrm{O}, \mathrm{E}) \;\&\; \mathrm{F} \notin \mathfrak{N}(\mathrm{O}, \mathrm{E}).$$

We consider the pattern of such extensions for $\mathrm{p} = 2, 3, 5$ in Excursus 3.1.

The case of an extension of $\mathfrak{F}_\mathrm{R}$ into a sub-field of the surdic field requires a little more consideration. Taking h to be a non-square in $\mathfrak{F}_\mathrm{R}$ we have to consider two cases according to the sign of h. If $\mathrm{h} > 0$, then the corresponding point F satisfies betweenness relations among the points of $\mathfrak{N}_\mathrm{R}$, and $\mathfrak{N}_\mathrm{R}(\mathrm{O}, \mathrm{E}, \mathrm{F})$ is a sub-field of $\mathfrak{N}_\mathrm{S}(\mathrm{O}, \mathrm{E})$. If $\mathrm{h} < 0$, then the corresponding point F cannot consistently with the axioms $A\mathscr{S}$ be related to O and E by any of the three relations $\mathrm{F}\,\mathscr{B}\,\mathrm{OE}$, $\mathrm{E}\,\mathscr{B}\,\mathrm{OF}$ or $\mathrm{O}\,\mathscr{B}\,\mathrm{EF}$.

(ii) Algebraic extensions

There is nothing in the geometry which requires that $\mathscr{M}_\mathrm{F}\mathrm{F} \in \mathfrak{N}(\mathrm{O}, \mathrm{E}, \mathrm{F})$, but it might be prescribed for $\mathrm{F}^{(\mathrm{n})} = (\mathscr{M}_\mathrm{F})^{\mathrm{n}-1}\mathrm{F}$, that $\mathrm{F}^{(\mathrm{n})} \in \mathfrak{N}(\mathrm{O}, \mathrm{E}, \mathrm{F}, \mathrm{F}^{(2)}, \ldots, \mathrm{F}^{(\mathrm{n}-1)})$, that is, to the net built by $\mathscr{D}$ transformers on the basis $\mathrm{O}, \mathrm{E}, \mathrm{F}, \ldots, \mathrm{F}^{(\mathrm{n}-1)}$. Then $\mathfrak{f}$ satisfies the algebraic condition

$$\phi : \mathfrak{f}^\mathrm{n} = \rho_0 + \rho_1\mathfrak{f} + \ldots + \rho_{\mathrm{n}-1}\mathfrak{f}^{\mathrm{n}-1},$$

where $\{\rho_\mathrm{i}\} \subset \mathfrak{F}(1)$. We can modify the relation ϕ to give

$$\phi' : \mathfrak{f}^{-1} = -\rho_0^{-1}(\rho_1 + \rho_2\mathfrak{f} + \ldots + \rho_{\mathrm{n}-1}\mathfrak{f}^{\mathrm{n}-2} - \mathfrak{f}^{\mathrm{n}-1})$$

so that the point F^* given by $\mathscr{M}_\mathrm{F}\mathrm{F}^* = \mathrm{E}$ also lies in $\mathfrak{N}(\mathrm{O}, \mathrm{E}, \mathrm{F}, \mathrm{F}^{(2)}, \ldots, \mathrm{F}^{(\mathrm{n}-1)})$. (The construction for F^* is derived in 3.2 *T* 6.)

The elements of the field $\mathfrak{F}(1, \mathfrak{f}, \mathfrak{f}^2, \ldots, \mathfrak{f}^{n-1})$ are

$$a_0 + a_1\mathfrak{f} + a_2\mathfrak{f}^2 + \ldots + a_{n-1}\mathfrak{f}^{n-1},$$

and, using the relations ϕ and ϕ' above, we may express every positive and negative integral power of $\mathfrak{f}$ in this form. Thus the net $\mathfrak{N}(O, E, F, F^{(1)}, \ldots, F^{(n-1)})$ is closed under the transformers $\mathscr{D}$ and $\mathscr{M}$.

In the case when the primary field is $\mathfrak{F}_R$, if the equation

$$x^n = a_0 + a_1 x + \ldots + a_{n-1} x^{n-1}$$

has a solution in the real number system, and we take $\mathfrak{f}$ to be that solution, then the corresponding point F may be related by betweenness to the members of $\mathfrak{N}_R(O, E)$ by an iterative process of which the simplest example was given in 2.5*C*4.

The set of solutions, in real numbers, to the set of algebraic equations of unrestricted orders constitute the field of *algebraic numbers*. The corresponding net of points could be called the *algebraic net* $\mathfrak{N}_A(O, E)$. Two types of subnets of $\mathfrak{N}_A(O, E)$ are: nets $\mathfrak{N}_R(O, E, F, \ldots, F^{(n-1)})$ for specified relations ϕ, and the surdic net $\mathfrak{N}_S(O, E)$.

(iii) Transcendental extension

Finally the specification of the point F could be such that for no value of n is $F^{(n)}, = (\mathscr{M}_F)^{n-1}F$, contained in the net $\mathfrak{N}(O, E, F, F^{(2)}, \ldots, F^{(n-1)})$, that is, for no value of n is there a relation such as ϕ among powers of the unit $\mathfrak{f}$. At this stage we verge on quite difficult algebraic territory, and the geometric equivalent is even more remote. When the primary field is $\mathfrak{N}_R$, a number which can be defined and calculated to any degree of approximation, but which does not satisfy any relation of the type ϕ, is called *transcendental*.

For example, the number t defined by the series

$$t = 1 - \tfrac{1}{2} + \tfrac{1}{3} - \tfrac{1}{4} + \ldots$$

is one of the simplest of the transcendentals (this does not mean that there is a simple proof that no algebraic relation $t^n = a_0 + a_1 t + \ldots + a_{n-1} t^{n-1}$ exists). In a geometric setting, we could suppose the corresponding point to be delimited thus:

Assume $t = s(OT/OE)$, and construct the two sequences

$$\mathscr{O}|H_1 H_2 \ldots, \quad \mathscr{O}|K_1 K_2 \ldots,$$

where

$$H_1 = O \qquad K_1 = E$$
$$s(OH_2/OE) = 1 - \tfrac{1}{2} \qquad s(OK_2/OE) = 1 - \tfrac{1}{2} + \tfrac{1}{3}$$
$$s(OH_3/OE) = 1 - \tfrac{1}{2} + \tfrac{1}{3} - \tfrac{1}{4} \qquad s(OK_3/OE) = 1 - \tfrac{1}{2} + \tfrac{1}{3} - \tfrac{1}{4} + \tfrac{1}{5}.$$

Then $\mathscr{O}|H_1H_2H_3 \dots H_r \dots T \dots K_r \dots K_3K_2K$.

$$s(H_rK_r/OE) = 1/(2r-1)$$

and can be made as small as we please. The betweenness relation of T to points of $\mathfrak{N}_R(O, E)$ are determinable from the relations $\mathscr{O}|H_rTK_r$.

The totality of algebraic and transcendental numbers constitute the *Real Field*, which we shall denote by $\mathfrak{F}_{\mathfrak{R}}$ and the corresponding geometric system by $\mathfrak{N}_{\mathfrak{R}}(O, E)$.

It will have been noticed that for $\mathfrak{N}_S$, $\mathfrak{N}_A$ and $\mathfrak{N}_{\mathfrak{R}}$ we have not needed to specify any basis beyond O, E (and the ideal point), so that strictly all these nets are primary nets, even although, to derive their definition it was necessary to extend the original net $\mathfrak{N}_{\mathfrak{R}}$ by some specified point. There are problems in which the extension of the net $\mathfrak{N}_R$ by adjoining a point and prescribing a relation plays a useful part, but, in general, geometry when not restricted to extensions of $\mathfrak{N}_p$ is concerned with the Real Net $\mathfrak{N}_{\mathfrak{R}}$.

A quadratic extension of $\mathfrak{N}_{\mathfrak{R}}$ is obtained by adjoining a point F such that $\mathscr{M}_F F \in \mathfrak{N}_{\mathfrak{R}}$ while $F \notin \mathfrak{N}_{\mathfrak{R}}$. That is, the unit $\mathfrak{f}$ is such that while $\mathfrak{f}^2 \in \mathfrak{F}_{\mathfrak{R}}$, $\mathfrak{f} \notin \mathfrak{F}_{\mathfrak{R}}$, and therefore $\mathfrak{f}^2 < 0$. Effectively therefore the quadratic extension of $\mathfrak{F}_{\mathfrak{R}}$ is the extension from the reals to the complexes.

The emphasis so far has been on the algebraic aspects of the first extension of a net; the first extension has however a geometrical property that marks it out from the higher extensions. In proving the uniqueness of the construction for $\mathscr{M}_A B$ we made use of the full Desargues axiom $A\mathscr{I}3$. For the first extension (as for the primary net itself) we shall prove that only the simply-special Desargues $A\mathscr{I}3^*$ is necessary.

3.4*T*3 The point constructed as $\mathscr{M}_A A$ can be proved to be unique using only the axiom $A\mathscr{I}3^*$ (cf. 3.2*T*1, 2, 3).

G $\quad \mathscr{L}\# \wp|OEF,\ F \notin \mathfrak{N}(O, E)$
S $\quad \wp|L : L \notin OE$
*C*1 $\quad M : M \in OL,\ FM \parallel EL$
*C*2 $\quad F^{(2)} : F^{(2)} \in OE,\ MF^{(2)} \parallel LF$
T $\quad F^{(2)} = \mathscr{M}_F F$
(i) *S* $\quad \wp|L' : L' \in EL\ \&\ \#|ELL'$
*C*3 $\quad M' : M' = FM \cap OL'$
T $\quad F^{(2)}M' \parallel FL'$
P

$$O\begin{Bmatrix} FF^{(2)} \\ LM \\ L'M' \end{Bmatrix} \Rightarrow \mathscr{L}|\langle FL \cap F^{(2)}M,\ FL' \cap F^{(2)}M',\ LL' \cap MM' \rangle$$

$$\Rightarrow F^{(2)}M' \parallel AL'.$$

Also $\mathscr{L}|FMM'$ and F, M, M′ and the ideal point on

FM are four of the ten points in the figure. I.e. only the axiom $A\mathscr{I}\,3^*$ is needed.

(ii) S $\quad \wp|\mathrm{L}'' : \mathrm{L}'' \in \mathrm{FL}\ \&\ \#|\mathrm{FLL}''$

C $\quad \mathrm{M}'' : \mathrm{M}'' \in \mathrm{OL}''\ \&\ \mathrm{FM}'' \parallel \mathrm{EL}''$

$T(X)$ $\quad \mathscr{L}|\mathrm{M}''\mathrm{MF}^{(2)}$

(iii) $T(X)$ $\quad \mathrm{F}^{(2)}$ is independent of the ideal line through the ideal point on OE and the proof depends only on $A\mathscr{I}\,3^*$.

3.4*T*4*X* Prove, by an inductive method, that the proof that the construction for

$$\mathrm{F}^{(r)} = (\mathscr{M}_{\mathrm{A}})^{r-1}\mathrm{F}$$

results in a unique point depends only on $A\mathscr{I}\,3^*$.

3.4*T*5*X* The proof that the construction of F*, where $\mathscr{M}_{\mathrm{F}}\mathrm{F}^* = \mathrm{E}$, is unique depends only on $A\mathscr{I}\,3^*$.

In consequence of these theorems

3.4*T*6 All constructions in the first extension of a primary net depend for their validity only on $A\mathscr{I}\,1, 2$ and $A\mathscr{I}\,3^*$.

CHAPTER 3.5

GENERAL EXTENSIONS, COMMUTATIVITY AND PAPPUS: QUATERNIONS

Let us assume that we have devised means of constructing any designated point of the net $\mathfrak{N}(O, E, F, F^{(2)}, \ldots)$ and that there is a point G in the collinear set OE that does not belong to this net. Corresponding to G we introduce a new unit

$$\mathfrak{g} = s(OG/OE).$$

As a simple example we might suppose that successive extensions of $\mathfrak{N}_R$ are defined by $\mathfrak{f}^2 = 2$, $\mathfrak{g}^2 = 3$ (this is not typical since F and G both belong to $\mathfrak{N}_S$, and satisfy betweenness relations). From F and G we can construct, using the transformers $\mathscr{D}$, points P given by

$$s(OP/OE) = r + s\mathfrak{f} + t\mathfrak{g} = r + s\sqrt{2} + t\sqrt{3}$$

where $(r, s, t) \in \mathfrak{F}_{\mathfrak{N}}$. This system is closed under the transformers $\mathscr{D}$. But using the transformers $\mathscr{M}$ we could generate points Q for which

$$s(OQ/OE) = r + s\sqrt{2} + t\sqrt{3} + u\sqrt{6}.$$

The resulting system, closed under transformers $\mathscr{M}$ as well as $\mathscr{D}$, is therefore $\mathfrak{N}(O, E, F, G, H)$, where $s(OH/OE) = \mathfrak{f}\mathfrak{g}$.

Let us return now to the general case. If

$$s(OA/OE) = a + a'\mathfrak{f} + a''\mathfrak{g} \quad \text{and} \quad s(OB/OE) = b + b'\mathfrak{f} + b''\mathfrak{g},$$

where $(a, a', a'', b, b', b'') \in \mathfrak{F}(1)$ and $P = \mathscr{M}_A B$ and $Q = \mathscr{M}_B A$,

then $s(OP/OE) = ab + (a'b + ab')\mathfrak{f} + (a''b + ab'')\mathfrak{g}$

$$+ a'b''\mathfrak{f}\mathfrak{g} + a''b'\mathfrak{g}\mathfrak{f} + a'b'\mathfrak{f}^2 + a''b''\mathfrak{g}^2$$

and $\qquad s(OP/OE) - s(OQ/OE) = (a'b'' - a''b')(\mathfrak{f}\mathfrak{g} - \mathfrak{g}\mathfrak{f}).$

Thus, unless $\mathfrak{f}\mathfrak{g} = \mathfrak{g}\mathfrak{f}$, $\mathscr{M}_A B \neq \mathscr{M}_B A$. So far we have not devised any geometric proof that $\mathscr{M}_A B = \mathscr{M}_B A$; if therefore we can devise an algebraic system which satisfies all the conditions prescribed in Chapter 3.3 and in which $\mathfrak{f}\mathfrak{g} \neq \mathfrak{g}\mathfrak{f}$, then there can be no geometric theorem, deducible from $A\mathscr{I}$ 1, 2, 3, which asserts that, for every pair of points {A, B} in the system, $\mathscr{M}_A B = \mathscr{M}_B A$ in relation to a basis on AB. The simplest of these algebraic systems is the quaternion algebra which is described in the next Chapter.

The first part of this Chapter is to be devoted to geometries in which, for every pair of points {A, B}, $\mathscr{M}_A B = \mathscr{M}_B A$ in relation to a basis on AB, but before we consider them further we consider the implications of this condition. In the next two theorems we are to prove that the commutativity of the transformer $\mathscr{M}$ is equivalent to the Pappus condition of collinearity, and that, as a consequence of $A\mathscr{I}$ 1 and 2 and the Pappus condition, the Desargues condition is always satisfied. That is, geometries in which $\mathscr{M}$ is commutative require as axioms only $A\mathscr{I}$ 1, 2 and the Pappus condition.

3.5*T*1 The condition $\mathscr{M}_A B = \mathscr{M}_B A$ for all pairs {A, B} and any basis on AB is equivalent to the Pappus condition:

$$\left.\begin{array}{r}\mathscr{L}|\text{PQR}\\ \&\ \mathscr{L}|\text{P}'\text{Q}'\text{R}'\end{array}\right\} \Rightarrow \mathscr{L}|\langle \text{QR}' \cap \text{Q}'\text{R}, \text{RP}' \cap \text{R}'\text{P}, \text{PQ}' \cap \text{P}'\text{Q}\rangle$$

G $\mathscr{L}\#\wp|$OEAB, ideal|u fixed

S $\wp|$L, L$\notin$OE

C M : M$\in$OL & BM$\parallel$EL
C : C$\in$OE & MC$\parallel$LA
(i.e. $\mathscr{M}_A B = C$).
N : N$\in$OL & AN$\parallel$EL

T (i) $\mathscr{M}_A B = \mathscr{M}_B A \Leftrightarrow$ NC$\parallel$LB

P Definition of $\mathscr{M}_B A$

T (ii) The Pappus condition:

$$\left.\begin{array}{r}\mathscr{L}|\text{ABC}\\ \&\ \mathscr{L}|\text{MNL}\end{array}\right\} \Rightarrow \mathscr{L}|\langle \text{AN} \cap \text{BM}, \text{AL} \cap \text{CM}, \text{BL} \cap \text{CN}\rangle \Rightarrow \text{BL}\parallel\text{CN}.$$

Thus, BL$\parallel$CN:
(i) if and only if $\mathscr{M}_A B = \mathscr{M}_B A$, or
(ii) if and only if the Pappus condition is satisfied.

We can ensure therefore that $\mathscr{M}_A B = \mathscr{M}_B A$ for all pairs in every collinear set OE (in relation to the basis O, E and the ideal point) by assuming:

$A\mathscr{I}$4 THE PAPPUS AXIOM

G $\mathscr{L}\#|$ABC
$\mathscr{L}\#|$A′B′C′ & #|AB, A′B′

C A″ : A″ = BC′ ∩ B′C
B″ : B″ = CA′ ∩ C′A
C″ : C″ = AB′ ∩ A′B

A $\mathscr{L}|$A″B″C″.

We now prove, by constructing a figure with nine of the Desargues incidences and proving that the tenth is a consequence of $\mathcal{AI}$4:

3.5*T*2 $\mathcal{AI}$1 & 2 & 4 $\Rightarrow$ $\mathcal{AI}$3.

*G*1 $\mathcal{L}_3|$KABC

*G*2 {A′, B′, C′} : A′ ∈ KA, B′ ∈ KB, C′ ∈ KC
& ≠|KAA′, ≠|KBB′, ≠|KCC′

C {L, M, N} : L = BC ∩ B′C′, M = CA ∩ C′A′, N = LM ∩ A′B′

T (as a consequence of $\mathcal{AI}$1 & 2 & 4)
$\mathcal{L}|$ABN

P S : S = A′B′ ∩ CC′
X : X = SB ∩ LM
Y : Y = C′X ∩ BB′
N* : N* = CY ∩ A′B′
T : T = SB ∩ CM

*t*1 $\mathcal{L}|$BB′Y & $\mathcal{L}|$C′CS $\Rightarrow$ $\mathcal{L}|$N*XL

*t*2 $\mathcal{L}|$SCC′ & $\mathcal{L}|$MXN* $\Rightarrow$ $\mathcal{L}|$YA′T

*t*3 $\mathcal{L}|$YA′T & $\mathcal{L}|$SCK $\Rightarrow$ $\mathcal{L}|$ABN*
$\Rightarrow$ N* = N

T $\mathcal{P}|$AB ∩ A′B′ ∩ LM = N.

The logical situation now is this: we assume always $\mathcal{AI}$1 and 2. Then

$$\mathcal{M}_{\mathrm{A}}\mathrm{B} = \mathcal{M}_{\mathrm{B}}\mathrm{A} \Leftrightarrow \text{Pappus } (\mathcal{AI}4)$$

and

$$\text{Pappus} \Rightarrow \text{Desargues } (\mathcal{AI}3),$$

but we have established that Pappus neither is nor is not a consequence of Desargues.

In the second half of this Chapter we shall show that there exist algebraic systems which satisfy all the conditions laid down in Chapter 3.3, but which are such that the system includes pairs of elements for which $x \times y \neq y \times x$. We have seen that this implies that the corresponding geometry does not conform to $\mathcal{AI}$4 (Pappus); we shall prove that in one such system the Desargues condition ($\mathcal{AI}$3) is satisfied. That is, the implication sign in the statement '$\mathcal{AI}$4 $\Rightarrow$ $\mathcal{AI}$3' cannot be reversed.

One other aspect of the Desargues and Pappus conditions: we may replace $\mathcal{AI}$3 by axioms which prescribe that there are points not in the coplanar set and that sets of points have certain additional combinatorial relations (as in Excursus 1.4). In other words, if the space is three-dimensional with suitable axioms of incidence, then in all planes in any geometry satisfying these axioms the Desargues condition is fulfilled. The Pappus condition is different: to spaces of any numbers of dimensions an algebra can be applied in which in general $x \times y \neq y \times x$, and the corresponding geometry will therefore not conform to the Pappus condition.

We finish this part of the Chapter by considering briefly what is involved in the study of planar nets satisfying $A\mathscr{I}$ 1, 2 and 4.

We have reached the stage at which the structure of the geometry depends very much on the structure of the corresponding algebraic system, and this lies beyond the scope of this book, although some of the possibilities are illustrated in Excursus 3.2. If the geometry is based on the rational net $\mathfrak{N}_R$ and extensions of $\mathfrak{N}_R$ into the surdic net $\mathfrak{N}_S$ and the algebraic net $\mathfrak{N}_A$, then the whole of the area of study covered by the titles 'theory of numbers', 'theory of equations' and 'theory of algebraic functions' is involved.

All the theorems that have been proved up to this point in Books 1, 2 and 3 are valid however far $\mathfrak{N}_R$ may be extended into $\mathfrak{N}_S$ and $\mathfrak{N}_A$. If the net is extended so that the underlying field is that of the reals or the complexes, the geometry is that made familiar in the standard texts.

The finite nets offer nothing of the same order of complexity, but while the three principal results can be stated in relatively simple terms, the proofs of these results lie quite deep. Briefly these results are:

(i) If the planar net consists of a finite number of points and conforms to $A\mathscr{I}$ 1, 2, 4, then

(*a*) the net is completely determined by two numbers:

p: the prime determining the primary field $\mathfrak{F}_p(1)$,
n: the degree of the extension $\mathfrak{F}_p(1, \mathfrak{f}, \mathfrak{f}^2, \ldots, \mathfrak{f}^{n-1})$.

That is, the net is determined by adjoining one point to a given primary net and prescribing the smallest value of n for which

$$F^{(n)} \in \mathfrak{N}(O, E, F, F^{(2)}, \ldots, F^{(n-1)}).$$

The field, however synthesized, is the Galois field $GF(p^n)$.

(*b*) the net can be generated cyclically, that is, a three-by-three matrix **M** with elements in $GF(p^n)$ can be found which is such that every point of the planar net is contained in the set

$$\{\mathbf{M}^r \epsilon_1 : r \in \{0, 1, \ldots, q^2+q\},\ q = p^n\} \quad \text{and} \quad \mathbf{M}^{q^2+q+1} = \kappa\mathbf{1}.$$

The indices, r, of the $q+1$ collinear sets of $q+1$ points $P_r = \mathbf{M}^r \epsilon_1$ which contain P_0 provide the rows of a $(q+1) \times (q+1)$ difference table.

(ii) There is no finite geometry conforming to $A\mathscr{I}1$, 2 and 3 (Desargues) in which there is any pair of points for which $\mathscr{M}_B A \neq \mathscr{M}_A B$. In algebraic terms, if any pair in a finite system is such that $a \times b \neq b \times a$ then the system cannot satisfy all the conditions for an algebraic field laid down in Chapter 3.3. For example, in Excursus 3.3 a system is described in which, as a consequence of relations such as $\alpha \times \beta = -\beta \times \alpha$, it is proved that for some sets of three elements the relation $\gamma \times (\alpha + \beta) = (\gamma \times \alpha) + (\gamma \times \beta)$ does not hold. The corresponding geometric property, $\mathscr{M}_C(\mathscr{D}_A B) = \mathscr{D}_{\mathscr{M}_C A}(\mathscr{M}_C B)$, was shown in 3.2$T$9 to be a consequence of $A\mathscr{I}3$. A good account of the algebraic aspect of these theorems will be found in I. N. Herstein *Topics in Algebra.*

Now let us assume for an algebra all the properties set out for the operators $+$ and $\times$ in Chapter 3.3 but not that, for every pair $\{a, b\}$ of elements, the additional relation $a \times b = b \times a$ holds. First we have to show that such algebraic systems exist, a problem of which the solution is not entirely trivial, since there are in fact no such finite systems.

The classical example is the system of quaternions, and, in this context we may assume them to be based on the primary field, $\mathfrak{F}_R$, of the rationals. We have already seen, in Chapter 3.4, that a single additional unit cannot lead to a non-commutative field; we adjoin therefore two new units i, j and prescribe for them the simplest possible non-commutative multiplicative relations, say

$$i^2 = -1, \quad j^2 = -1,$$

$$ij = -ji.$$

The field $\mathfrak{F}_R(1, i, j, ij)$ is then the set of elements

$$\{\mathfrak{g}\} = \{a + bi + (c + di)j\}, \quad (a, b, c, d) \in \mathfrak{F}_R.$$

The new units are required to be associative in multiplication, so that we have

$$(ij)i = -(ji)i = -j(ii) = j,$$

$$i(ij) = (ii)j = -j.$$

Similarly

$$j(ij) = i = -(ij)j,$$

$$(ij)(ij) = -i(ij)j = -1.$$

We are not going to prescribe any polynomial identity among i, j, ij, which, since $i^2 = j^2 = (ij)^2 = -1$, means that we are not going to prescribe a set $\{\alpha, \beta, \gamma, \delta\}$ of members of $\mathfrak{F}_R$, not all zero, such that $\alpha + \beta i + \gamma j + \delta ij = 0$; that is, we assume $ij \notin \mathfrak{F}_R(1, i, j)$. There is therefore no problem with the distributive property of multiplication. (Compare the system Ω in Excursus 3.3 in which we have a set of units similar in multiplication properties to i, j, ij, but connected by linear identities. Suppose that we had prescribed in the present case that $ij = i + j$ and that multiplication should be distributive over addition. We should have $(i+j)j = ij - 1$; also $(i+j)j = (ij)j = -i$ so that $ij = 1 - i$. Similarly $j(i+j)$ gives $-ij - 1 = i$, that is, $ij = -i - 1$. It follows that we cannot have both a relation $ij = i + j$, and a full distributive property.)

For a product $\mathbf{q}\mathbf{q}'$ we have:

$$\begin{aligned}\mathbf{q}\mathbf{q}' &= (a + bi + cj + dij)(a' + b'i + c'j + d'ij)\\ &= (aa' - bb' - cc' - dd') + (ab' + ba' + cd' - dc')i\\ &\quad + (ac' + ca' - bd' + db')j + (ad' + da' + bc' - cb')ij.\end{aligned}$$

Thus the product is uniquely definable, and in general $\mathbf{q}\mathbf{q}' \neq \mathbf{q}'\mathbf{q}$. Take

$$\bar{\mathbf{q}} = a - bi - cj - dij,$$

then from the above expression

$$\mathbf{q}\bar{\mathbf{q}} = a^2 + b^2 + c^2 + d^2 + 0i + 0j + 0ij.$$

This relation enables us to complete the system, since now we have:

$$\text{additive inverse of } \mathbf{q} : -\mathbf{q} = -a - bi - cj - dij,$$
$$\text{multiplicative inverse of } \mathbf{q} : \bar{\mathbf{q}}/(a^2 + b^2 + c^2 + d^2),$$

with of course the proviso, $a^2 + b^2 + c^2 + d^2 \neq 0$. This condition is satisfied for all non-zero sets $(a, b, c, d) \in \mathfrak{F}_R$.

Until this last property, no specific use was made of the restriction of the underlying primary field to the field of rationals, $\mathfrak{F}_R$. We could replace $\mathfrak{F}_R$ by any field ($\mathfrak{F}_S$ or $\mathfrak{F}_A$ for example) in which the sum of a set of (non-zero) squares is never zero, but we could not replace $\mathfrak{F}_R$ by any finite primary field, because in any such field sets of four squares can be easily found whose sum is zero. (In fact, in Excursus 2.4 we proved that in $\mathfrak{F}_p$ there are always exactly $p + 1$ solutions to an equation such as $x^2 + y^2 + 1 = 0$. *A fortiori* no compound finite field would have the required property.)

In this system of quaternions we have specified completely a field which satisfies all the conditions laid down in Chapter 3.3, and which is not commutative in multiplication. Now that we are assured that such fields exist, we can proceed to discuss the geometry of a planar net over a non-commutative field.

Take points to be the vectors

$$\boldsymbol{\xi} = \left\{\begin{bmatrix} x \\ y \\ 1 \end{bmatrix}, \begin{bmatrix} x \\ 1 \\ 0 \end{bmatrix}, \begin{bmatrix} 1 \\ 0 \\ 0 \end{bmatrix}\right\}$$

and the collinear sets to be the vectors

$$\boldsymbol{\alpha} = \left\{ \begin{bmatrix} u \\ v \\ 1 \end{bmatrix}, \begin{bmatrix} u \\ 1 \\ 0 \end{bmatrix}, \begin{bmatrix} 1 \\ 0 \\ 0 \end{bmatrix} \right\}.$$

When the algebra is non-commutative there are two different ways in which we may specify that a point belongs to a collinear set: we may take either

$$\boldsymbol{\alpha}^T\boldsymbol{\xi} = 0 \quad \text{or} \quad \boldsymbol{\xi}^T\boldsymbol{\alpha} = 0.$$

It does not matter which, but once the choice has been made, only the chosen form may be used. We shall take $\boldsymbol{\alpha}^T\boldsymbol{\xi} = 0$.

For any elements λ, μ of the field, the set of points $\boldsymbol{\xi}$ which satisfy $\boldsymbol{\alpha}^T\boldsymbol{\xi} = 0$ is the same as the set which satisfies

$$\lambda(\boldsymbol{\alpha}^T\boldsymbol{\xi})\mu = 0, \quad \text{i.e. } (\lambda\boldsymbol{\alpha}^T)(\boldsymbol{\xi}\mu) = 0.$$

We may therefore replace at any time:

a point $\boldsymbol{\xi}$ by $\boldsymbol{\xi}\mu = (x\mu, y\mu, \mu)$ or (x', y', z')

and the collinear set $\boldsymbol{\alpha}^T$ by $\lambda\boldsymbol{\alpha}^T = [\lambda u, \lambda v, \lambda]$ or $[u', v', w']$.

Alternatively any linear relation

$$u'x' + v'y' + w'z' = 0$$

may be reduced to standard form by replacing it by

$$\lambda(u'x' + v'y' + w'z')\mu = 0$$

where, if $w' \neq 0$, then $\lambda w' = 1$ and if $z' \neq 0$, then $z'\mu = 1$, and similar replacements are made in the other cases. We *cannot* replace $\boldsymbol{\alpha}^T$ by $\boldsymbol{\alpha}^T\rho$ nor $\boldsymbol{\xi}$ by $\rho\boldsymbol{\xi}$, because the set satisfying $\boldsymbol{\alpha}^T\rho\boldsymbol{\xi} = 0$ is different from that satisfying $\boldsymbol{\alpha}^T\boldsymbol{\xi} = 0$. For example, take $\boldsymbol{\alpha}^T = [ij, 1, 1]$. One point of the collinear set $\boldsymbol{\alpha}^T$ is $(ij, 0, 1)$. Now replace $\boldsymbol{\alpha}^T$ by $\boldsymbol{\alpha}^T j = [-i, j, j]$. The condition that $(ij, 0, 1)$ should belong to this collinear set is that $[-i, j, j](ij, 0, 1) = 0$, but the value of the product is $j + 0 + j = 2j \neq 0$. The standard form of $[-i, j, j]$ is $-j[-i, j, j] = [-ij, 1, 1]$.

We have now to verify that the system satisfies the axioms $A\mathscr{I}$ 1, 2, and shall prove first that, given two distinct points $\boldsymbol{\pi} = (p, q, 1)$, $\boldsymbol{\pi}' = (p', q', 1)$ there is a single vector $\boldsymbol{\alpha}$ such that $\boldsymbol{\alpha}^T\boldsymbol{\pi} = 0$ and $\boldsymbol{\alpha}^T\boldsymbol{\pi}' = 0$. We consider first some special cases:

(i) If $p = 0$, $q \neq 0$, then $v = -q^{-1}$ and u is determined uniquely from the other equation unless $p' = 0$.

If $p = 0$ & $p' = 0$ the only line containing $\boldsymbol{\pi}$ and $\boldsymbol{\pi}'$ is $[1, 0, 0]$.

If $p = 0$ & $q = 0$ the line passes through $(0, 0, 1)$ and either it has co-ordinates $[u, 1, 0]$, u being determined by $up' + q' = 0$, or, if $p' = 0$, it has co-ordinates $[1, 0, 0]$.

(ii) If there is an element k, $k \notin \{0, 1\}$, such that

$$p' = pk, \quad q' = qk,$$

then the two conditions are

$$up+vq+1 = 0, \quad (up+vq)k+1 = 0$$

and therefore
$$up+vq = 0$$

and the line is
$$[-qp^{-1}, 1, 0] \quad \text{or} \quad [1, 0, 0].$$

We shall assume therefore that

$$pq \neq 0 \quad \text{and} \quad q^{-1}q' \neq p^{-1}p'.$$

Then,
$$\begin{aligned} up+vq+1 = 0 \;\&\; up'+vq'+1 = 0 \\ \Rightarrow v(qp^{-1}p'-q')+(p^{-1}p'-1) = 0 \\ \Rightarrow v = -(p^{-1}p'-1)(p^{-1}p'-q^{-1}q')^{-1}q^{-1}. \end{aligned}$$

Similarly
$$u = -(q^{-1}q'-1)(q^{-1}q'-p^{-1}p')^{-1}p^{-1}.$$

Thus, given two distinct points there is always at least one collinear set $\boldsymbol{\alpha}^T\boldsymbol{\xi} = 0$ containing them both. Moreover since the argument is in the form 'if there is such a collinear set, then $\boldsymbol{\alpha}$ satisfies certain conditions', and the final condition is in the form '$wr = s$' with $r \neq 0$, so that there is one and only one value of w, namely, $w = sr^{-1}$, it follows that the collinear set is unique.

3.5*X*1 Prove that any two collinear sets, $\boldsymbol{\alpha}^T\boldsymbol{\xi} = 0$, $\boldsymbol{\alpha}'^T\boldsymbol{\xi} = 0$, have one and only one common member.

3.5*X*2

$$\mathbf{M} = \begin{bmatrix} \rho & 0 & 0 \\ 0 & \sigma & 0 \\ h & k & 1 \end{bmatrix}, \rho \neq 0, \sigma \neq 0 \quad \text{and} \quad \mathbf{N} = \begin{bmatrix} \rho^{-1} & 0 & 0 \\ 0 & \sigma^{-1} & 0 \\ -h\rho^{-1} & -k\sigma^{-1} & 1 \end{bmatrix}.$$

Prove that (i) $\mathbf{MN} = \mathbf{NM} = \mathbf{1}$ and (ii) if $\{\boldsymbol{\xi}\}$ is a set of collinear points, so is $\{\mathbf{M}\boldsymbol{\xi}\}$.

The form introduced in Chapter 2.4 for a general point of the collinear set $\langle \boldsymbol{\pi}, \boldsymbol{\pi}' \rangle$ determined by the two points $\boldsymbol{\pi} = (p, q, 1)$ and $\boldsymbol{\pi}' = (p', q', 1)$ is still available, namely,

$$\begin{bmatrix} p & p' \\ q & q' \\ 1 & 1 \end{bmatrix} \begin{bmatrix} \kappa \\ 1 \end{bmatrix} (\kappa+1)^{-1}$$

(the standardisation factor $(\kappa+1)^{-1}$ may be omitted), since

$$\boldsymbol{\alpha}^T\boldsymbol{\pi} = 0 \;\&\; \boldsymbol{\alpha}^T\boldsymbol{\pi}' = 0 \;\Rightarrow\; \boldsymbol{\alpha}^T[\boldsymbol{\pi}, \boldsymbol{\pi}'] \begin{bmatrix} \kappa \\ 1 \end{bmatrix} = 0,$$

for every κ. That is, every point of the set determined by $\{\kappa\}$ satisfies the condition that it shall belong to the collinear set $\langle \boldsymbol{\pi}, \boldsymbol{\pi}' \rangle$.

In the same way collinear sets containing the point common to $\boldsymbol{\alpha}^{\mathrm{T}}\boldsymbol{\xi} = 0$ and $\boldsymbol{\alpha}'^{\mathrm{T}}\boldsymbol{\xi} = 0$ form the system

$$(\rho+1)^{-1}[\rho,1]\begin{bmatrix}\boldsymbol{\alpha}^{\mathrm{T}}\\ \boldsymbol{\alpha}'^{\mathrm{T}}\end{bmatrix}\boldsymbol{\xi} = 0$$

(and the standardising factor may be omitted).

Finally we prove that the geometry defined in this way by a non-commutative algebraic field is Desarguesian. We have already in fact given in Excursus 2.1.2 a proof which could be adapted, but we repeat it here to emphasize the order in which the elements in products have to be written.

3.5*T*3 The Desargues condition holds in a non-commutative field. In the Desargues figure take:

$$\mathrm{K} = \begin{bmatrix}1\\1\\1\end{bmatrix}, \quad \mathrm{A} = \begin{bmatrix}1\\0\\0\end{bmatrix}, \quad \mathrm{B} = \begin{bmatrix}0\\1\\0\end{bmatrix}, \quad \mathrm{C} = \begin{bmatrix}0\\0\\1\end{bmatrix}.$$

Since $\mathrm{A}' \in \mathrm{KA}$, A' may be taken to be

$$\mathrm{A}' = \begin{bmatrix}1 & 1\\0 & 1\\0 & 1\end{bmatrix}\begin{bmatrix}\alpha-1\\1\end{bmatrix} = \begin{bmatrix}\alpha\\1\\1\end{bmatrix},$$

where $\alpha \neq 1$, since $\# |\mathrm{KAA}'$. Similarly B' and C' may be taken as

$$\mathrm{B}' = \begin{bmatrix}1\\\beta\\1\end{bmatrix}, \quad \mathrm{C}' = \begin{bmatrix}1\\1\\\gamma\end{bmatrix}, \quad \beta \neq 1, \gamma \neq 1.$$

BC is the collinear set $[1, 0, 0]$ and a general point of $\mathrm{B}'\mathrm{C}'$ is

$$\begin{bmatrix}1 & 1\\\beta & 1\\1 & \gamma\end{bmatrix}\begin{bmatrix}\rho\\1\end{bmatrix},$$

so that L is given by the value of ρ for which

$$[1, 0, 0]\begin{bmatrix}1 & 1\\\beta & 1\\1 & \gamma\end{bmatrix}\begin{bmatrix}\rho\\1\end{bmatrix} = 0,$$

that is, $\rho+1 = 0$, and therefore

$$\mathrm{L} = \begin{bmatrix} 1 & 1 \\ \beta & 1 \\ 1 & \gamma \end{bmatrix} \begin{bmatrix} -1 \\ 1 \end{bmatrix} = \begin{bmatrix} 0 \\ \beta-1 \\ 1-\gamma \end{bmatrix}.$$

Similarly

$$\mathrm{M} = \begin{bmatrix} 1-\alpha \\ 0 \\ \gamma-1 \end{bmatrix}, \quad \mathrm{N} = \begin{bmatrix} \alpha-1 \\ 1-\beta \\ 0 \end{bmatrix}.$$

Since $\alpha \neq 1, \beta \neq 1, \gamma \neq 1$, these three points belong to the collinear set $[(\alpha-1)^{-1}, (\beta-1)^{-1}, (\gamma-1)^{-1}]$.

Thus whilst $A\mathscr{I}\,1, 2, 4$ (Pappus) $\Rightarrow$ $A\mathscr{I}\,3$ (Desargues), (3.5 T 2), it is not true that $A\mathscr{I}\,1, 2, 3 \Rightarrow A\mathscr{I}\,4$, since the Desargues condition can be satisfied whether or not the underlying field is commutative. On the other hand we have proved (3.5 T 1) that the Pappus condition is fulfilled only when the field is commutative. Moreover, unlike the Desargues condition, the Pappus condition does not depend on the restriction of the geometry to a plane: the elements of a non-commutative field can be used as co-ordinates in space of any number of dimensions.

The result that Pappus $\Rightarrow$ Desargues, but Desargues $\not\Rightarrow$ Pappus is in many respects a surprising one, since there is actually an apparently greater freedom of choice in the Desargues figure. In Desargues, after fixing $\mathscr{P}_3|$KABC, we may choose each of the three points A′, B′, C′ arbitrarily from a collinear set. On the other hand if we name the Pappus points according to the scheme:

$$\left.\begin{matrix} \mathscr{L}|\mathrm{ABC} \\ \& \ \mathscr{L}|\mathrm{A'B'C'} \end{matrix}\right\} \Rightarrow \mathscr{L}|\mathrm{A''B''C''},$$

we may begin with $\mathscr{P}_3|$AB′CB″, then may choose arbitrarily {A′, B}, A′ ∈ CB″ and B ∈ AC, and the rest of the figure is determined. The points {A, B′, C, B″} determine completely a primary net. Both A′ and B can provide extensions to the net, that is, between them they can require the adjunction of two new units to the field. This is the minimum number required for a non-commutative field. On the other hand, referring back to 3.5 T 3, we could assume that in the Desargues figure {K, A, B, C} determine the primary net and that α, β, γ are all three of them new units adjoined to the primary field. In the course of the proof however the only algebraic operations to which α, β, γ are subjected are: (i) the addition of -1 to each of them, (ii) the multiplication of each of the expressions $\alpha-1, \beta-1, \gamma-1$ by its own inverse. In neither of these operations is the commutativity of multiplication involved.

3.5*X*3 Take in the Pappus figure:

$$A = \begin{bmatrix}1\\0\\0\end{bmatrix}, \quad B' = \begin{bmatrix}0\\1\\0\end{bmatrix}, \quad C = \begin{bmatrix}0\\0\\1\end{bmatrix}, \quad B'' = \begin{bmatrix}1\\1\\1\end{bmatrix},$$

$$A' = \begin{bmatrix}0 & 1\\0 & 1\\1 & 1\end{bmatrix}\begin{bmatrix}i-1\\1\end{bmatrix}, \quad B = \begin{bmatrix}j\\0\\1\end{bmatrix},$$

where i, j are quaternion units, and prove that $\mathscr{L}|A''B''C''$.

3.5*X*4 Prove analytically for the 'simply-special' Pappus figure, in which $\mathscr{P}|AB \cap A'B' \cap A''B''$, that the theorem '$\mathscr{L}|A''B''C''$' can be deduced from $A\mathscr{I}$ 1, 2, 3, namely,

*G*1 $\mathscr{L}_3|ACB'B''$
*G*2 $V : V \in AC \ \& \ \#|AC$
*C*1 $\{A', C'\} : A' = CB'' \cap VB', C' = AB'' \cap VB'$
*C*2 $\{A'', C''\} : A'' = CB' \cap VB'', C'' = AB' \cap VB''$
*C*3 $B : B = AC \cap A'C''$
T $\mathscr{L}|BA''C'$.

In this case, after the primary net has been fixed by {A, C, B′, B″}, only one point, V, is selected randomly in a collinear set, and the rest of the figure then obtained by uniquely constructed points. The argument may therefore be adapted to prove that the 'simply-special' Pappus condition is a consequence of $A\mathscr{I}$ 1, 2, 3*.

EXCURSUS ON BOOK 3

EXCURSUS 3.1

SOME EXTENSIONS OF FINITE PRIMARY FIELDS

The main theme of this excursus is the quadratic extension of $\mathfrak{F}_3$ and $\mathfrak{F}_5$ by the adjunction of the square root of a non-square, but we discuss first the extension of $\mathfrak{F}_2$, to which, since there are no non-squares, this method does not apply.

Exc. 3.1.1 $\mathfrak{F}_2$ to $\mathfrak{F}_4$ and $\mathfrak{F}_8$

In geometric terms we are extending the net $\mathfrak{N}_2(\mathrm{O, E})$ by adjoining a point F, and then assigning a new unit $\mathfrak{f} = \mathrm{s(OF/OE)}$ to the corresponding algebra, so as to obtain the field $\mathfrak{F}_2(1, \mathfrak{f})$. Assume first that the extension is to be quadratic, that is, geometrically $\mathscr{M}_{\mathrm{F}}\mathrm{F} \in \mathfrak{N}(\mathrm{O, E, F})$, and algebraically $\mathfrak{f}^2 \in \mathfrak{F}(1, \mathfrak{f})$. The elements in $\mathfrak{F}(1, \mathfrak{f})$ are $0, 1, \mathfrak{f}, 1+\mathfrak{f}$, and the only one of these which leads to a non-trivial extension of the field is $1+\mathfrak{f}$. Assume then

$$1+\mathfrak{f}+\mathfrak{f}^2 = 0.$$

There are then four proper points in the net $\mathfrak{N}_2(\mathrm{O, E, F})$, namely, O, E, F, $\mathrm{F}^{(2)}$, where $\mathrm{F}^{(2)} = \mathscr{M}_{\mathrm{F}}\mathrm{F} = \mathscr{D}_{\mathrm{F}}\mathrm{E}$.

Assume next that $\mathrm{F}^{(2)} \notin \mathfrak{N}(\mathrm{O, E, F})$ but that $\mathrm{F}^{(3)} \in \mathfrak{N}(\mathrm{O, E, F, F^{(2)}})$. The new unit $\mathfrak{f}$ satisfies some relation

$$\mathfrak{f}^3+\alpha\mathfrak{f}^2+\beta\mathfrak{f}+\gamma = 0, \quad (\alpha, \beta, \gamma) \in \{0, 1\},$$

which, by a slight change of basis ($\mathfrak{f}+\alpha$ for $\mathfrak{f}$), can be reduced to

$$\mathfrak{f}^3+\beta'\mathfrak{f}+\gamma' = 0.$$

This relation is to be irreducible, so that clearly $\gamma' \neq 0$; also $\beta' \neq 0$, since $\mathfrak{f}^3+1 = (\mathfrak{f}+1)(\mathfrak{f}^2+\mathfrak{f}+1)$, so that the only non-trivial relation is

$$\mathfrak{f}^3+\mathfrak{f}+1 = 0.$$

We have also therefore

$$(\mathfrak{f}^4+\mathfrak{f}^2+\mathfrak{f}+1)(\mathfrak{f}^3+\mathfrak{f}+1) = 0,$$

that is
$$\mathfrak{f}^7 = 1.$$

We have obtained therefore the eight proper points of $\mathfrak{N}_2(\mathrm{O}, \mathrm{E}, \mathrm{F}, \mathrm{F}^{(2)})$, namely, in addition to the basis,

$$\begin{aligned} &\mathscr{M}_{\mathrm{F}}^{3}\mathrm{E} = \mathrm{F}^{(3)} = \mathscr{D}_{\mathrm{F}}\mathrm{E} && (\mathfrak{f}^3 = \mathfrak{f}+1)\\ &\mathscr{M}_{\mathrm{F}}^{4}\mathrm{E} = \mathrm{F}^{(4)} = \mathscr{D}_{\mathrm{F}^{(2)}}\mathrm{F} && (\mathfrak{f}^4 = \mathfrak{f}^2+\mathfrak{f})\\ &\mathscr{M}_{\mathrm{F}}^{5}\mathrm{E} = \mathrm{F}^{(5)} = \mathscr{D}_{\mathrm{F}^{(2)}}\mathscr{D}_{\mathrm{F}}\mathrm{E} && (\mathfrak{f}^5 = \mathfrak{f}^3+\mathfrak{f}^2 = \mathfrak{f}^2+\mathfrak{f}+1)\\ &\mathscr{M}_{\mathrm{F}}^{6}\mathrm{E} = \mathrm{F}^{(6)} = \mathscr{D}_{\mathrm{F}^{(2)}}\mathrm{E} && (\mathfrak{f}^6 = \mathfrak{f}^4+\mathfrak{f}^3 = \mathfrak{f}^2+1)\\ \text{while}\quad &\mathscr{M}_{\mathrm{F}}^{7}\mathrm{E} = \mathrm{F}^{(7)} = \mathrm{E} && (\mathfrak{f}^7 = \mathfrak{f}^5+\mathfrak{f}^4 = 1). \end{aligned}$$

Exc. 3.1.2 $\mathfrak{F}_3(1)$ to $\mathfrak{F}_3(1, \mathfrak{f})$

In $\mathfrak{F}_3(1)$ there is one non-square, namely, -1, so that to obtain the first extension into the surdic net we adjoin to the field a new unit $\mathfrak{f}$ with $\mathfrak{f}^2 = -1$. The nine elements of $\mathfrak{F}_3(1, \mathfrak{f})$ are $\{0, \pm 1, \pm\mathfrak{f}, \pm 1 \pm \mathfrak{f}\}$. Now take

$$\mathrm{j} = 1-\mathfrak{f}$$

so that

$$\mathrm{j}^2 = \mathfrak{f}, \quad \mathrm{j}^3 = \mathfrak{f}\mathrm{j} = 1+\mathfrak{f}, \quad \mathrm{j}^4 = -1;$$

the elements may be expressed as $\{0, \pm 1, \pm\mathrm{j}, \pm\mathrm{j}^2, \pm\mathrm{j}^3\}$ or as $\{0, 1, \mathrm{j}, \ldots, \mathrm{j}^7\}$. Each set of three elements is linearly related; the relations may be determined from the relations

$$\mathrm{j}^4 = -1, \quad \mathrm{j}^{\mathrm{r}}+\mathrm{j}^{\mathrm{r}-1}-\mathrm{j}^{\mathrm{r}-2} = 0,$$

derived from

$$\mathrm{j}^2+\mathrm{j} = \mathfrak{f}+(1-\mathfrak{f}) = 1.$$

In the geometry we may therefore name the proper points of $\mathfrak{N}_3(\mathrm{O}, \mathrm{E}, \mathrm{F})$ as $\mathrm{O}, \mathrm{J}_0 = \mathrm{E}, \mathrm{J}_1, \mathrm{J}_2 = \mathrm{F}, \mathrm{J}_3, \ldots, \mathrm{J}_7$, where

$$\mathscr{M}_{\mathrm{J}_{\mathrm{r}}}\mathrm{J}_{\mathrm{s}} = \mathrm{J}_{\mathrm{r+s}} \quad\text{and}\quad \mathscr{D}_{\mathrm{J}_{\mathrm{r}}}\mathrm{J}_{\mathrm{s}} = \mathrm{J}_{\mathrm{t}} \quad\text{and}\quad \mathscr{D}_{\mathrm{J}_{\mathrm{r}}}\mathrm{J}_{\mathrm{r}+4} = \mathrm{O},$$

where t is determined from r, s by the relations [illegible] the powers of j above.

The squares in $\mathfrak{F}_3(1, \mathfrak{f})$ are $\{0, 1, \mathrm{j}^2, \mathrm{j}^4 = -1, \mathrm{j}^6 = -\mathrm{j}^2\}$ so that the next extension into the surdic net would be obtained by adjoining a new unit $\mathfrak{g}$ with $\mathfrak{g}^2 = \mathrm{j}$. The 81 elements of the net are then

$$\mathrm{a}+\mathrm{b}\mathfrak{g}+\mathrm{c}\mathfrak{g}^2+\mathrm{d}\mathfrak{g}^3, \quad (\mathrm{a}, \mathrm{b}, \mathrm{c}, \mathrm{d}) \in \{0, 1, -1\}$$

with $\mathfrak{g}^4 = -1$.

There is an extension to a set of 27 elements which cannot be formed by adjoining a new element to $\mathfrak{F}(1, \mathfrak{f})$, but which is derived by adjoining a single new unit $\mathfrak{h}$, such that $\mathfrak{h}^3 \in \mathfrak{F}(1, \mathfrak{h}, \mathfrak{h}^2)$.

Following the pattern set in $\mathfrak{F}_2$ we may avail ourselves of the cyclic generation of the planar net $\mathfrak{N}_3$ established in Excursus 2.3.3, based on a matrix $\mathbf{K}$ such that $\mathbf{K}^{13} = \mathbf{1}$. Take $\mathfrak{h} = -\mathbf{K}$, so that $\mathfrak{h}^{13} = -1$, $-\mathfrak{h}^{\mathrm{r}} = \mathfrak{h}^{\mathrm{r}+13}$, $\mathfrak{h}^{26} = 1$. For the matrices $\mathbf{K}$ we have the relation

$$-\mathbf{1}+\mathbf{K}+\mathbf{K}^2+\mathbf{K}^3 = \mathbf{0},$$

and, therefore, among the units $\mathfrak{h}$, there is the relation

$$\mathfrak{h}^3-\mathfrak{h}^2+\mathfrak{h}+1 = 0.$$

From this we obtain

$$\mathfrak{h}^4 = \mathfrak{h}^3-\mathfrak{h}^2-\mathfrak{h} = \mathfrak{h}-1$$

corresponding to the three-term relation $\mathbf{1}+\mathbf{K}+\mathbf{K}^4 = \mathbf{0}$. We may therefore adapt the cyclic table of points $\mathscr{K}^r P_0$ to find the sum of any two powers of $\mathfrak{h}$. In the following table

.	+	+	+	1	...	$\mathfrak{h}^r$	...	$\mathfrak{h}^{26}$
+	.	+	−	$\mathfrak{h}$	...	$\mathfrak{h}^{r+1}$	...	1
+	−	.	+	$\mathfrak{h}^4$	...	$\mathfrak{h}^{r+4}$	...	$\mathfrak{h}^3$
−	−	+	.	$\mathfrak{h}^6$	...	$\mathfrak{h}^{r+6}$	...	$\mathfrak{h}^5$

the sum of any three of the terms in any column, with the signs listed in front of the table, is zero, a term $-\mathfrak{h}^r$ being replaced by $\mathfrak{h}^{r+13}$.

Exc. 3.1.3 A quadratic extension of $\mathfrak{F}_5$

In $\mathfrak{F}_5(1)$, $1^2 = 4^2 = 1$, $2^2 = 3^2 = 4$, and 2 and 3 are the non-squares. The first extension into the surdic net is therefore obtained by adjoining a unit $\mathfrak{f}$ with $\mathfrak{f}^2 = 2$. Then $\mathfrak{f}^4 = 4$, $\mathfrak{f}^6 = 3$, $\mathfrak{f}^8 = 1$. The twenty-five elements of the extended field $\mathfrak{F}_5(1, \mathfrak{f})$ are $a+b\mathfrak{f}$, $(a, b) \in \{0, 1, 2, 3, 4\}$.

In order to express the elements as powers of a single unit, $\mathfrak{j}$, we have to find a function of $\mathfrak{f}$ such that $\mathfrak{j}^3 =$ (an odd power of $\mathfrak{f}$) and therefore $\mathfrak{j}^{24} = 1$. We are led towards a suitable choice by considering the squares in the extended field. We have $(1+\mathfrak{f})^2 = 3+2\mathfrak{f} = \mathfrak{f}^3(1+\mathfrak{f}^3)$. $\mathfrak{f}^3$ is a non-square and therefore $(1+\mathfrak{f}^3)$ is a non-square; thus the squares in $\mathfrak{F}_5(1, \mathfrak{f})$ are

$$\mathfrak{f}^{2r}, \quad \mathfrak{f}^{2r+1}(1+\mathfrak{f}^3), \quad \mathfrak{f}^{2r+1}(1+\mathfrak{f}^3)^{-1}, \quad r \in \{0, 1, 2, 3\}.$$

We take then

$$\mathfrak{j} = 1+\mathfrak{f}^3.$$

We find

$$\mathfrak{j}^3 = 1+3\mathfrak{f}^3+3\mathfrak{f}^6+\mathfrak{f}^9 = 3\mathfrak{f}^3+\mathfrak{f}$$
$$= 2\mathfrak{f} = \mathfrak{f}^3,$$

and can easily compute other powers of $\mathfrak{j}$ in the form $a+b\mathfrak{f}$. They form the following table:

$a+b\mathfrak{f}$ a \ b	0	1	2	3	4
0	0	$\mathfrak{j}^9$	$\mathfrak{j}^3$	$\mathfrak{j}^{15}$	$\mathfrak{j}^{21}$
1	1	$\mathfrak{j}^{14}$	$\mathfrak{j}$	$\mathfrak{j}^5$	$\mathfrak{j}^{22}$
2	$\mathfrak{j}^{18}$	$\mathfrak{j}^{23}$	$\mathfrak{j}^8$	$\mathfrak{j}^{16}$	$\mathfrak{j}^{19}$
3	$\mathfrak{j}^6$	$\mathfrak{j}^7$	$\mathfrak{j}^4$	$\mathfrak{j}^{20}$	$\mathfrak{j}^{11}$
4	$\mathfrak{j}^{12}$	$\mathfrak{j}^{10}$	$\mathfrak{j}^{17}$	$\mathfrak{j}^{13}$	$\mathfrak{j}^2$

From this table we may compute immediately $\mathfrak{j}^r + \mathfrak{j}^{r+s}$, since $1 + \mathfrak{j}^s$ is in the same column as and in the row below $\mathfrak{j}^s$ in the table. For example,

$$1 + (1 + 3\mathfrak{f}) = 2 + 3\mathfrak{f} \Rightarrow 1 + \mathfrak{j}^5 = \mathfrak{j}^{16},$$

and, therefore, since $\mathfrak{j}^{12} = -1$,

$$\mathfrak{j}^r + \mathfrak{j}^{r+4} + \mathfrak{j}^{r+5} = 0.$$

We may now name the 25 proper points of the linear net $\mathfrak{N}_5(O, E, F)$ as $O, J_0(=E), J_1, \ldots, J_9(=F), \ldots, J_{24}$, in such a way that

$$\mathscr{M}_{J_r} J_s = J_{r+s} \quad \text{and} \quad \mathscr{D}_{J_l} J_m = J_n,$$

where n is deducible from l and m by using the table of powers.

Let us look briefly at an extension in which $\mathfrak{f}^2 \notin \mathfrak{F}_5(1, \mathfrak{f})$. Suppose, for example

$$\mathfrak{g}^3 \in \mathfrak{F}_5(1, \mathfrak{g}, \mathfrak{g}^2)$$

and is given by

$$\mathfrak{g}^3 = \mathfrak{g} + 2.$$

This is a possible relation, since $x^3 - x - 2$ is irreducible in $\mathfrak{F}_5(1)$—that is, $x^3 - x - 2 = 0$ has no roots in the field. From this relation we could express every power of $\mathfrak{g}$ as $a + b\mathfrak{g} + c\mathfrak{g}^2$, where $(a, b, c) \in \{0, 1, 2, 3, 4\}$. It can be verified, for example, that

$$\mathfrak{g}^4 = \mathfrak{g}^2 + 2\mathfrak{g}, \quad \mathfrak{g}^5 = 2\mathfrak{g}^2 + \mathfrak{g} + 2, \quad \mathfrak{g}^6 = \mathfrak{g}^2 - \mathfrak{g} - 1,$$

$$\mathfrak{g}^7 = 2 - \mathfrak{g}^2, \quad \mathfrak{g}^{31} = 2, \quad \mathfrak{g}^{62} = 4, \quad \mathfrak{g}^{93} = 3, \quad \mathfrak{g}^{124} = 1.$$

EXCURSUS 3.2

VARIOUS EXTENSIONS OF FINITE PLANAR PRIMARY NETS

In this Excursus we use some of the extended fields for which we have constructed cyclic representations in Excursus 3.1 to build up the corresponding geometries.

Exc. 3.2.1 $\Gamma^{(2)}$ to $\Gamma^{(4)}$

We take the elements of $\mathfrak{F}_2(1,\omega)$ to be $\{0, 1, \omega, \omega^2\}$ with $1+\omega+\omega^2 = 0$, and seek a matrix $\mathbf{M}$ over this field which is such that for $n = 21$ and for no smaller value, $\mathbf{M}^n = \mathbf{1}$ or $\omega\mathbf{1}$ or $\omega^2\mathbf{1}$. We have therefore to find a matrix such that the equation $\det(\rho\mathbf{1}-\mathbf{M}) = 0$ has no solutions in the field, and which is not such that its cube or seventh power is a multiple of the identity matrix. For the convenience in manipulation that such matrices present, we try to choose a matrix of the pattern

$$\begin{bmatrix} 0 & 1 & 0 \\ 1 & 0 & \alpha \\ 0 & \alpha & \beta \end{bmatrix} \quad \text{and find fairly easily that} \quad \mathbf{M} = \begin{bmatrix} 0 & 1 & 0 \\ 1 & 0 & \omega^2 \\ 0 & \omega^2 & \omega \end{bmatrix}$$

fulfils all the requirements.†

Starting from $P_0 = \boldsymbol{\epsilon}_1 = (1, 0, 0)$, the powers of this matrix produce the following sequence of points P_r:

r				r				r				r			
0	1	0	0												
1	0	1	0	6	ω^2	ω	1	11	1	0	1	16	1	ω	0
2	1	0	ω^2	7	ω	0	ω^2	12	0	ω	ω	17	ω	1	1
3	0	ω^2	1	8	0	0	1	13	ω	1	ω	18	1	1	1
4	ω^2	ω^2	0	9	0	ω^2	ω	14	1	ω^2	0	19	1	ω	1
5	ω^2	ω^2	ω	10	ω^2	1	1	15	ω^2	1	ω	20	ω	ω	ω^2
												21	ω	0	0

By the same argument as was used in Excursus 2.3, the linear relations among the above vectors are also the linear relations among the powers of the matrices. We see, then, that

$$\mathbf{1}+\mathbf{M}+\omega\mathbf{M}^4 = \mathbf{0},$$
$$\mathbf{1}+\omega\mathbf{M}+\mathbf{M}^{16} = \mathbf{0},$$
$$\mathbf{1}+\omega^2\mathbf{M}+\mathbf{M}^{14} = \mathbf{0}.$$

† Notice that because $\omega^3 = 1$ we cannot adjust the matrix with a scalar multiplier such that $\det(k\mathbf{M}) = 1$.

These relations lead to the difference square

0	1	4	14	16.	Collinear set	0	0	1
20	0	3	13	15		0	ω	1
17	18	0	10	12		0	1	1
7	8	11	0	2		0	1	0
5	6	9	19	0		0	ω^2	1

which is the 'reciprocal' of the arrangement used in Excursus 1.1.4.

One of the systems $\Gamma^{(2)}$ which forms a subset of $\Gamma^{(4)}$ is provided by the points whose co-ordinates take the values 0 and 1 only, namely, $P_0, P_1, P_4, P_8, P_{11}, P_{12}, P_{18}$, but one more directly related to $\Gamma^{(4)}$ consists of the points P_{3r}, $= (\mathbf{M}^3)^r \epsilon_1$. Associated with this set are the two sets $\mathbf{M}^{3r+1}\epsilon_1$, and $\mathbf{M}^{3r+2}\epsilon_1$, each of which also forms a system $\Gamma^{(2)}$.

The pattern can be exhibited more clearly by rearranging the table (so that it is no longer quite a difference table) in the following way:

	3r		3r + 1	3r + 2	
0	3	15	13	20	[0, ω, 1]
18	0	12	10	17	[0, 1, 1]
6	9	0	19	5	[0, ω^2, 1]
1	4	16	7	(0)	[0, 0, 1]
8	11	2	(0) 14		[0, 1, 0]

The arrangement of the points in collinear sets is naturally the same for all three systems $\Gamma^{(2)}$, namely,

r = 0	1	2	3	4	5	6
2	3	4	5	6	0	1
3	4	5	6	0	1	2

Exc. 3.2.2 $\Gamma^{(2)}$ to $\Gamma^{(8)}$

We take the elements of the underlying field to be $0, 1, \alpha, \ldots, \alpha^6$ with $\alpha^7 = 1$, the set of relations among the powers of α being given by the cyclic pattern

$$\alpha^r + \alpha^{r+1} + \alpha^{r+3} = 0.$$

The number of points in $\Gamma^{(8)}$ is 73 so that unless a projectivity matrix $\mathbf{M}$ leaves a point of $\Gamma^{(8)}$ fixed it will have the required property that $\mathbf{M}^{73} = \alpha^r \mathbf{1}$. In this case it is of some advantage to investigate systematically the matrices of a standard pattern in order to pick out those which are suitable.

Take

$$\mathbf{M} = \begin{bmatrix} 0 & 1 & 0 \\ 1 & 0 & \alpha^r \\ 0 & \alpha^r & \alpha^s \end{bmatrix}.$$

The equation
$$\det(\rho\mathbf{1}-\mathbf{M}) = 0$$
is
$$\rho^3+(\alpha^{2r}+1)\rho+\alpha^s(\rho^2+1) = 0.$$

Omit to begin with the cases in which $r = 0$ and write $\alpha^u = \alpha^{2r}+1$; this relation gives the following pairs of values

$$\begin{matrix} r = 1 & 2 & 3 & 4 & 5 & 6, \\ u = 6 & 5 & 2 & 3 & 1 & 4. \end{matrix}$$

Substitute for ρ in the function $\rho^3+\alpha^u\rho+\alpha^s(\rho^2+1)$ the various powers of α in turn. For $\rho = \alpha$ we obtain $\alpha^3+\alpha^{u+1}+\alpha^{s+6}$, so that any pair of values (u, s) for which this vanishes is not suitable. In this way we obtain six sets of restrictive conditions, namely,

$$1+\alpha^{u+a}+\alpha^{s+b} = 0$$

for the pairs of values (a, b) = (5, 3), (3, 6), (1, 0), (6, 5), (4, 0), (2, 0).

Since $\det \mathbf{M} = \alpha^s$, it is desirable if possible to choose $s = 0$. The equation then becomes
$$(\alpha^{2r}+1)\rho = \rho^3+\rho^2+1.$$

Regarding this as a relation between r and ρ, we obtain the table of values:

$$\begin{matrix} \rho = 1 & \alpha & \alpha^2 & \alpha^3 & \alpha^4 & \alpha^5 & \alpha^6 \\ \rho^3+\rho^2+1 = 1 & \alpha^4 & \alpha & 0 & \alpha^2 & 0 & 0 \\ \alpha^{2r}+1 = 1 & \alpha^3 & \alpha^6 & 0 & \alpha^5 & 0 & 0 \\ \alpha^r = & \alpha^4 & \alpha & 1 & \alpha^2 & 1 & 1 \end{matrix}$$

The permissible set of values of r is therefore {3, 5, 6}; we choose $\alpha^r = \alpha^6$ $(=\alpha^{-1})$, as being likely to be the simplest for manipulation; that is, we choose

$$\mathbf{M} = \begin{bmatrix} 0 & 1 & 0 \\ 1 & 0 & \alpha^6 \\ 0 & \alpha^6 & 1 \end{bmatrix},$$

for which $\det(\rho\mathbf{1}-\mathbf{M}) = \rho^3+\alpha^4\rho+(\rho^2+1)$ and is irreducible. From $\mathbf{M}^r\boldsymbol{\epsilon}_1 = \mathrm{P}_r$ we find

$$\mathrm{P}_0 = (1, 0, 0),\quad \mathrm{P}_1 = (0, 1, 0),\quad \mathrm{P}_2 = (1, 0, \alpha^6),\quad \mathrm{P}_3 = (0, \alpha^4, \alpha^6),$$

so that the basic relation among the powers of the matrix is

$$\mathbf{M}^3 = \mathbf{M}^2+\alpha^4\mathbf{M}+\mathbf{1}.$$

(If we had approached this problem differently, we should have taken a cubic extension of $\mathfrak{F}_2$ using the relation

$$j^3 = j^2+\alpha^4 j+1.)$$

By actual computation it will be found that this matrix gives the difference table

0	1	5	12	18	21	49	51	59	line [0, 0, 1]
72	0	4	11	17	20	48	50	58	$[0, \alpha^6, 1]$
68	69	0	7	13	16	44	46	54	$[0, \alpha^4, 1]$
61	62	66	0	6	9	37	39	47	[0, 1, 1]
55	56	60	67	0	3	31	33	41	$[0, \alpha^2, 1]$
52	53	57	64	70	0	28	30	38	$[0, \alpha^3, 1]$
24	25	29	36	42	45	0	2	10	[0, 1, 0]
22	23	27	34	40	43	71	0	8	$[0, \alpha^5, 1]$
14	15	19	26	32	35	63	65	0	$[0, \alpha, 1]$.

The 'basic' $\Gamma^{(2)}$ in this system, consisting of points with co-ordinates {0, 1} only, is:

$$\begin{matrix} 100 & 010 & 001 & 111 & 011 & 101 & 110 \\ P_0 & P_1 & P_{25} & P_{61} & P_{37} & P_{42} & P_{49} \end{matrix}$$

Any set derived from this by a projectivity is of course also a set $\Gamma^{(2)}$, in particular every power of **M** operating on this set gives such a set. Further, $\mathbf{M}^2$, $\mathbf{M}^4$, $\mathbf{M}^6$, $\mathbf{M}^8$, $\mathbf{M}^{10}$ and possibly others give sets which are mutually disjoint and disjoint from the basic set. It would appear to be quite a problem to find out what is the greatest possible number of disjoint sets, $\Gamma^{(2)}$.

Exc. 3.2.3 $\Gamma^{(3)}$ to $\Gamma^{(9)}$

Take the elements of the underlying field $\mathfrak{F}(1, \mathfrak{f})$ (that is, GF(3^2)) to be $\{0, \pm 1, \pm \mathrm{j}, \pm \mathrm{j}^2, \pm \mathrm{j}^3\}$, where

$$\mathrm{j}^2 + \mathrm{j} - 1 = 0 \;\&\; \mathrm{j}^4 = -1,$$

and consider matrices of the form

$$\mathbf{M}_{rs} = \begin{bmatrix} 0 & 1 & 0 \\ 1 & 0 & \mathrm{j}^r \\ 0 & \mathrm{j}^r & \mathrm{j}^s \end{bmatrix}.$$

By an investigation similar to that carried out for $\Gamma^{(8)}$, we find that among the polynomials $\det(\rho\mathbf{1} - \mathbf{M}_{rs})$ that are irreducible are $\mathbf{M}_{02}$ and $\mathbf{M}_{12}$, but that $\mathbf{M}_{02}^{13} = \mathrm{j}^2\mathbf{1}$, while $\mathbf{M}_{12}^{n} = \mathrm{j}^t\mathbf{1}$ for n = 91 and for no lower value. We select initially, therefore, as the matrix to generate the cycle of 91 points in $\Gamma^{(9)}$:

$$\mathbf{M} = -\mathrm{j}^2 \begin{bmatrix} 0 & 1 & 0 \\ 1 & 0 & \mathrm{j} \\ 0 & \mathrm{j} & \mathrm{j}^2 \end{bmatrix},$$

where the factor $-j^2$ is inserted so that $\det \mathbf{M} = 1$. From this matrix, by calculating the vectors $\mathbf{M}^r \boldsymbol{\epsilon}_1$, we derive the difference table based on

$$0 \quad 1 \quad 27 \quad 33 \quad 49 \quad 63 \quad 72 \quad 74 \quad 84 \quad 87$$

The core system, $\Gamma^{(3)}$, of points with co-ordinates $\{0, \pm 1\}$, arranged in the order of the points $\mathbf{K}^r \boldsymbol{\epsilon}_1$, where

$$\mathbf{K} = \begin{bmatrix} 0 & 1 & 0 \\ 1 & 0 & 1 \\ 0 & 1 & -1 \end{bmatrix}$$

(Excursus 2.3.3), is

$$P_0 \quad P_1 \quad P_{82} \quad P_3 \quad P_{27} \quad P_8 \quad P_{84} \quad P_{80} \quad P_{10} \quad P_{12} \quad P_{20} \quad P_{13} \quad P_{56}.$$

Our next objective is to rearrange the system $\Gamma^{(9)}$ so that the points of this core system occupy the positions 0, 7, 14, ..., 84 in the sequence. First we observe that if, in relation to any geometry $\Gamma^{(q)}$, $\mathbf{R}$ is any matrix with elements in $\mathfrak{F}_q$ which is such that the least index for which $\mathbf{R}^n = \kappa \mathbf{1}$ is $n = q^2+q+1$, and $\mathbf{S}$ is any non-singular matrix in $\mathfrak{F}_q$, then $\bar{\mathbf{R}} = \mathbf{S}^{-1}\mathbf{R}\mathbf{S}$ is a matrix with the same property, since $\bar{\mathbf{R}}^t = \mathbf{S}^{-1}\mathbf{R}^t\mathbf{S}$ and therefore $\mathbf{R}^n = \kappa\mathbf{1} \Leftrightarrow \bar{\mathbf{R}}^n = \kappa\mathbf{1}$. The set $\{(\mathbf{S}^{-1}\mathbf{R}\mathbf{S})^r \boldsymbol{\epsilon}_1\}$ is the set $\{\mathbf{R}^r \boldsymbol{\epsilon}_1\}$ arranged in a different order.

Next, if $\mathbf{H} = \mathbf{M}^7$, then the points $\mathbf{H}^r \boldsymbol{\epsilon}_1$ form a set $\Gamma^{(3)}$, and consequently we shall have achieved our aim if we can find a matrix $\mathbf{S}$ such that

$$\mathbf{H} = \mathbf{S}^{-1}\mathbf{K}^a\mathbf{S} = \mathbf{S}^{-1}\mathbf{L}\mathbf{S} \quad \text{(say)}$$

for some value of a.

We have therefore to investigate the conditions satisfied by matrices $\mathbf{H}$ and $\mathbf{L}$ related in this way. We have

$$\begin{aligned} \det(\rho\mathbf{1} - \mathbf{H}) &= \det(\rho\mathbf{1} - \mathbf{S}^{-1}\mathbf{L}\mathbf{S}) = \det[\mathbf{S}^{-1}(\rho\mathbf{1} - \mathbf{L})\,\mathbf{S}] \\ &= \det\mathbf{S}^{-1}\det\mathbf{S}\det(\rho\mathbf{1} - \mathbf{L}) \\ &= \det(\rho\mathbf{1} - \mathbf{L}). \end{aligned}$$

That is, given $\mathbf{H}$, $\mathbf{L}$, we cannot find $\mathbf{S}$ unless the two polynomials $\det(\rho\mathbf{1} - \mathbf{H})$, $\det(\rho\mathbf{1} - \mathbf{L})$ are identical (but we have not proved, nor is it always true, that if the polynomials are identical then a matrix such as $\mathbf{S}$ can be formed).

$$\mathbf{H} = \mathbf{M}^7 = -j\begin{bmatrix} j^2 & 1 & 1 \\ 1 & 1 & 1 \\ 1 & 1 & 0 \end{bmatrix},$$

$$\begin{aligned} \eta = \det(\rho\mathbf{1} - \mathbf{H}) &= \begin{vmatrix} \rho+j^3 & j & j \\ j & \rho+j & j \\ j & j & \rho \end{vmatrix} \\ &= \rho^3 - \rho^2 - \rho - 1. \end{aligned}$$

We therefore look at the powers of $\mathbf{K}$ to see for which of them, if any, $\det(\rho\mathbf{1}-\mathbf{K}^{a}) = \eta$; we find that it is so for $a \in \{4, 12, 10\}$. We may choose any one of these; take

$$\mathbf{H} = \mathbf{S}^{-1}\mathbf{K}^{10}\mathbf{S},$$

so that $$\mathbf{M}^{28} = \mathbf{H}^{4} = \mathbf{S}^{-1}\mathbf{K}^{40}\mathbf{S} = \mathbf{S}^{-1}\mathbf{K}\mathbf{S}.$$

We have now to find a matrix $\mathbf{S}$ such that

$$\mathbf{S}\mathbf{M}^{7} = \mathbf{K}^{10}\mathbf{S},$$

that is, we have to solve, if they are consistent, the equations

$$\begin{bmatrix} a_1 & b_1 & c_1 \\ a_2 & b_2 & c_2 \\ a_3 & b_3 & c_3 \end{bmatrix}\begin{bmatrix} -j^3 & -j & -j \\ -j & -j & -j \\ -j & -j & 0 \end{bmatrix} = \begin{bmatrix} 0 & 0 & -1 \\ 0 & -1 & 1 \\ -1 & 1 & -1 \end{bmatrix}\begin{bmatrix} a_1 & b_1 & c_1 \\ a_2 & b_2 & c_2 \\ a_3 & b_3 & c_3 \end{bmatrix}.$$

In fact, there are many solutions, we may choose $a_1 = 1$, $b_1 = c_1 = 0$ and find then

$$\mathbf{S} = \begin{bmatrix} 1 & 0 & 0 \\ -1 & -1 & -j \\ j^3 & j & j \end{bmatrix}.$$

Finally, take $$\mathbf{N} = \mathbf{S}\mathbf{M}^{4}\mathbf{S}^{-1},$$

So that, $$\mathbf{N}^{7} = \mathbf{S}\mathbf{M}^{28}\mathbf{S}^{-1} = \mathbf{S}\mathbf{H}^{4}\mathbf{S}^{-1} = \mathbf{K}.$$

That is, the matrix

$$\mathbf{N} = \begin{bmatrix} 1 & 0 & 0 \\ -1 & -1 & -j \\ j^3 & j & j \end{bmatrix}\begin{bmatrix} j^3 & -1 & j^3 \\ -1 & j^2 & j \\ j^3 & j & j^2 \end{bmatrix}\begin{bmatrix} 1 & 0 & 0 \\ 1 & j^2 & j^2 \\ -j^3 & -j^2 & -j \end{bmatrix}$$

$$= -j^2\begin{bmatrix} 1 & -j & j^3 \\ -j & -j & 1 \\ j^3 & 1 & 0 \end{bmatrix}.$$

It can be checked that $\mathbf{N}^{7} = \mathbf{K}$. (We have in fact found, by a somewhat devious method, the 'seventh root' of $\mathbf{K}$ over GF(9)!)

Since the matrices $\mathbf{N}^{r}$ and $\mathbf{M}^{r}$ arrange the points of $\Gamma^{(9)}$ in different orders we have to calculate a new difference table for the set $\mathbf{N}$. We find one of the generating lines of the table to be

0 6 7 25 28 30 38 42 71 82

We rearrange this line to bring the numbers 7r together and take it to be the first line of the table; the structure is much more clearly displayed if we write the sequence of numbers in the form r^s, where

$$r^s = 7r+s, \quad r \in \{0, 1, \ldots, 12\}, \quad s \in \{0, 1, \ldots, 6\}.$$

0^0	1^0	4^0	6^0	0^6	11^5	3^4	5^3	4^2	10^1
12^0	0^0	3^0	5^0	12^6	10^5	2^4	4^3	3^2	9^1
9^0	10^0	0^0	2^0	9^6	7^5	12^4	1^3	0^2	6^1
7^0	8^0	11^0	0^0	7^6	5^5	10^4	12^3	11^2	4^1
12^1	0^1	3^1	5^1	0^0	10^6	2^5	4^4	3^3	9^2
1^2	2^2	5^2	7^2	2^1	0^0	4^6	6^5	5^4	11^3
9^3	10^3	0^3	2^3	10^2	8^1	0^0	1^6	0^5	6^4
7^4	8^4	11^4	0^4	8^3	6^2	11^1	0^0	11^6	4^5
8^5	9^5	12^5	1^5	9^4	7^3	12^2	1^1	0^0	5^6
2^6	3^6	6^6	8^6	3^5	1^4	6^3	8^2	7^1	0^0

The superscripts, s (the remainder on division by 7), pick out a set of seven disjoint systems $\Gamma^{(3)}$ of which $\Gamma^{(9)}$ may be compounded. The set [K] is in the top left corner.

In Excursus 3.1.2 we saw that we could organise the elements of $\mathfrak{F}_3(1, \mathfrak{f})$ as 'reals', 0, 1, -1, and pairs of 'conjugate complexes' (j, j^3), (j^2, j^6), (j^5, j^7). The pattern of the co-ordinates of the points of $\Gamma^{(9)}$ is displayed much more clearly if we pick out these pairs of conjugates and write:

$$\lambda = j, \quad \bar{\lambda} = j^3; \qquad \lambda+\bar{\lambda} = -1, \quad \lambda\bar{\lambda} = -1;$$
$$\mu = j^6, \quad \bar{\mu} = j^2; \qquad \mu+\bar{\mu} = 0, \quad \mu\bar{\mu} = 1;$$
$$\nu = j^7, \quad \bar{\nu} = j^5; \qquad \nu+\bar{\nu} = 1, \quad \nu\bar{\nu} = -1.$$

(In the tables of co-ordinates we shall also write, but only as a matter of convenience, $\bar{1} = -1$.)

The 91 points of $\Gamma^{(9)}$ split into seven sets of 13 points, each forming a system $\Gamma^{(3)}$, namely, [K], consisting of all the 'real' points, and three pairs {[L], [L′]}, {[M], [M′]}, {[N], [N′]} in which we shall find that the co-ordinates of the points of one member of a pair, when written in standard form, are the 'conjugate complexes' of those of the points of the other. The whole system can be displayed in the table on page 262. In this table (in which the two halves are to be considered together) the points in the same row, r, lie on the line k_r and are named K_r, $L_r, \ldots, N'_r$. The serial number, t, to the right of each column of co-ordinates give the expressions for the points as $\mathbf{N}^t\epsilon_1$. The first row corresponds to the first row of the difference table.

The columns headed '*' display a method of naming the lines in the six systems [L], …, [N′]. The lines are named in this way in Table 1 (i) $\Gamma^{(9)}$ of Excursus 1.3.1. The line of each system through $K_0 = (1, 0, 0)$, that is, in fact, the lines $K_0L_{10}, \ldots, K_0N'_{10}$ are named $l_0, \ldots, n'_0$ in such a way that:

$$l_0 = [0, \mu, 1], \quad m_0 = [0, \bar{\lambda}, 1], \quad n_0 = [0, \nu, 1],$$
$$l'_0 = [0, \bar{\mu}, 1], \quad m'_0 = [0, \lambda, 1], \quad n'_0 = [0, \bar{\nu}, 1].$$

Then $l_r = [0, \mu, 1] N^{-r}$, etc. The other lines in the table are then named $l_0, \ldots, n'_0$ by reference to the middle co-ordinate of the corresponding point. Thus $l_0 \supset \{K_0, L_2, L_3, L_6, L_8, L'_{10}, M_1, M'_{11}, N_4, N'_5\}$, for all of which the middle co-ordinate is μ, so that all lie on $\mu y + 1 = 0$.

Lines		Points								
		r	$\{K_r\}$	t	$\{L_r\}$	t	*	$\{L'_r\}$	t	*
k_0	$0\ 0\ 1$	0	$1\ 0\ 0$	0	$\lambda\ 1\ 0$	6		$\bar{\lambda}\ 1\ 0$	71	
k_1	$\bar{1}\ 0\ 1$	1	$0\ 1\ 0$	7	$1\ \lambda\ 1$	13	m	$1\ \bar{\lambda}\ 1$	78	m′
k_2	$\bar{1}\ \bar{1}\ 1$	2	$1\ 0\ 1$	14	$\bar{\lambda}\ \mu\ 1$	20	l	$\lambda\ \bar{\mu}\ 1$	85	l′
k_3	$1\ \bar{1}\ 1$	3	$0\ 1\ 1$	21	$\nu\ \mu\ 1$	27	l	$\bar{\nu}\ \bar{\mu}\ 1$	1	l′
k_4	$\bar{1}\ 1\ 0$	4	$1\ 1\ 0$	28	$\nu\ \nu\ 1$	34	n	$\bar{\nu}\ \bar{\nu}\ 1$	8	n′
k_5	$1\ 1\ 1$	5	$1\ 1\ 1$	35	$\mu\ \bar{\nu}\ 1$	41	n′	$\bar{\mu}\ \nu\ 1$	15	n
k_6	$1\ 1\ 0$	6	$\bar{1}\ 1\ 0$	42	$\bar{\mu}\ \mu\ 1$	48	l	$\mu\ \bar{\mu}\ 1$	22	l′
k_7	$0\ 1\ 1$	7	$1\ \bar{1}\ 1$	49	$\bar{\nu}\ \bar{1}\ 1$	55		$\nu\ \bar{1}\ 1$	29	
k_8	$1\ 0\ 1$	8	$\bar{1}\ \bar{1}\ 1$	56	$\bar{1}\ \mu\ 1$	62	l	$\bar{1}\ \bar{\mu}\ 1$	36	l′
k_9	$0\ 1\ 0$	9	$\bar{1}\ 0\ 1$	63	$\nu\ 0\ 1$	69		$\bar{\nu}\ 0\ 1$	43	
k_{10}	$1\ 0\ 0$	10	$0\ 0\ 1$	70	$0\ \bar{\mu}\ 1$	76	l′	$0\ \mu\ 1$	50	l
k_{11}	$\bar{1}\ 1\ 1$	11	$0\ \bar{1}\ 1$	77	$\bar{\nu}\ \bar{\lambda}\ 1$	83	m′	$\nu\ \lambda\ 1$	57	m
k_{12}	$0\ \bar{1}\ 1$	12	$\bar{1}\ 1\ 1$	84	$\lambda\ 1\ 1$	90		$\bar{\lambda}\ 1\ 1$	64	

r	Points											
	$\{M_r\}$	t	*	$\{M'_r\}$	t	*	$\{N_r\}$	t	*	$\{N'_r\}$	t	*
0	$\mu\ 1\ 0$	25		$\bar{\mu}\ 1\ 0$	38		$\nu\ 1\ 0$	30		$\bar{\nu}\ 1\ 0$	82	
1	$1\ \mu\ 1$	32	l	$1\ \bar{\mu}\ 1$	45	l′	$1\ \nu\ 1$	37	n	$1\ \bar{\nu}\ 1$	89	n′
2	$\nu\ \bar{\nu}\ 1$	39	n′	$\bar{\nu}\ \nu\ 1$	52	n	$\mu\ \bar{\lambda}\ 1$	44	m′	$\bar{\mu}\ \lambda \cdot 1$	5	m
3	$\bar{\mu}\ \bar{\lambda}\ 1$	46	m′	$\mu\ \lambda\ 1$	59	m	$\lambda\ \nu\ 1$	51	n	$\bar{\lambda}\ \bar{\nu}\ 1$	12	n′
4	$\lambda\ \lambda\ 1$	53	m	$\bar{\lambda}\ \bar{\lambda}\ 1$	66	m′	$\mu\ \mu\ 1$	58	l	$\bar{\mu}\ \bar{\mu}\ 1$	19	l′
5	$\bar{\lambda}\ \lambda\ 1$	60	m	$\lambda\ \bar{\lambda}\ 1$	73	m′	$\nu\ \bar{\mu}\ 1$	65	l′	$\bar{\nu}\ \mu\ 1$	26	l
6	$\bar{\lambda}\ \nu\ 1$	67	n	$\lambda\ \bar{\nu}\ 1$	80	n′	$\bar{\nu}\ \lambda\ 1$	72	m	$\nu\ \bar{\lambda}\ 1$	33	m′
7	$\mu\ \bar{1}\ 1$	74		$\bar{\mu}\ \bar{1}\ 1$	87		$\bar{\lambda}\ \bar{1}\ 1$	79		$\lambda\ \bar{1}\ 1$	40	
8	$\bar{1}\ \lambda\ 1$	81	m	$\bar{1}\ \bar{\lambda}\ 1$	3	m′	$\bar{1}\ \bar{\nu}\ 1$	86	n′	$\bar{1}\ \nu\ 1$	47	n
9	$\bar{\lambda}\ 0\ 1$	88		$\lambda\ 0\ 1$	10		$\bar{\mu}\ 0\ 1$	2		$\mu\ 0\ 1$	54	
10	$0\ \lambda\ 1$	4	m	$0\ \bar{\lambda}\ 1$	17	m′	$0\ \nu\ 1$	9	n	$0\ \bar{\nu}\ 1$	61	n′
11	$\bar{\lambda}\ \bar{\mu}\ 1$	11	l′	$\lambda\ \mu\ 1$	24	l	$\mu\ \nu\ 1$	16	n	$\bar{\mu}\ \bar{\nu}\ 1$	68	n′
12	$\bar{\nu}\ 1\ 1$	18		$\nu\ 1\ 1$	31		$\mu\ 1\ 1$	23		$\bar{\mu}\ 1\ 1$	75	

Although 91 is not the power of a prime it still has relatively simple numerical properties. In particular there is only one pair of numbers $\pm c$ $(2 \leqslant c \leqslant 89)$ such that $c^2 \equiv 1 \pmod{91}$, namely, $c = 27, 64$. Thus, with any number r $(0 \leqslant r \leqslant 90)$, there is paired another number s, such that $s \equiv 27r$, $r \equiv 27s$, and $r \equiv s$ only when $r = 7t$, $t \in \{0, \ldots, 12\}$. (For $c = 64$, $7tc \equiv 7(13-t)c$.) In fact, the two points $N^r \epsilon_1$ and $N^{27r} \epsilon_1$ are complex conjugates; for example, from the first row of the difference table, M_0 is given by N^{25} and M_0 by N^{38}, and $25 \times 27 \equiv 38 \pmod{91}$.

Exercise 1 (i) Verify that the only cubic polynomials in $\mathfrak{F}_3(1, \mathfrak{f})$ with coefficients in $\mathfrak{F}_3(1)$ which have no zeros are:

$$\begin{aligned}
f_0(x) &= f(x) &&= x^3+x^2+x-1\\
f_1(x) &= -f(-x) &&= x^3-x^2+x+1\\
f_2(x) &= -x^3f(1/x) &&= x^3-x^2-x-1\\
f_{12}(x) &= -x^3f(-1/x) &&= x^3+x^2-x+1\\
g_0(x) &= g(x) = -f(1-x) &&= x^3-x^2+1\\
g_1(x) &= -g(-x) &&= x^3+x^2-1\\
g_2(x) &= x^3g(1/x) &&= x^3-x+1\\
g_{12}(x) &= x^3g(-1/x) &&= x^3-x-1.
\end{aligned}$$

(ii) Prove that if $\mathbf{X}$ is any matrix such that

$$\det \mathbf{X} = 1 \text{ and } \det(\rho\mathbf{1}-\mathbf{X}) = h(\rho),$$

then $\det(\rho\mathbf{1}-\mathbf{X}^{-1}) = \rho^3 h(1/\rho)$, and that for the matrix $\mathbf{K}$

$$\det[(\rho+1)\mathbf{1}-\mathbf{K}] = \det[\rho\mathbf{1}-(-\mathbf{1}+\mathbf{K})] = \det(\rho\mathbf{1}-\mathbf{K}^6).$$

(iii) Prove that $\det(\rho\mathbf{1}-\mathbf{K}^r)$ gives the following functions:

$$r = 1, 3, 9\colon f_0(\rho), \quad r = 12, 10, 4\colon f_2(\rho),$$
$$r = 2, 6, 5\ (=18 \text{ mod. } 13)\colon g_1(\rho), \quad r = 11, 7, 8\colon g_{12}(\rho).$$

(This seems to imply that it was rather a lucky accident that the function $\det(\rho\mathbf{1}-\mathbf{M}^7)$ coincided with one of the functions $\det(\rho\mathbf{1}-\mathbf{K}^a)$!)

Exercise 2 Establish a relation between $\Gamma^{(2)}$ generated by

$$\mathbf{M} = \begin{bmatrix} 0 & 1 & 0\\ 1 & 0 & 1\\ 0 & 1 & 1 \end{bmatrix},$$

as in Excursus 2.3.2 and $\Gamma^{(4)}$ corresponding to that between $\Gamma^{(3)}$ and $\Gamma^{(9)}$ and the matrices $\mathbf{K}$ and $\mathbf{N}$. That is, find a matrix $\mathbf{G}$ such that $\mathbf{G}^3 = \mathbf{M}$.

EXCURSUS 3.3

$\Omega^{(9)}$ AND $\Omega^{(9)*}$ AND THEIR RELATIONS TO $\Gamma^{(9)}$†

Exc. 3.3.1 The underlying algebra

Take again the elements of $\mathfrak{F}_3(1, \mathfrak{f})$ to be $0, \pm 1, \lambda, \bar{\lambda}, \mu, \bar{\mu}, \nu, \bar{\nu}$. Among λ, μ, ν we have the additive relations:

(A) $$\lambda-\mu = 1, \qquad \nu-\lambda = 1;$$
$$\mu-\nu = 1, \qquad \lambda+\mu+\nu = 0,$$

a set of relations in which (with coefficients in $\mathfrak{F}_3$) the first two imply the others; and the multiplicative relations

(MΓ) $$\lambda^2 = -\mu, \qquad \mu\nu = -\lambda,$$
$$\mu^2 = -1, \qquad \nu\lambda = 1,$$
$$\nu^2 = \mu, \qquad \lambda\mu = \nu.$$

Because of the relation of λ, μ, ν to the powers of j we know that multiplication in this system is commutative, associative and distributive over addition.

To obtain the new geometric systems $\Omega^{(9)*}$ and $\Omega^{(9)}$ we define a new algebraic system Ω in which we replace (but only for convenience) $\{\lambda, \mu, \nu\}$ by a new set of units $\{\alpha, \beta, \gamma\}$ and define an algebraic system in which there is a set of nine elements $\Phi = \{0, \pm 1, \pm\alpha, \pm\beta, \pm\gamma\}$ and in which the coefficients $\{0, 1, -1\}$ are to be treated as elements of $\mathfrak{F}_3(1)$. For Ω we assume the same *additive* relations among the units as those above, namely,

(A) $$\alpha-\beta = \beta-\gamma = \gamma-\alpha = 1,$$
$$\alpha+\beta+\gamma = 0.$$

In effect, we assume α as an unspecified unit, and define β and γ by $\beta = \alpha-1$, $\gamma = \alpha+1$, in a system in which addition is commutative and associative and the coefficients are remainders modulo 3.

The *multiplication* relations are replaced by the symmetrical but *non-commutative* set

(MΩ) $$\alpha^2 = \beta^2 = \gamma^2 = -1,$$
$$\beta\gamma = -\gamma\beta = \alpha,$$
$$\gamma\alpha = -\alpha\gamma = \beta,$$
$$\alpha\beta = -\beta\alpha = \gamma.$$

† The geometries $\Omega^{(9)*}$ and $\Omega^{(9)}$ were the first finite non-Desarguesian geometries to be invented. (Veblen and Wedderburn: Non-Desarguesian and non-Pascalian geometries, *Trans. Am. Math. Soc.* **8** (1907) 379–88.)

We assume that *multiplication is associative*, but not that it is fully distributive over addition, in fact we prove easily that it cannot be; thus

$$\left.\begin{aligned}\gamma(\alpha+\beta) = \gamma(-\gamma) &= -\gamma^2 = 1,\\ \text{whilst}\qquad \gamma\alpha+\gamma\beta = \beta-\alpha &= -1\end{aligned}\right\}$$

$$\left.\begin{aligned}\text{and}\qquad \gamma(\alpha+1) = \gamma\gamma &= -1,\\ \text{whilst}\qquad \gamma\alpha+\gamma = \beta+\gamma &= -\alpha.\end{aligned}\right\}$$

The system Ω is therefore *not left-distributive.*

On the other hand

$$\left.\begin{aligned}(\alpha+\beta)\gamma &= -\gamma\gamma = 1,\\ \alpha\gamma+\beta\gamma &= -\beta+\alpha = 1,\end{aligned}\right\}$$

$$\left.\begin{aligned}(\beta+\gamma)\gamma &= -\alpha\gamma = \beta,\\ \beta\gamma+\gamma^2 &= \alpha-1 = \beta,\end{aligned}\right\}$$

$$\left.\begin{aligned}(\alpha+1)\gamma = \gamma\gamma &= -1,\\ \alpha\gamma+\gamma = -\beta+\gamma &= -1,\end{aligned}\right\}\quad \text{etc.}$$

The system Ω therefore is *right-distributive.*

The sum and product of every two members of Φ is defined by the relations (A) and (MΩ) so that the system Ω is closed under these operations. There are moreover two immediate deductions from the relations:

Theorem (I) For any fixed element ρ and variable elements x,

(i) $\rho\in\Phi \Rightarrow \{\rho+\mathrm{x} : \mathrm{x}\in\Phi\} = \Phi$,
(ii) $\rho\in\Phi \;\&\; \rho \neq 0 \Rightarrow \{\rho\mathrm{x} : \mathrm{x}\in\Phi\} = \Phi$.

Theorem (II) $\rho\in\Phi \;\&\; \sigma\in\Phi \;\&\; \rho\sigma = 0 \Rightarrow$ either $\rho = 0$ or $\sigma = 0$.

(*Proof*) Assume $\sigma \neq 0$, so that $\sigma^2 = 1$ or -1. Then

$$\rho\sigma = 0 \Rightarrow (\rho\sigma)\sigma = \rho\sigma^2 = \pm\rho = 0.$$

That is,

$$\rho\sigma = 0 \;\&\; \sigma \neq 0 \Rightarrow \rho = 0.$$

Exc. 3.3.2 The geometry $\Omega^{(9)*}$ based on the algebra Ω

Because the system Ω is not distributive in both senses any algebraic operation involving both multiplication and addition has to be carried out with extreme caution. Thus, we must use, at least initially, the prescribed forms for the vectors corresponding to points and lines, namely,

$$\xi = \left\{ \begin{bmatrix} x \\ y \\ 1 \end{bmatrix}, \begin{bmatrix} x \\ 1 \\ 0 \end{bmatrix}, \begin{bmatrix} 1 \\ 0 \\ 0 \end{bmatrix} \right\} \quad \text{and} \quad \alpha^T = \left\{ \begin{matrix} [u, v, 1] \\ [u, 1, 0] \\ [1, 0, 0] \end{matrix} \right\}.$$

Consider now the equation

$$\alpha^T \xi \equiv ux + vy + 1 = 0;$$

this is the same, for any $\rho \neq 0$, as

$$\rho(ux + vy + 1) = 0$$

by Theorem (I), but, unless $\rho = \pm 1$, *not* as

$$(\rho u)x + (\rho v)y + \rho = 0,$$

since the system is not left-distributive, so that the co-ordinates $[u, v, 1]$ of a line cannot be replaced by $[\rho u, \rho v, \rho]$.

On the other hand, the equation

$$(ux + vy + 1)\sigma = 0$$

is satisfied by the same set of points as

$$u(x\sigma) + v(y\sigma) + \sigma = 0,$$

because the system is right-distributive. Thus, the point represented by the vector $(x, y, 1)$ may also be represented by $(x\sigma, y\sigma, \sigma)$ for any $\sigma \neq 0$. It is in fact convenient, sometimes, to use the form $ux + vy + z = 0$ for the condition that a set of points $\{(xz^{-1}, yz^{-1}, 1)\}$ is collinear. But the set of points satisfying

$$ux + vy + 1 = 0$$

(which is to be a collinear set) is *not* the same as the set

$$(wu)x + (wv)y + w = 0$$

which may well not be collinear.

With these considerations in mind, our first task is to prove that the sets of points $\{\xi\}$ satisfying the relations $\alpha^T\xi = 0$ have all the properties of collinear sets, namely,

(1) there is a single vector α determined by any two distinct points,

(2) any two different vectors α, α' are such that there is a single point satisfying the two conditions $\alpha^T\xi = 0$, $\alpha'^T\xi = 0$.

It will be sufficient to prove only one of these statements, since there are 91 points ξ and 91 vectors α. We shall prove (1), and do in fact also investigate the problem of finding the point ξ satisfying both $\alpha^T\xi = 0$ and $\alpha'^T\xi = 0$.

We are to prove that, given (a, b, 1) and (a′, b′, 1), there is a single relation $\boldsymbol{\alpha}^T\boldsymbol{\xi} = 0$ satisfied by both points, and we consider two special cases:

(i) a = 0 & a′ = 0. We have

$$[v \quad 1]\begin{bmatrix} b \\ 1 \end{bmatrix} = 0 \ \& \ [v \quad 1]\begin{bmatrix} b' \\ 1 \end{bmatrix} = 0$$

$$\Rightarrow vb = vb'$$

$$\Rightarrow v = 0 \quad \text{since} \quad b \neq b'.$$

If we assume that the collinear set is [u, 0, 1] we reach a contradiction, so that, the only collinear set containing both (0, b, 1) and (0, b′, 1) is [1, 0, 0].

(ii) a′ = ak & b′ = bk & k ≠ 0 & k ≠ 1 & a ≠ 0.

If we assume that the collinear set containing (a, b, 1) and (ak, bk, 1) is [u, v, 1], then

$$ua + vb + 1 = 0 \quad \text{and} \quad (ua + vb)k + 1 = 0,$$

since the algebra is right-distributive. But we cannot have simultaneously

$$(ua + vb) = -1 \quad \text{and} \quad (ua + vb) = -k^{-1}.$$

Assuming then that the collinear set is [u, 1, 0], we find that the collinear set is $[-ba^{-1}, 1, 0]$ and only this set contains both points.

Thus both these exceptional cases lead to collinear sets which contain the point (0, 0, 1)—the 'origin'.

Turning now to the general case, from which pairs of points satisfying the conditions (i) or (ii) are excluded, we assume that (a, b, 1), (a′, b′, 1) belong to the collinear set [u, v, 1], so that

$$ua + vb + 1 = 0 \quad \text{and} \quad ua' + vb' + 1 = 0.$$

For any pair $\{\sigma, \sigma'\}$ we have in consequence

$$(ua + vb + 1)\sigma + (ua' + vb' + 1)\sigma' = 0,$$

and therefore in particular

$$(ua + vb + 1)a^{-1} - (ua' + vb' + 1)a'^{-1} = 0.$$

Since the algebra is right-distributive we may reduce this to

$$vba^{-1} - vb'a'^{-1} + a^{-1} - a'^{-1} = 0$$

or say

$$vp + vq = r$$

(but *not* v(p + q) = r!), where p, q, r are members of Φ determined by a, b, a′, b′. For the given p, q, r we have to find which of the nine elements of Φ satisfies the equation, and have first to see that there is at most one such. Assume that

$$vp+vq = r \quad \text{and} \quad v'p+v'q = r$$

so that (since the system is right-distributive),

$$(v-v')p+(v-v')q = 0,$$

say $v''p+v''q = 0$, where $v'' \neq 0$. From Theorem I (ii) we infer that $p = -q$. But $p = -q \Rightarrow ba^{-1} = b'a'^{-1}$. This is equivalent to the condition (ii), $a' = ak$ & $b' = bk$, which we have excluded. It follows that there is at most one value v satisfying the equation $vp+vq = r$ for given p, q, r. Thus on substituting in $vp+vq$ the nine possible values of v, we shall obtain in turn all members of Φ. One of these is r.

The system $\Omega^{(9)*}$ has the property that any two points determine a single vector $\boldsymbol{\alpha}$, such that the relation $\boldsymbol{\alpha}^T\boldsymbol{\xi} = 0$ is satisfied by them both. We have proved that: the system $\Omega^{(9)*}$, consisting of the set of 91 points

$$\boldsymbol{\xi} = \left\{\begin{bmatrix} x \\ y \\ 1 \end{bmatrix}, \begin{bmatrix} x \\ 1 \\ 0 \end{bmatrix}, \begin{bmatrix} 1 \\ 0 \\ 0 \end{bmatrix}\right\} \quad \text{and 91 vectors} \quad \boldsymbol{\alpha}^T = \left\{\begin{matrix} [u, v, 1] \\ [u, 1, 0] \\ [1, 0, 0] \end{matrix}\right\},$$

in which a collinear set is a set of points satisfying one of the 91 relations $\boldsymbol{\alpha}^T\boldsymbol{\xi} = 0$, satisfies axioms $A\mathscr{I}$ 1 and 2.

We consider now the problem of determining the point common to two collinear sets, say

$$ux+vy+1 = 0, \quad u'x+v'y+1 = 0.$$

From these two equations, we can deduce, since addition is commutative and associative, that

$$ux-u'x = v'y-vy.$$

If we assume $y \neq 0$, since the system Ω is right-distributive, this relation implies

$$(u-u')xy^{-1} = v'-v.$$

Thus xy^{-1} is determined (assuming $u \neq u'$) as

$$xy^{-1} = (u-u')^{-1}(v'-v) = k, \text{ say.}$$

Now replace the first equation of this sequence by

$$uxy^{-1}+v+y^{-1} = 0,$$

that is, if

$$v+uk \neq 0,$$

$$y = -(v+uk)^{-1},$$

then

$$x = -k(v+uk)^{-1}.$$

That is, if none of the exceptional conditions excluded in the course of the argument holds, the point is completely determined. The exceptional conditions present no difficulties.

We have now to show that the geometry is non-Desarguesian. One example of a non-Desarguesian figure is

points				collinear sets			
K	1	1	1				
A	1	0	0	KAA′	0	-1	1
B	0	1	0	KBB′	-1	0	1
C	0	0	1	KCC′	-1	1	0
A′	α	1	1	BCL	1	0	0
B′	1	β	1	CAM	0	1	0
C′	$-\gamma$	$-\gamma$	1	ABN	0	0	1
L	0	$-\beta$	1	B′C′L	1	$-\beta$	1
M	β	0	1	C′A′M	β	α	1
N	$-\alpha$	1	0	A′B′N	$-\alpha$	1	1
				LM	β	$-\beta$	1

The condition for $N \in LM$ is

$$[\beta, -\beta, 1](-\alpha, 1, 0) = 0,$$

that is, $\gamma - \beta = 0$, which is clearly not satisfied.

In $\Omega^{(9)*}$ we cannot use projectivities as we have done in all other geometrical systems so far, even if the elements of the matrix determining the projectivity all belong to the set $\{0, 1, -1\}$, because collinear sets usually do not transform into collinear sets. Suppose that $(\boldsymbol{\varkappa}\}$ is the collinear set of points satisfying $\boldsymbol{\alpha}^T\boldsymbol{\xi} = 0$, where $\boldsymbol{\alpha}^T = [u, v, 1]$. Then the points of the set also satisfy $\boldsymbol{\alpha}^T(\mathbf{M}^{-1}\mathbf{M})\boldsymbol{\xi} = 0$, but no further significant statement of relations can be made, because

(i) matrix multiplication is no longer necessarily associative, that is, $\boldsymbol{\alpha}^T(\mathbf{M}^{-1}\mathbf{M})\boldsymbol{\xi} \neq (\boldsymbol{\alpha}^T\mathbf{M}^{-1})(\mathbf{M}\boldsymbol{\xi})$,

(ii) the vector $\boldsymbol{\alpha}^T\mathbf{M}^{-1}$ is not generally of one of the standard forms which corresponds to lines, namely, one in which the last-written non-zero component is 1.

Take for example the collinear set

$$\boldsymbol{\delta}^T = [\alpha, 1, 1],$$

and the three points

$$\boldsymbol{\varkappa}_1 = (-\alpha, 1, 1), \quad \boldsymbol{\varkappa}_2 = (\beta, -\beta, 1), \quad \boldsymbol{\varkappa}_3 = (-\beta, \alpha, 1)$$

which belong to it, and consider its transform by the matrix $\mathbf{K}$, namely,

$$(\boldsymbol{\delta}^T\mathbf{K}^{-1})(\mathbf{K}\boldsymbol{\xi}) \equiv [\alpha, 1, 1]\begin{bmatrix} -1 & 1 & 1 \\ 1 & 0 & 0 \\ 1 & 0 & -1 \end{bmatrix}\begin{bmatrix} 0 & 1 & 0 \\ 1 & 0 & 1 \\ 0 & 1 & -1 \end{bmatrix}\boldsymbol{\xi} = 0.$$

The transforms $\mathbf{K}\boldsymbol{\varkappa}_i$ of the points are

$$\boldsymbol{\varkappa}'_1 = (\beta, 1, 0)(-\beta), \quad \boldsymbol{\varkappa}'_2 = (\gamma, -1, 1)(-\alpha), \quad \boldsymbol{\varkappa}'_3 = (-\gamma, -\alpha, 1)\beta.$$

The transform $\boldsymbol{\delta}^{\mathrm{T}}\mathbf{K}^{-1}$ of $\boldsymbol{\delta}^{\mathrm{T}}$ is

$$\boldsymbol{\delta}'^{\mathrm{T}} = [-\gamma, \alpha, \beta].$$

This is not one of the standard forms determining collinear sets; moreover, whilst

$$\boldsymbol{\delta}'^{\mathrm{T}}\boldsymbol{\varkappa}'_2 = 1-\alpha+\beta = 0,$$

contrarily $$\boldsymbol{\delta}'^{\mathrm{T}}\boldsymbol{\varkappa}'_1 = \alpha+\alpha+0 \neq 0$$

and $$\boldsymbol{\delta}'^{\mathrm{T}}\boldsymbol{\varkappa}'_3 = -1+1+\beta \neq 0.$$

The line joining $\boldsymbol{\varkappa}'_2$ and $\boldsymbol{\varkappa}'_3$ is $[-\beta, -\beta, 1]$ and this set of collinear points clearly does not contain $\boldsymbol{\varkappa}'_1 = (\beta, 1, 0)$.

Summarizing: $\boldsymbol{\delta}'^{\mathrm{T}} = \boldsymbol{\delta}^{\mathrm{T}}\mathbf{K}^{-1}$ is not a vector determining a collinear set, only one of the three expressions $\boldsymbol{\delta}'^{\mathrm{T}}\boldsymbol{\varkappa}'_i$ vanishes, and the three points $\boldsymbol{\varkappa}'_i = \mathbf{K}\boldsymbol{\varkappa}_i$ are not collinear!

Exercise 1 Prove that (i) if $\boldsymbol{\Lambda}$ is the matrix of a projectivity, that is, if, for all vectors $\boldsymbol{\alpha}$, $(\boldsymbol{\alpha}\boldsymbol{\Lambda}^{-1})(\boldsymbol{\Lambda}\boldsymbol{\xi}) = 0$ is the equation of a collinear set, then the third column of $\boldsymbol{\Lambda}^{-1}$ is $(0, 0, \pm 1)$,

(ii) the point (0, 0, 1) is invariant under all projectivities,

(iii) $\begin{bmatrix} 0 & 1 & 0 \\ 1 & 0 & 0 \\ 0 & 0 & 1 \end{bmatrix}$ is the matrix of a projectivity,

(iv) no line is invariant under all projectivities,

(v) $\Omega^{(9)*}$ is not a self-dual system.

Exercise 2 Prove that, (i) for all values h, k, and all non-zero values of ρ, σ the matrices

$$\begin{bmatrix} 1 & 0 & 0 \\ 0 & 1 & 0 \\ \mathrm{h} & \mathrm{k} & 1 \end{bmatrix}, \quad \begin{bmatrix} \rho & 0 & 0 \\ 0 & \sigma & 0 \\ 0 & 0 & 1 \end{bmatrix}, \quad \begin{bmatrix} 1 & 1 & 0 \\ 1 & -1 & 0 \\ 0 & 0 & 1 \end{bmatrix} \text{ define projectivities,}$$

(ii) these matrices generate the complete group of projectivities.

Exercise 3 $\mathscr{T}_1$ is the transformation in which, in the vectors for points and lines, α, β, γ are interchanged respectively with $-\alpha, -\gamma, -\beta$. $\mathscr{T}_2$ and $\mathscr{T}_3$ are defined symmetrically. Prove that (i) $\mathscr{T}_1\mathscr{T}_2\mathscr{T}_1 = \mathscr{T}_3$, (ii) $\mathscr{T}_1$ and $\mathscr{T}_2$ generate the group of substitutions on three elements, and (iii) if $\{\boldsymbol{\xi}\}$ is a set of collinear points, so also is $\{\mathscr{T}_1\boldsymbol{\xi}\}$. (iv) What points are invariant under $\mathscr{T}_1$?

Exercise 4 Find the vertices of the diagonal triangle of the quadrangle of which the vertices are (1, 0, 0), (1, 0, 1), (0, 1, 1) and (1, α, 1). Prove that the set of seven vertices forms a geometry $\Gamma^{(2)}$.

Exc. 3.3.3 The geometry $\Omega^{(9)}$, a hybrid between $\Gamma^{(9)}$ and $\Omega^{(9)*}$

The geometry $\Omega^{(9)}$ described in Excursus 1.3 is not the same as $\Omega^{(9)*}$, although it is based initially on the same algebraic system Ω, and the points have the same co-ordinates. The difference between the two systems is this:

in $\Omega^{(9)*}$ the points of a collinear set are those whose co-ordinates satisfy a linear relation $\boldsymbol{\alpha}^T\boldsymbol{\xi} = 0$,

in $\Omega^{(9)}$ a basic set of 13 collinear sets of points satisfy the relations $ux+vy+wz = 0$ where $(u, v, w) \in \{0, 1, -1\}$, and the points of the remaining collinear sets are the transforms of these by the matrices $\mathbf{K}^r$, where

$$\mathbf{K} = \begin{bmatrix} 0 & 1 & 0 \\ 1 & 0 & 1 \\ 0 & 1 & -1 \end{bmatrix}.$$

The geometry [K] therefore forms the core of $\Omega^{(9)}$, giving the thirteen points $K_0, \dots K_{12}$ and thirteen lines $k_0, \dots, k_{12}$. The other 78 points are named in the following way:

First, we introduce symbols $\bar{\alpha}$, $\bar{\beta}$, $\bar{\gamma}$ corresponding to $\bar{\lambda}$, $\bar{\mu}$, $\bar{\nu}$ in Excursus 3.2.3, thus

$$\bar{\alpha} = -\gamma, \quad \bar{\beta} = -\beta, \quad \bar{\gamma} = -\alpha;$$

then take

$$A_0 = \begin{bmatrix} \alpha \\ 1 \\ 0 \end{bmatrix}, \quad A_1 = \mathbf{K}\begin{bmatrix} \alpha \\ 1 \\ 0 \end{bmatrix} = \begin{bmatrix} 1 \\ \alpha \\ 1 \end{bmatrix}, \quad A_2 = \mathbf{K}^2\begin{bmatrix} \alpha \\ 1 \\ 0 \end{bmatrix} = \begin{bmatrix} \bar{\alpha} \\ \beta \\ 1 \end{bmatrix}\bar{\beta}, \quad \dots,$$

$$A_{12} = \mathbf{K}^{12}\begin{bmatrix} \alpha \\ 1 \\ 0 \end{bmatrix} = \begin{bmatrix} \alpha \\ 1 \\ 1 \end{bmatrix},$$

$$A_0' = \begin{bmatrix} \bar{\alpha} \\ 1 \\ 0 \end{bmatrix}, \quad B_0 = \begin{bmatrix} \beta \\ 1 \\ 0 \end{bmatrix}, \quad B_0' = \begin{bmatrix} \bar{\beta} \\ 1 \\ 0 \end{bmatrix}, \quad C_0 = \begin{bmatrix} \gamma \\ 1 \\ 0 \end{bmatrix}, \quad C_0' = \begin{bmatrix} \bar{\gamma} \\ 1 \\ 0 \end{bmatrix}$$

and $A_r' = \mathbf{K}^r A_0$, etc. In the table below the co-ordinates are given in standard form, although in this system the reduction is irrelevent except for easy identification.

It will be noticed that in this table there are regular cyclic progressions across the rows in the first co-ordinates and second co-ordinates, and the names of the lines, namely, either $\alpha\bar{\alpha}\beta\bar{\beta}\gamma\bar{\gamma}$ or $\bar{\alpha}\alpha\bar{\gamma}\gamma\bar{\beta}\beta$ (or aa′bb′cc′ or a′ac′cb′b).

	A	A′	B	B′	C	C′
0	$\alpha\ 1\ 0$	$\bar{\alpha}\ 1\ 0$	$\beta\ 1\ 0$	$\bar{\beta}\ 1\ 0$	$\gamma\ 1\ 0$	$\bar{\gamma}\ 1\ 0$
1	$1\ \alpha\ 1$ b′	$1\ \bar{\alpha}\ 1$ b	$1\ \beta\ 1$ a′	$1\ \bar{\beta}\ 1$ a	$1\ \gamma\ 1$ c′	$1\ \bar{\gamma}\ 1$ c
2	$\bar{\alpha}\ \beta\ 1$ a′	$\alpha\ \bar{\beta}\ 1$ a	$\bar{\gamma}\ \gamma\ 1$ c′	$\gamma\ \bar{\gamma}\ 1$ c	$\bar{\beta}\ \alpha\ 1$ b′	$\beta\ \bar{\alpha}\ 1$ b
3	$\bar{\gamma}\ \bar{\beta}\ 1$ a	$\gamma\ \beta\ 1$ a′	$\bar{\beta}\ \bar{\alpha}\ 1$ b	$\beta\ \alpha\ 1$ b′	$\bar{\alpha}\ \bar{\gamma}\ 1$ c	$\alpha\ \gamma\ 1$ c′
4	$\gamma\ \gamma\ 1$ c′	$\bar{\gamma}\ \bar{\gamma}\ 1$ c	$\alpha\ \alpha\ 1$ b′	$\bar{\alpha}\ \bar{\alpha}\ 1$ b	$\beta\ \beta\ 1$ a′	$\bar{\beta}\ \bar{\beta}\ 1$ a
5	$\bar{\beta}\ \gamma\ 1$ c′	$\beta\ \bar{\gamma}\ 1$ c	$\bar{\alpha}\ \alpha\ 1$ b′	$\alpha\ \bar{\alpha}\ 1$ b	$\bar{\gamma}\ \beta\ 1$ a′	$\gamma\ \bar{\beta}\ 1$ a
6	$\bar{\beta}\ \beta\ 1$ a′	$\beta\ \bar{\beta}\ 1$ a	$\bar{\alpha}\ \gamma\ 1$ c′	$\alpha\ \bar{\gamma}\ 1$ c	$\bar{\gamma}\ \alpha\ 1$ b′	$\gamma\ \bar{\alpha}\ 1$ b
7	$\bar{\gamma}\ \bar{1}\ 1$	$\gamma\ \bar{1}\ 1$	$\bar{\beta}\ \bar{1}\ \bar{1}$	$\beta\ \bar{1}\ 1$	$\bar{\alpha}\ \bar{1}\ 1$	$\alpha\ \bar{1}\ 1$
8	$\bar{1}\ \bar{\beta}\ 1$ a	$\bar{1}\ \beta\ 1$ a′	$\bar{1}\ \bar{\alpha}\ 1$ b	$\bar{1}\ \alpha\ 1$ b′	$\bar{1}\ \bar{\gamma}\ 1$ c	$\bar{1}\ \gamma\ 1$ c′
9	$\gamma\ 0\ 1$	$\bar{\gamma}\ 0\ 1$	$\alpha\ 0\ 1$	$\bar{\alpha}\ 0\ 1$	$\beta\ 0\ 1$	$\bar{\beta}\ 0\ 1$
10	$0\ \bar{\beta}\ 1$ a	$0\ \beta\ 1$ a′	$0\ \bar{\alpha}\ 1$ b	$0\ \alpha\ 1$ b′	$0\ \bar{\gamma}\ 1$ c	$0\ \gamma\ 1$ c′
11	$\gamma\ \alpha\ 1$ b′	$\bar{\gamma}\ \bar{\alpha}\ 1$ b	$\alpha\ \beta\ 1$ a′	$\bar{\alpha}\ \bar{\beta}\ 1$ a	$\beta\ \gamma\ 1$ c′	$\bar{\beta}\ \bar{\gamma}\ 1$ c
12	$\bar{\alpha}\ 1\ 1$	$\alpha\ 1\ 1$	$\bar{\gamma}\ 1\ 1$	$\gamma\ 1\ 1$	$\bar{\beta}\ 1\ 1$	$\beta\ 1\ 1$

The basic collinear sets $a_0, \ldots, c'_0$ are:

a_0: points with second co-ordinate $\bar{\beta}$,

$a'_0, b_0, b'_0, c_0, c'_0$.: points with second co-ordinates respectively $\beta, \bar{\alpha}, \alpha, \bar{\gamma}, \gamma$, so that

$$a_0 \supset K_0, A_3, A_8, A_{10}, A'_2, A'_6, B'_1, B'_{11}, C'_4, C'_5,$$

$$a'_0 \supset K_0, A'_3, A'_8, A'_{10}, A_2, A_6, B_1, B_{11}, C_4, C_5, \text{etc.}$$

The remaining collinear sets are then typified by

$$a_r \supset K_r, A_{3+r}, A_{8+r}, A_{10+r}, A'_{2+r}, A'_{6+r}, B'_{1+r}, B'_{11+r}, C'_{4+r}, C'_{5+r}.$$

The whole scheme is that described in Excursus 1.3.2, Table 1 (ii) $\Omega^{(9)}$. The pattern looks so similar to that of $\Gamma^{(9)}$ as tabulated in Excursus 3.2.3 that it is worth comparing the two systems more closely. We do this by equating α, β, γ with respectively λ, μ, ν, and then we notice that, for example, the co-ordinates of seven points A_r are the same as those of the corresponding points L_r, while those of the other six A_s are the same as those of the corresponding L'_s, and in each of the pairs of sets there is a similar interchange of six of the pairs of complex conjugates.

BOOK 4

CONGRUENCE

The critical problem that has faced the F-G since his discovery of a mathematical model for 'lengths' on the same line has been that of finding a method for comparing 'lengths' on non-parallel lines. If he has a straight stick represented, when it is in a certain position, by a segment ⌜AB⌝, and a unit measuring rod which, when placed ⁑parallel to the stick, is represented by ⌜OE⌝, then he can assign a ⁑length to the stick in that or any ⁑parallel position, namely, the number s(AB/OE). It seems to be a property of his physical world that, in relation to the same unit measuring rod, the stick should preserve its ⁑length in whatever position it is placed. He should therefore be able to devise some mathematical model for the physical process of moving the stick from one place to another.

He has already some clue to the method of attack in his representation of ⁑lengths on the same ⁑line by means of his mathematical system of transformers $\mathscr{D}$. In the physical system he considers the stick as being moved from one position to another keeping it ⁑parallel to its original direction; in the mathematical system he considers two separate point-pairs, related to each other by a certain geometrical construction (2.1*D*1), namely,

G $\mathscr{L}\wp$|ABP
C Q : PQ || AB & BQ || AP
D $\mathrm{Q} = \mathscr{D}_{\mathrm{AB}}\mathrm{P}$.

He now tries to express his second problem (of the relation between the lengths of the stick in non-parallel positions) in physical terms which will correspond to this. He can move the rod from ⁑⌜AB⌝ to ⁑⌜CD⌝ by first moving it to the parallel position ⁑⌜CK⌝ and then '⁑rotating' it about the point ⁑C until ⁑K coincides with ⁑D. Then the question is, can he find a mathematical model for this physical operation of ⁑rotation? Rotation is apparently a ⁑continuous operation, which would seem to require information about all intervening positions. This is not necessarily an insuperable difficulty—the same difficulty is encountered in the transition from the physical operation of 'moving the stick parallel to itself' to formulating the construction for the transformer $\mathscr{D}$. But in this case there seems to be no corresponding simple geometric operation.

The F-G next recalls that he has observed a physical phenomenon which might provide him with just the model he seeks, although, to begin with, it appears to require the use of objects above and below as well as on his paddock. He observes that the image of a stick in a still sheet of water, however the stick may be placed above the surface of the water, appears identical with the stick itself, and further seems to stand in exactly the same geometrical relation to the surface, but beneath it instead of above. He finds in fact that he can derive a mathematical model, the 'Reflection Transformation' similar to the Displacement Transformation, which gives all the properties he requires, and which, moreover, can be specified completely in terms of the geometrical relations he has already investigated, namely, incidence, parallelism and perpendicularity.

Throughout Book 4 we assume that, if the field is finite and of order q, then q is odd.

CHAPTER 4.1

THE REFLECTION TRANSFORMER, $\mathscr{R}$

With the physical picture of 'reflection' in mind we define a geometric transformation $\mathscr{R}$ in the following way:

4.1 *D*1 The reflection transformation.

G $\wp|$l
S $\wp|$P
*C*1 L : L$\in$l & PL $\perp$ l
*C*2 Q : Q = $\mathscr{D}_{PL}$L
D Q = $\mathscr{R}_l$P.

We shall read this relation as 'Q is the reflection of P in l', and refer to $\mathscr{R}_l$ as a *reflection transformer*, and l as the *axis of reflection*. It should be noted that the restriction P$\notin$l is not imposed; the definition holds formally unchanged if P$\in$l. Thus P$\in$l $\Rightarrow$ P = L $\Rightarrow$ P = Q. Conversely, since, with the notation of *D*1, $\mathscr{R}_l P = \mathscr{D}_{LQ}\mathscr{D}_{PL}P = \mathscr{D}^2_{PL}P$, if P is a proper point, then P = $\mathscr{R}_l$P only if P = L, i.e. P$\in$l, thus

4.1 *T*1 $\wp|$P & P = $\mathscr{R}_l$P $\Leftrightarrow$ P$\in$l.

In terms of the ideal line and the orthogonality involution the definition *D*1 becomes:

4.1 *D*2 Reflection transformation with assigned orthogonality involution.

*G*1 ideal|u & orthogonality involution $\mathscr{I}$
*G*2 l : #|lu
S P : P$\notin$u
C U : U = l $\cap$ u
V : V = $\mathscr{I}$U
L : L = VP $\cap$ l
Q : Q = $\mathscr{H}_{LV}$P
D Q = $\mathscr{R}_{l(u,\mathscr{I})}$P.

There is no reflection transformer determined by the ideal line.

If ideal|P & #|PUV, then L = U and Q = $\mathscr{H}_{UV}$P. Thus under the transformation $\mathscr{R}_l$ the ideal line is over-all invariant and the points

U and V are such that $P = U \Rightarrow P = L \Rightarrow P = Q$, and $P = V \Rightarrow P = L \Rightarrow P = Q$ so that

4.1*T*2

G # |lu & ideal|u
T (i) $P \in u \Rightarrow \mathscr{R}_l P \in u$
T (ii) $\mathscr{R}_l(u \cap l) = u \cap l$, $\mathscr{R}_l[\mathscr{I}(u \cap l)] = \mathscr{I}(u \cap l)$.

By constructing the point $\mathscr{R}_l Q$ we prove that

4.1*T*3 $\mathscr{R}$ is involutory, i.e., $\mathscr{R} = \mathscr{R}^{-1}$.

G ℘|l
S ℘|Q
C $L : L \in l$ & $LQ \perp l$
$P : P = \mathscr{D}_{QL} L$
T $\mathscr{R}_l Q = P$.

We may write the relation as

$$\mathscr{R}_l^2 = \mathscr{I} \quad \text{or} \quad \mathscr{R}_l^{-1} = \mathscr{R}_l.$$

Consider next the resultants of pairs of transformations $\mathscr{R}$.

4.1*T*4 The resultant of reflections in two parallel lines is a displacement.

G ||#|l, m
S ℘|P
C $Q = \mathscr{R}_l P$
$R = \mathscr{R}_m Q$
$L : L = PQ \cap l$
$M : M = QR \cap m$
T (i) $\mathscr{L}$|PQR
T (ii) $\mathscr{R}_m \mathscr{R}_l P = \mathscr{D}_{LM}^2 P$
P (ii) $Q = \mathscr{D}_{LQ}^2 P \quad (\mathscr{D}_{PL} = \mathscr{D}_{LQ})$
$R = \mathscr{D}_{QM}^2 Q$
$= \mathscr{D}_{QM}^2 \mathscr{D}_{LQ}^2 P = (\mathscr{D}_{QM} \mathscr{D}_{LQ})^2 P$
$= \mathscr{D}_{LM}^2 P.$
Since P is arbitrarily selected:
T $\mathscr{R}_m \mathscr{R}_l = \mathscr{D}_{LM}^2.$

In the same way $\mathscr{R}_l \mathscr{R}_m = \mathscr{D}_{ML}^2$, so that, at least for parallel lines, two reflections do not commute.

Consider next the case $l \perp m$.

4.1*T*5 The resultant of reflections in two perpendicular lines.

G $l, m : l \perp m \ \& \ l \cap m = G$
S $\wp | P$
C $Q, R : Q = \mathscr{R}_l P, R = \mathscr{R}_m Q$
T (i) $\mathscr{L} | PRG \ \& \ s(PG/GR) = 1$
(ii) $\mathscr{R}_m \mathscr{R}_l = \mathscr{R}_l \mathscr{R}_m$
P *c* $L, M : L = PQ \cap l, M = QR \cap m$
*t*1 $l \perp m \ \& \ PL \perp l \Rightarrow m \| PL$
& similarly $l \| QM$
*t*2 $LQ \| GM \ \& \ GL \| MQ \ \&$ *C*
$\Rightarrow \mathscr{D}_{LG} = \mathscr{D}_{QM} = \mathscr{D}_{MR} \ \& \ \mathscr{D}_{GM} = \mathscr{D}_{LQ} = \mathscr{D}_{PL}$
$\Rightarrow \mathscr{D}_{LM} = \mathscr{D}_{LQ} \mathscr{D}_{QM} = \mathscr{D}_{GM} \mathscr{D}_{MR} = \mathscr{D}_{GR}$
& $\mathscr{D}_{LM} = \mathscr{D}_{PG}$
T (i) $\mathscr{D}_{PG} = \mathscr{D}_{GR}$
T (ii) *X* $\mathscr{R}_m \mathscr{R}_l = \mathscr{R}_l \mathscr{R}_m$

Now take $\# | lm \ \& \ l \not\perp m \ \& \ \wp | P \in l \ \& \ P \notin m$, then

$$\mathscr{R}_m \mathscr{R}_l P = \mathscr{R}_m P = Q, \quad \text{say,} \quad \text{while } \mathscr{R}_l \mathscr{R}_m P = \mathscr{R}_l Q.$$

Thus

$$\wp | Q \ \& \ Q = \mathscr{R}_l Q \Rightarrow Q \in l \quad \text{and} \quad Q \in l \Rightarrow m \perp l.$$

4.1*T*6*X* Two distinct reflections commute if and only if they are reflections in perpendicular lines.

G $\# \wp | l, m$
T $\mathscr{R}_l \mathscr{R}_m = \mathscr{R}_m \mathscr{R}_l \Leftrightarrow l \perp m.$

4.1*T*7 The transformers $\mathscr{R}$ are associative.

G $\# \wp | lmn$
S $\wp | P$
C $Q = \mathscr{R}_l P, R = \mathscr{R}_m Q, S = \mathscr{R}_n R$
T $\mathscr{R}_n(\mathscr{R}_m \mathscr{R}_l) P = \mathscr{R}_n R = S$
$= (\mathscr{R}_n \mathscr{R}_m) Q = (\mathscr{R}_n \mathscr{R}_m) \mathscr{R}_l P$
i.e. $\mathscr{R}_n(\mathscr{R}_m \mathscr{R}_l) = (\mathscr{R}_n \mathscr{R}_m) \mathscr{R}_l.$

4.1*T*8 Given any two distinct points P, Q there is a single transformer $\mathscr{R}$ such $\mathscr{R}P = Q$.

G $\# \wp | PQ$
C $L : L = \text{mid} - \ulcorner PQ \urcorner$
$l : l \supset L \ \& \ l \perp PQ$
T $\mathscr{R}_l P = Q.$

l is unique, since it is the only line satisfying the two conditions in the construction. The transformers $\mathscr{R}$ therefore determine an inverse construction based on 4.1*T*8.

4.1*C*1 The line in which each of two given points is the reflection of the other.

G $\# \wp|\mathrm{PQ}$
C $\mathrm{l}: \mathscr{R}_1\mathrm{P} = \mathrm{Q}$ (as in 4.1*T*8).

4.1*T*9 The reflection of a collinear set is a collinear set.

G $\wp|\mathrm{l}$
S $\mathscr{L}\wp|\{\mathrm{P_i}\}$ $(\mathrm{i} = 1, 2, \ldots)$
C $\{\mathrm{Q_i}\}: \mathrm{Q_i} = \mathscr{R}_1\mathrm{P_i}$
T (i) $\mathscr{L}|\{\mathrm{Q_i}\}$
(ii) $\mathscr{P}|\mathrm{l} \cap \mathrm{P_i P_j} \cap \mathrm{Q_i Q_j}$
P *c*1 $\mathrm{K}: \mathrm{K} = \mathrm{l} \cap \mathrm{P_i P_j}$
*c*2 $\mathrm{L_i}: \mathrm{L_i} = \mathrm{l} \cap \mathrm{P_i Q_i}$
*t*1 $\mathrm{s(KL_i/KL_j)} = \mathrm{s(P_i L_i/P_j L_j)}$
*t*2 $\mathrm{Q_j} \in \mathrm{KQ_i}$
p $\mathrm{s(Q_i L_i/Q_j L_j)} = \mathrm{s(P_i L_i/P_j L_j)}$
$= \mathrm{s(KL_i/KL_j)}$
T $\{\mathrm{Q_j}\} \subset \mathrm{KQ_i}$.

4.1*T*10 The reflection of parallel lines.

T (i) $\mathrm{p} \| \mathrm{q} \Rightarrow \mathscr{R}_1\mathrm{p} \| \mathscr{R}_1\mathrm{q}$
(ii) $\mathscr{D}_{\mathrm{AB}}\mathrm{P} = \mathrm{Q} \Rightarrow \mathscr{D}_{\mathscr{R}_1\mathrm{A}\mathscr{R}_1\mathrm{B}}\mathscr{R}_1\mathrm{P} = \mathscr{R}_1\mathrm{Q}$
P *T*9 & *T*2.

4.1*T*11 The reflection of perpendicular lines.

*G*1 $\wp|\mathrm{l}$
*G*2 $\wp|\{\mathrm{A, B}\} \subset \mathrm{l}$
$\mathrm{P}: \mathrm{AP} \perp \mathrm{PB}$
*G*3 $\mathrm{Q}: \mathrm{Q} = \mathscr{R}_1\mathrm{P}$
T $\mathrm{AQ} \perp \mathrm{QB}$
P 2.6*T*7.

4.1*T*12 A line associated with a given point in relation to two reflection transformations.

*G*1 $\# \wp|\mathrm{lm}, \mathrm{l} \# \mathrm{m}$
*G*2 $\wp|\mathrm{P}$

*C*1	$Q : Q = \mathscr{R}_l P$
	$R : R = \mathscr{R}_m Q$
*C*2	$h : \mathscr{R}_h P = R$
T	$h \supset l \cap m$ (i.e. $\mathscr{P}\|l \cap m \cap h$)
P	2.6*T*8 with A B C HD HE HF
	replaced by P Q R m h l

It is to be noticed particularly that this is a theorem for a selected point in relation to two reflections: given l and m, for different selections of P we obtain different lines h. It follows that the resultant transformation $\mathscr{R}_m \mathscr{R}_l$ is not a reflection.

4.1*T*13*X* A line, which in relation to reflections in two lines meeting in G, is associated with a given line through G.

*G*1	$\# \wp\|lm \;\&\; l \nparallel m \;\&\; l \cap m = G$
*G*2	$\# \wp\|PP', PP' \supset G$
*C*1	$Q, Q' : Q = \mathscr{R}_l P, Q' = \mathscr{R}_l P'$
	$R, R' : R = \mathscr{R}_m Q, R' = \mathscr{R}_m Q'$
*C*2	$h : \mathscr{R}_h P = R$
T	$\mathscr{R}_h P' = R'$.

This theorem establishes, in relation to l, m, a one-to-one correspondence between lines p = GP, and h, (h ⊃ G) but it is not an involution.

4.1*T*14*X* The equivalent reflection for a given point of a sequence of reflections in concurrent lines.

G	$\mathscr{P}\wp\|l_1 \cap l_2 \cap \ldots \cap l_r = G \;\&\; \wp\|G$
S	$\wp\|P$
*C*1	$P' : P' = \mathscr{R}_{l_r} \mathscr{R}_{l_{r-1}} \ldots \mathscr{R}_{l_1} P$
*C*2	$k : \mathscr{R}_k P = P'$
T	$k \supset G$.

The stage is now set for the theorem on which the process of compounding reflections depends, namely, 'the resultant of reflections in three concurrent lines is a reflection in a line concurrent with them'. Stated explicitly the theorem is

G	$\mathscr{P}\#\wp\|l \cap m \cap n = G \;\&\; \wp\|G$
S	$\wp\|A$
*C*1	$\{B, C, D\} : B = \mathscr{R}_l A, C = \mathscr{R}_m B, D = \mathscr{R}_n C$
*C*2	$k : \mathscr{R}_k A = D$
T(i)	$k \supset G$ (this is a consequence of *T* 14)
T(ii)	k is independent of the choice of A, i.e. $\mathscr{R}_k A = \mathscr{R}_n \mathscr{R}_m \mathscr{R}_l A$ for all A so that the transformers themselves related by

$$\mathscr{R}_k = \mathscr{R}_n \mathscr{R}_m \mathscr{R}_l.$$

Join all points concerned to G and take the section by the ideal line u, with orthogonality involution $\mathscr{J}$. Write

$$u \cap \{^{o}GA, GB, GC, GD, l, m, n\} = \{^{o}P, Q, R, S, L, M, N\}$$

and

$$\mathscr{J}\{^{o}L, M, N\} = \{^{o}L', M', N'\}.$$

Then, assuming T(i), if the required theorem, T(ii), is valid, we shall have the following theorem for collinear points in involution:

G	$\mathscr{L}\#\|LMNL'M'$
$C1$	$N' : N' = \mathscr{J}_{L,L';M,M'}$
S	$P : P \in LM$
$C2$	$\{Q, R, S\} : Q = \mathscr{H}_{LL'}P,\ R = \mathscr{H}_{MM'}Q,\ S = \mathscr{H}_{NN'}R.$

$T\mathscr{L}$(i) There exists K, $K \in LM$, such that

$$\mathscr{H}_{PS}K = \mathscr{J}_{L,L';M,M'}K.$$

$T\mathscr{L}$(ii) K is independent of the choice of P.

It is to be noticed that we had already proved $T\mathscr{L}$(i) when we stated the theorem in terms of reflections. In this second form of the theorem, which is a theorem for three pairs of points in involution in a collinear set, we should prove $T\mathscr{L}$(i) independently. In fact we shall obtain a rather more striking result, namely:

$T\mathscr{L}$(iii) The pairs {P, S} form an involution, which is always hyperbolic; the double-points are linearly constructable, and form a pair in the involution $\mathscr{J}_{L,L';M,M'}$. Again the clause in the theorem which states that the points are linearly constructable is a consequence of the construction in the statement of the theorem in terms of reflections, but an independent proof is desirable.

This theorem, $T\mathscr{L}$(iii), is of such significance in the development of the properties of systems of reflections, that we should much like to have a completely geometric proof (using the quadrangle construction for the involution and so on), but in fact there does not seem to be any proof of reasonable length on such a basis—a reason for expecting that this should be so will be found at the conclusion of the algebraic proof.

The theorem to be proved is:

4.1 *T*15 The basic lemma in the proof of the theorem of the three reflections

$G1$	$\mathscr{L}\#\|LMNL'M'$
$C1$	$N' : N' = \mathscr{J}_{L,L';M,M'}N$
S	$P : P \in LM$

*C*2 $Q:Q = \mathscr{H}_{LL'}P$
$R:R = \mathscr{H}_{MM'}Q$
$S:S = \mathscr{H}_{NN'}R$

T(i) The pairs {P, S} belong to an involution,
(ii) this involution always has a pair of double points, each of which can be finitely constructed from L, L′, M, M′, N,
(iii) the pair of double points is a pair of the involution $\mathscr{J}$.

P *s* Basis on LM:
M = O, M′ = U, L = E

*c*1 $s(ON/OE) = c$, $s(ON'/OE) = c'$
so that, for a pair $\{X, X' = \mathscr{J}X\}$,
$s(OX/OE)s(OX'/OE) = cc'$
and $s(OL'/OE) = cc'$

*c*2 $\lambda = \frac{1}{2}(1+cc')$, $\nu = \frac{1}{2}(c+c')$

*c*3 $s(OP/OE) = p$, etc.

*t*1 PQ $\mathscr{H}$ LL′ $\Rightarrow \mathrm{R}(LL'/PQ) = -1$
$\Rightarrow pq-\lambda(p+q)+cc' = 0$
$\Rightarrow q(p-\lambda)+cc'-\lambda p = 0$

*t*2 QR $\mathscr{H}$ MM′ i.e. QR $\mathscr{H}$ OU
$\Rightarrow q+r = 0$

*t*3 RS $\mathscr{H}$ NN′
$\Rightarrow rs-\nu(r+s)+cc' = 0$
$\Rightarrow q(s-\nu)-(cc'-\nu s) = 0$

*t*4 *t*1 & *t*3
$\Rightarrow (p-\lambda)(\nu s-cc')+(s-\nu)(\lambda p-cc') = 0$
$\Rightarrow ps(\nu+\lambda)-(p+s)(cc'+\lambda\nu)+(\lambda+\nu)cc' = 0$

T(i) P and S are pairs in an involution (*t*4)

*t*5 The involution has double points if the equation

$$p^2-2p\frac{cc'+\lambda\nu}{\lambda+\nu}+cc' = 0$$

has roots

*t*6 $$\frac{cc'+\lambda\nu}{\lambda+\nu} = \frac{c(1+c')^2+c'(1+c)^2}{2(1+c)(1+c')} \quad (c2)$$

*t*7 $$p^2-2p\frac{cc'+\lambda\nu}{\lambda+\nu}+cc'$$
$$\equiv \left(p-\frac{c(1+c')}{1+c}\right)\left(p-\frac{c'(1+c)}{1+c'}\right)$$

T(ii) The involution has double-points given by

$$s(OK/OE) = \frac{c(1+c')}{1+c}, \quad s(OK'/OE) = \frac{c'(1+c)}{1+c'}$$

T(iii) $s(OK/OE)s(OK'/OE) = cc'$
$\Rightarrow K' = \mathscr{J}K$.

It should be noticed that the involution $\mathscr{J}$ is restricted neither to have nor not to have double-points; the involution of pairs {P, S}, always has double-points, they are linearly constructable and form a pair in $\mathscr{J}$. The reason for expecting that a geometric proof is likely to be highly intractable is now clear: the result of the construction on which the proof is based must lead from the five points O, E, U, N, N′, where

$$s(ON/OE) = c, \quad s(ON'/OE) = c'$$

to the point

$$s(OK/OE) = c(1+c')/(1+c).$$

We can now state and prove the 'theorem of the three reflections', which, in relation to a given orthogonality involution, provides a construction for a unique fourth line from a set of three given concurrent lines.

4.1*T*16 The theorem of the three reflections.

G	$\mathscr{P}\#\wp\lvert l \cap m \cap n = G \;\&\; \wp\lvert G$	
S	$\wp\lvert A, \#\lvert AG$	
*C*1	$D : D = \mathscr{R}_n\mathscr{R}_m\mathscr{R}_l A$	
*C*2	$k : D = \mathscr{R}_k A$	(4.1*C*1)
T(i)	$k \supset G$	(4.1*T*14)
T(ii)	$\mathscr{R}_k = \mathscr{R}_n\mathscr{R}_m\mathscr{R}_l$	
P	*c* ideal|PQRSLMNL′M′N′KK′ ∊ u as in the introduction to 4.1*T*15	
	*t*1 $u \cap \{^{\mathscr{O}}AB, BC, CD\} = \{^{\mathscr{O}}L', M', N'\}$	
	*t*2 $u \cap \{^{\mathscr{O}}k, PS\} = \{^{\mathscr{O}}K, K'\}$	
T(ii)	$\mathscr{R}_k = \mathscr{R}_n\mathscr{R}_m\mathscr{R}_l$.	

This last relation can be written as

$$\mathscr{R}_m\mathscr{R}_l = \mathscr{R}_n\mathscr{R}_n\mathscr{R}_m\mathscr{R}_l = \mathscr{R}_n\mathscr{R}_k$$

so that the resultant, $\mathscr{R}_m$, $\mathscr{R}_l$, of reflections in non-parallel lines l and m is the same as the resultant, $\mathscr{R}_n\mathscr{R}_k$, of reflections in an arbitrary line n through $l \cap m$ and a line k uniquely determined by l, m, and n.

In terms of physical Euclidean geometry, $\mathscr{R}_m\mathscr{R}_l$ is the ⁑rotation about the point $l \cap m$ through twice the ⁑angle between l and m; it is therefore determined by the point (centre), $l \cap m$, and the ⁑measure of the angle, and can be generated as $\mathscr{R}_n\mathscr{R}_k$, where n is an arbitrary line through $l \cap m$ and k is ⁑inclined to it at the required ⁑angle. For this reason we shall use the notation:

4.1*N*1 *Rotation, centre*: the resultant transformer $\mathscr{R}_m\mathscr{R}_l$ of reflections in two non-parallel lines l, m is a *rotation* of which the *centre* is $l \cap m$.

4.1*T*17*X* The only proper point invariant under a rotation is the centre.

In 4.1*C*1 we defined a construction of a line determined, by reflection transformations, from two points. We have now found a construction which determines a line concurrent with three concurrent lines:

4.1*C*2 Line determined by the three-reflections theorem.

G $\mathscr{P}\wp|\mathrm{l} \cap \mathrm{m} \cap \mathrm{n} = \mathrm{G}\ \&\ \wp|\mathrm{G}$

C $\mathrm{k}: \mathscr{R}_\mathrm{k} = \mathscr{R}_\mathrm{n}\mathscr{R}_\mathrm{m}\mathscr{R}_\mathrm{l}$ (as in 4.1*T*16).

4.1*T*18*X*

G $\mathscr{P}\wp|\mathrm{l} \cap \mathrm{m} \cap \mathrm{n}$

T $\mathscr{R}_\mathrm{l}\mathscr{R}_\mathrm{m}\mathscr{R}_\mathrm{n} = \mathscr{R}_\mathrm{n}\mathscr{R}_\mathrm{m}\mathscr{R}_\mathrm{l}$,

that is, while reflections are not commutative, a sequence of reflections in three concurrent lines has a property similar to but rather weaker than commutativity.

While we have stated the three reflections theorem explicitly for non-parallel lines, the theorem is trivially true for the case

$$\mathrm{l} \cap \mathrm{m} \cap \mathrm{n} = \mathrm{G}\ \&\ \mathrm{ideal}|\mathrm{G},$$

'rotations' being replaced by 'displacements'. Also there is one special case of the theorem for which the proof is direct.

4.1*T*19 The resultant of reflections in a sequence of three concurrent lines of which the first two are perpendicular.

*G*1 $\#\wp\mathscr{P}|\mathrm{l} \cap \mathrm{m} \cap \mathrm{n} = \mathrm{G}\ \&\ \wp|\mathrm{G}$

*G*2 $\mathrm{l} \perp \mathrm{m}$

C $\mathrm{k}: \mathrm{k} \supset \mathrm{G}\ \&\ \mathrm{k} \perp \mathrm{n}$

T $\mathscr{R}_\mathrm{n}\mathscr{R}_\mathrm{m}\mathscr{R}_\mathrm{l} = \mathscr{R}_\mathrm{k}$

P *c* ideal|LL′MM′NN′PQRS (as in *T*15)

*t*1 L′ = M, M′ = L (*G*2)

*t*2 P = R (*t*1)

*t*3 $\mathrm{S} = \mathscr{H}_{\mathrm{NN'}}\mathrm{P}$

*t*4 $\mathrm{N'} = \mathscr{J}\mathrm{N}\ \&\ \mathrm{PS}\ \mathscr{H}\ \mathrm{NN'} \Rightarrow (\mathrm{N}, \mathrm{N'}) = (\mathrm{K}, \mathrm{K'})$

*t*5 k: k = GN′

T $\mathscr{R}_\mathrm{n}\mathscr{R}_\mathrm{m}\mathscr{R}_\mathrm{l} = \mathscr{R}_\mathrm{k}$.

We defer until the next chapter the consideration of the construction of a line l with regard to which given lines a, b are such that $\mathscr{R}_\mathrm{l}\mathrm{a} = \mathrm{b}$, but we discuss below the possible points B on b, when a, b, l are given, which are the reflections (in lines other than l) of a given point A on a. It is to be remembered that

$$\mathscr{R}_\mathrm{l}\mathrm{a} = \mathrm{b} \Rightarrow \mathscr{P}|\mathrm{a} \cap \mathrm{b} \cap \mathrm{l} \quad (4.1T9).$$

4.1*T*20 Points on a given line which can be obtained as reflections of a fixed point on another given line under reflection transformations which transform one of the given lines into the other.

*G*1 $\mathscr{P}\wp|\mathrm{a}\cap\mathrm{b}\cap\mathrm{l}=\mathrm{G}\ \&\ \mathscr{R}_{\mathrm{l}}\mathrm{a}=\mathrm{b}$
S $\wp|\mathrm{A},\ \mathrm{A}\in\mathrm{a}\ \&\ \#|\mathrm{AG}$
C $\mathrm{B}:\mathrm{B}=\mathscr{R}_{\mathrm{l}}\mathrm{A}\quad(\Rightarrow\ \mathrm{B}\in\mathrm{b}\ \&\ \#|\mathrm{BG})$
*G*2 $\mathrm{m}:\mathscr{R}_{\mathrm{m}}\mathrm{a}=\mathrm{b}\ \&\ \mathscr{R}_{\mathrm{m}}\mathrm{A}=\mathrm{B}'\quad(\Rightarrow\ \mathrm{B}'\in\mathrm{b}\ \&\ \mathrm{m}\supset\mathrm{G})$
T(i) either $\mathrm{m}=\mathrm{l}\ \&\ \mathrm{B}'=\mathrm{B}$
or $\mathrm{m}\perp\mathrm{l}\ \&\ \mathscr{D}_{\mathrm{BG}}=\mathscr{D}_{\mathrm{GB}'}$
(ii) $\mathrm{m}\perp\mathrm{l}\Rightarrow\mathrm{lm}\ \mathscr{H}\ \mathrm{ab}$
P *c* $\mathrm{n}:\mathscr{R}_{\mathrm{n}}=\mathscr{R}_{\mathrm{l}}\mathscr{R}_{\mathrm{m}}\mathscr{R}_{\mathrm{b}}$
*t*1 $\mathrm{B}=\mathscr{R}_{\mathrm{l}}\mathrm{A}=\mathscr{R}_{\mathrm{l}}\mathscr{R}_{\mathrm{m}}\mathrm{B}'=\mathscr{R}_{\mathrm{n}}\mathscr{R}_{\mathrm{b}}\mathrm{B}'$
$=\mathscr{R}_{\mathrm{n}}\mathrm{B}'$
either $\mathrm{B}'=\mathrm{B}\ \&\ \mathrm{n}=\mathrm{b}\ \&\ \mathscr{R}_{\mathrm{l}}\mathscr{R}_{\mathrm{m}}=\mathscr{I}$
or $\mathrm{n}\perp\mathrm{b}\ \&\ \mathrm{l}\perp\mathrm{m}$
T(i) either $\mathrm{m}=\mathrm{l}\ \&\ \mathrm{B}'=\mathrm{B}$
or $\mathrm{m}\perp\mathrm{l}\ \&\ \mathrm{G}=\text{mid}-\ulcorner\mathrm{BB}'\urcorner$.
g $\mathrm{m}\perp\mathrm{l}\ \&\ \mathrm{R}_{\mathrm{l}}\mathrm{a}=\mathrm{b}\ \&\ \mathrm{R}_{\mathrm{l}}\mathrm{A}=\mathrm{B}\ \&\ \mathrm{A}\in\mathrm{a}\ \&\ \mathrm{R}_{\mathrm{m}}\mathrm{a}=\mathrm{b}$
*t*2 $\mathrm{l}\cap\mathrm{AB}=\text{mid}-\ulcorner\mathrm{AB}\urcorner$
*t*3 $\mathrm{m}\perp\mathrm{l}\Rightarrow\mathrm{m}\|\mathrm{AB}$
*t*4 $(\mathrm{AB}\cap\mathrm{l})=\mathscr{H}_{\mathrm{AB}}(\mathrm{AB}\cap\mathrm{m})$
T(ii) $\mathrm{lm}\ \mathscr{H}\ \mathrm{ab}$.

It follows that if we know that l is such that $\mathscr{R}_{\mathrm{l}}\mathrm{a}=\mathrm{b}$, then there is one and only one other line m such that $\mathscr{R}_{\mathrm{m}}\mathrm{a}=\mathrm{b}$ and $\mathrm{m}\perp\mathrm{l}$. In physical geometry l and m are the ⁑angle-bisectors of the ⁑angles formed by the lines a and b.

4.1*T*21 The resultant of reflections in three non-concurrent lines can be expressed uniquely as a reflection in a line combined with a displacement parallel to the line (the displacement and the reflection commute, 4.1*T*5).

G $\wp\#|\mathrm{abc}\ \&\ \mathrm{a}\cap\mathrm{b}\cap\mathrm{c}=\emptyset$
C $\mathrm{b}':\mathrm{b}'\supset\mathrm{b}\cap\mathrm{c}\ \&\ \mathrm{b}'\perp\mathrm{a}$
$\mathrm{c}':\mathscr{R}_{\mathrm{c}'}\mathscr{R}_{\mathrm{b}'}=\mathscr{R}_{\mathrm{c}}\mathscr{R}_{\mathrm{b}}$ (4.1*C*2)
$\mathrm{b}'':\mathrm{b}''\supset\mathrm{a}\cap\mathrm{b}'\ \&\ \mathrm{b}''\|\mathrm{c}'$
$\mathrm{d}:\mathscr{R}_{\mathrm{b}''}\mathscr{R}_{\mathrm{d}}=\mathscr{R}_{\mathrm{b}'}\mathscr{R}_{\mathrm{a}}$
*t*1 $\mathscr{R}_{\mathrm{c}}\mathscr{R}_{\mathrm{b}}\mathscr{R}_{\mathrm{a}}=\mathscr{R}_{\mathrm{c}'}\mathscr{R}_{\mathrm{b}''}\mathscr{R}_{\mathrm{d}}$
c $\{\mathrm{L},\mathrm{M}\}\subset\mathrm{d}\ \&\ \mathscr{D}_{\mathrm{LM}}=\mathscr{D}^{2}_{\mathrm{d}\cap\mathrm{b}'',\,\mathrm{d}\cap\mathrm{c}'}$
T $\mathscr{R}_{\mathrm{c}}\mathscr{R}_{\mathrm{b}}\mathscr{R}_{\mathrm{a}}=\mathscr{R}_{\mathrm{d}}\mathscr{D}_{\mathrm{LM}}=\mathscr{D}_{\mathrm{LM}}\mathscr{R}_{\mathrm{d}}$.

4.1*N*2 *Glide-reflection*: the resultant of the reflection in a line and a displacement parallel to the line.

For manipulative purposes it is often better to express the resultant of reflections in three non-concurrent lines in terms of some elements that can be chosen arbitrarily.

4.1*T*22 The resultant of three reflections in non-concurrent lines can be expressed as the resultant of a reflection in an arbitrary line followed by a uniquely determined rotation or displacement.

G $\wp\#|\text{abc}\ \&\ \text{a}\cap\text{b}\cap\text{c}=\emptyset$
S $\wp|\text{l}$
*C*1 $\text{b}':\text{b}'=\langle\text{b}\cap\text{c},\ \text{a}\cap\text{l}\rangle$
*C*2 $\text{n}:\mathscr{R}_\text{n}=\mathscr{R}_\text{c}\mathscr{R}_\text{b}\mathscr{R}_{\text{b}'}$
*C*3 $\text{m}:\mathscr{R}_\text{m}=\mathscr{R}_{\text{b}'}\mathscr{R}_\text{a}\mathscr{R}_\text{l}$
T $\mathscr{R}_\text{n}\mathscr{R}_\text{m}\mathscr{R}_\text{l}=\mathscr{R}_\text{c}\mathscr{R}_\text{b}\mathscr{R}_{\text{b}'}\mathscr{R}_{\text{b}'}\mathscr{R}_\text{a}\mathscr{R}_\text{l}\mathscr{R}_\text{l}$
$\quad=\mathscr{R}_\text{c}\mathscr{R}_\text{b}\mathscr{R}_\text{a}.$

The elements that can be selected arbitrarily are:
the line l: this determines the point G, $=\text{m}\cap\text{n}$, which, if proper, is the centre of the rotation and, if ideal, is the orthogonal conjugate of the point determining the displacement;
the line m: restricted only to pass through G.

4.1*T*23*X* The resultant of three reflections in non-concurrent lines can be expressed as the resultant of a rotation about an arbitrarily assigned centre followed by a uniquely determined reflection.

An immediate consequence of *T* 22 is:

4.1*T*24 The resultant of a sequence of four reflections (i.e., of two rotations) is a rotation or a displacement.

G $\wp|\text{a, b, c, d};\ \mathscr{T}=\mathscr{R}_\text{d}\mathscr{R}_\text{c}\mathscr{R}_\text{b}\mathscr{R}_\text{a}$
C $\text{d}',\text{c}':\mathscr{R}_{\text{d}'}\mathscr{R}_{\text{c}'}\mathscr{R}_\text{a}=\mathscr{R}_\text{d}\mathscr{R}_\text{c}\mathscr{R}_\text{b}$
T $\mathscr{T}=\mathscr{R}_{\text{d}'}\mathscr{R}_{\text{c}'}\mathscr{R}_\text{a}\mathscr{R}_\text{a}=\mathscr{R}_{\text{d}'}\mathscr{R}_{\text{c}'}.$

Thus, since any sequence of four reflections may be replaced by a sequence of two reflections, it follows that any sequence of an $\begin{cases}\text{even}\\ \text{odd}\end{cases}$ number of reflections may be replaced by a sequence of $\begin{cases}\text{two}\\ \text{three}\end{cases}$ reflections. The principal theorem for combinations of reflections now follows:

4.1*T*25 The resultant of a sequence of r reflections is (i) a rotation or a displacement (or identity) when r is even and (ii) a glide-reflection (or a simple reflection) when r is odd.

4.1*N*3 *Congruence transformation*: the resultant of any sequence of reflections.

4.1*T*26*X* The ideal line is over-all invariant under every congruence transformation; under the various types of congruence transformations the following points are invariant:

Congruence Transformation	Sets of invariant points	
	Proper	Ideal
Reflection	Every point of the line of reflection (Each line ⊥ the line of reflection is over-all invariant)	Each point of one pair of orthogonal conjugates
Displacement	$\emptyset$ (Each line ∥ the displacement is over-all invariant)	Each point of the ideal line
Rotation	1 point (the centre)	$\emptyset$
Glide-reflection	$\emptyset$	Each point of one pair of orthogonal conjugates

Congruence transformations have the properties:
the resultant of any two of them is a congruence transformation,
the operation of forming the resultant is associative,
identity is a congruence transformation,
every congruence transformation has an inverse.

The set of congruence transformations therefore forms a (non-commutative) group under the operation of forming the resultant. The group is said to be *generated* by the reflections. The members of the group (apart from identity) are (i) reflections, (ii) rotations and displacements and (iii) glide-reflections.

The resultant of two displacements is a displacement, and the resultant of two rotations or of a rotation and a displacement is a rotation or a displacement, so that displacements, and rotations together with displacements form subgroups of the group of congruence transformations. Moreover, for any congruence transformer $\mathscr{T}$, $\mathscr{T}\mathscr{R}_m\mathscr{R}_1\mathscr{T}^{-1}$ is the product of an even number of reflections and is therefore a rotation or displacement, so that rotations together with displacements form a *normal* (or *self-conjugate*) *subgroup* of the congruence group.

4.1*X*1 Do displacements form a normal subgroup of the congruence group?

We complete this Chapter with statements of the effect of reflections on 'sense' determined by sets of points in the rational, surdic or real net. The proofs of the theorems are immediate.

4.1 *T27X* Senses determined by the reflections of two points.

G $\wp|$l & $\# \wp|$AB
C A′, B′ : A′ = $\mathscr{R}_l$A, B′ = $\mathscr{R}_l$B
T (i) l$\|$AB $\Rightarrow$ ∂(AB/A′B′) = I (1.8*D*1)
(ii) l $\perp$ AB $\Rightarrow$ ∂(AB/A′B′) = J

4.1 *T28X* Reflections of a point in two perpendicular lines.

*G*1 $\wp|$lm & l $\perp$ m & l $\cap$ m = G
*G*2 $\wp|$A & A $\notin$ l & A $\notin$ m
C B, C : B = $\mathscr{R}_l$A, C = $\mathscr{R}_m$A
T $\mathscr{L}\mathscr{B}|$BGC.

(In fact, G = mid-⌜BC⌝.)

4.1 *T29X* Reflections of three points.

G $\mathscr{L}\wp|$ABC, $\wp|$l
C A′, B′, C′ : {^{o}A′, B′, C′} = $\mathscr{R}_l${^{o}A, B, C}
T ∂(ABC/A′B′C′) = J
P *s* g as in 1.8*D*3 & g $\perp$ l.

CHAPTER 4.2

THE MATRIX REPRESENTATION OF REFLECTIONS

We assume that a co-ordinate system $\mathfrak{C}(OE, OF)$ has been selected in the plane, with co-ordinates (x, y, 1) or (x, y, z) over an assigned field, which is such that the ideal points on OE, OF are conjugates in the orthogonality involution (i.e. OE $\perp$ OF). We assume further that a pair of ideal points $(\rho, 1, 0)$, $(\rho', 1, 0)$ are orthogonal conjugates if

$$\rho\rho' = h,$$

where h is a fixed non-square, that is, the two lines $ux+vy+1 = 0$, $u'x+v'y+1 = 0$ are perpendicular if and only if

$$huu' = vv'.$$

For the rational, surdic or real net we may assume $h = -1$, and shall find, as the work progresses, that we shall need to consider these nets separately from the finite nets. Further, nets $\mathfrak{N}_q$ for which $q = 2^m$ are to be excluded from consideration.

To obtain the equations representing reflection in the line

$$l:\boldsymbol{\alpha}^T\boldsymbol{\xi} = ux+vy+1 = 0,$$

we may proceed thus (assuming $uv \neq 0$):

the ideal point L on l is $\mathbf{l} = (-u^{-1}v, 1, 0)$,
the orthogonal conjugate M of L is $\mathbf{m} = (-huv^{-1}, 1, 0)$.

Take P to be

$$\mathbf{p} = (a, b, 1),$$

then any point of PM is $\mathbf{p}+\kappa\mathbf{m}$, and the point $N = PM \cap l$ is given by the value of κ for which

$$\boldsymbol{\alpha}^T(\mathbf{p}+\kappa\mathbf{m}) = 0.$$

If $\overline{P}$, $= \mathscr{R}_l P = \mathscr{H}_{NM} P$, has co-ordinate vector $\overline{\mathbf{p}}$, then, provided $\boldsymbol{\alpha}^T\mathbf{m} \neq 0$,

$$\overline{\mathbf{p}} = \mathbf{p}+2\kappa\mathbf{m} = \mathbf{p}-\frac{2\boldsymbol{\alpha}^T\mathbf{p}}{\boldsymbol{\alpha}^T\mathbf{m}}\mathbf{m}$$

$$= \left(1-\frac{2\mathbf{m}\boldsymbol{\alpha}^T}{\boldsymbol{\alpha}^T\mathbf{m}}\right)\mathbf{p},$$

say

$$\overline{\mathbf{p}} = \mathbf{R}_l\mathbf{p}.$$

4.2 *X*1 Verify that $\mathbf{R}_l^2 = \mathbf{1}$.

Writing $\mathbf{R}_l$ in full we shall find

$$\mathbf{R}_l = (hu^2 - v^2)^{-1}\begin{bmatrix} -(hu^2+v^2) & -2huv & -2hu \\ 2uv & hu^2+v^2 & 2v \\ 0 & 0 & hu^2-v^2 \end{bmatrix}.$$

Since h is a non-square, the condition $hu^2 - v^2 \neq 0$, i.e. $\boldsymbol{\alpha}^T\mathbf{m} \neq 0$, is always satisfied.

$\mathbf{R}_l$ may be expressed as the product

$$\mathbf{R}_l = \begin{bmatrix} 1 & 0 & -2hu\eta \\ 0 & 1 & 2v\eta \\ 0 & 0 & 1 \end{bmatrix}\begin{bmatrix} -(hu^2+v^2)\eta & -2huv\eta & 0 \\ 2uv\eta & (hu^2+v^2)\eta & 0 \\ 0 & 0 & 1 \end{bmatrix},$$

where $$\eta = (hu^2 - v^2)^{-1}, \quad \text{say}$$

$$\mathbf{R}_l = \mathbf{D}\mathbf{R}_0.$$

This expresses the reflection $\mathscr{R}_l$ as the resultant of a reflection in a parallel line through the origin followed by a displacement, so that, for most of the following investigation, we may confine our attention to reflections in lines through 0.

The equation for the reflection in the line $ux + y = 0$ is

$$\begin{bmatrix} \bar{a} \\ \bar{b} \end{bmatrix} = (hu^2 - 1)^{-1}\begin{bmatrix} -(hu^2+1) & -2hu \\ 2u & hu^2+1 \end{bmatrix}\begin{bmatrix} a \\ b \end{bmatrix}, \quad \text{say}$$

$$\bar{\mathbf{p}} = \mathbf{R}_u\mathbf{p}.$$

The 'three reflections' theorem can be verified by carrying out some rather heavy algebraic manipulation. Thus, if $\mathbf{R}_i$ is the matrix $\mathbf{R}_u$ above with u_i replacing u, it can be verified that

$$\mathbf{R}_2\mathbf{R}_1 = (hu'^2 - v'^2)^{-1}\begin{bmatrix} hu'^2+v'^2 & 2hu'v' \\ 2u'v' & hu'^2+v'^2 \end{bmatrix},$$

where $u' = u_2 - u_1$ and $v' = hu_1u_2 - 1$. This is the matrix form for a rotation about 0.

The next step is simplified if we write down the conditions for the validity of the 'three reflections' theorem, and show that they are satisfied. Assume, then, that

$$\mathbf{R}_4 = \mathbf{R}_3\mathbf{R}_2\mathbf{R}_1,$$

i.e. that $$\mathbf{R}_3\mathbf{R}_4 = \mathbf{R}_2\mathbf{R}_1.$$

Writing $u'' = u_3 - u_4$, $v'' = hu_3u_4 - 1$, we shall have to show that a unique value u_4 can be found such that

$$\frac{hu'^2+v'^2}{hu'^2-v'^2} = \frac{hu''^2+v''^2}{hu''^2-v''^2}$$

and
$$\frac{u'v'}{hu'^2-v'^2} = \frac{u''v''}{hu''^2-v''^2}.$$

These two equations are satisfied simultaneously if and only if
$$u'v'' = u''v',$$
a relation which leads to the unique solution
$$u_4 = \frac{u_1-u_2+u_3-hu_1u_2u_3}{1-h(u_2u_3-u_3u_1+u_1u_2)}.$$

This result verifies that the reflection corresponding to $\mathbf{R}_4$, namely, the reflection in $u_4x+y = 0$, is the resultant of the sequence of reflections in $u_1x+y = 0$, $u_2x+y = 0$, $u_3x+y = 0$. The resultant of the four reflections (applied in the order of the indices) is the identity operation, so that the relations may be written symmetrically as
$$\mathbf{R}_4\mathbf{R}_3\mathbf{R}_2\mathbf{R}_1 = \mathbf{1}$$
$$\Leftrightarrow u_1-u_2+u_3-u_4+h(u_2u_3u_4-u_3u_4u_1+u_4u_1u_2-u_1u_2u_3) = 0. \tag{4.2.1}$$

In the case of the rational, surdic or real nets we may take $h = -1$ and recover the familiar Cartesian forms. Take the line l to be†
$$x\cos\phi+y\sin\phi = p \quad \left(u = -\frac{1}{p}\cos\phi, \quad v = -\frac{1}{p}\sin\phi\right)$$
then
$$\mathbf{R}_1 = \begin{bmatrix} -\cos 2\phi & -\sin 2\phi & 2p\cos\phi \\ -\sin 2\phi & \cos 2\phi & 2p\sin\phi \\ 0 & 0 & 1 \end{bmatrix}.$$
The relation (4.2.1) connecting the four quantities $u_i = \cot\phi_i$ becomes
$$\tan(\phi_1-\phi_2+\phi_3-\phi_4) = 0.$$

Let us turn now to the inverse problem of finding the reflection matrices (i) which interchange two given points and (ii) which interchange two given lines through 0.

Take the points to be $\mathbf{p} = (a, b, 1)$ and $\mathbf{p}' = (a', b', 1)$, then, if we leave the geometry out of consideration, we are faced with the formidable task of finding u, v such that the matrix equation
$$\mathbf{p}' = \mathbf{R}_1\mathbf{p}$$
is satisfied. However from the geometrical point of view we know that the line l in which the reflection is to be made satisfies the conditions:
$$l \perp \langle \mathbf{p}, \mathbf{p}' \rangle, \quad \text{i.e.} \quad l \perp (b-b')x-(a-a')y = 0,$$

† Cos ϕ and sin ϕ denote any two elements in the field the sum of whose squares is 1.

and $\quad l \supset \text{mid-}\ulcorner \mathbf{p}, \mathbf{p}'\urcorner$, i.e. $l \supset (\tfrac{1}{2}(a+a'),\ \tfrac{1}{2}(b+b'),\ 1)$.

The equation of l is therefore

$$l : (a-a')\,x - h(b-b')\,y - \tfrac{1}{2}(a^2-a'^2) + \tfrac{1}{2}h(b^2-b'^2) = 0.$$

The matrix $\mathbf{R}_l$ can be constructed from the coefficients in this equation.

Before investigating the second inverse problem, namely, that of finding the lines which determine reflection transformations which interchange two given lines, we need to construct the matrix form for the line which is the reflection of a given line under a given reflection transformation. We may restrict our investigation to lines through 0, and can find the equation of the reflection of $tx+y=0$ in $ux+y=0$ by finding the reflection $(\lambda', -\lambda't', 1)$ of the point $(\lambda, -\lambda t, 1)$. We have

$$\begin{bmatrix} \lambda' \\ -\lambda't' \end{bmatrix} = (hu^2-1)^{-1}\begin{bmatrix} -(hu^2+1) & -2hu \\ 2u & hu^2+1 \end{bmatrix}\begin{bmatrix} \lambda \\ -\lambda t \end{bmatrix},$$

so that
$$\lambda't' = \lambda(hu^2-1)^{-1}\{-2u+(hu^2+1)\,t\},$$
$$\lambda' = \lambda(hu^2-1)^{-1}\{-(hu^2+1)+2hut\},$$

and therefore
$$t' = \frac{2u-(hu^2+1)\,t}{hu^2+1-2hut}.$$

That is,
$$2hutt' - (hu^2+1)(t+t') + 2u = 0 \qquad (4.2.2)$$

is the condition that $tx+y=0$ and $t'x+y=0$ should be reflections in $ux+y=0$. Regarding u as fixed and (t, t') as a variable pair, this is the equation of the involution in which the double lines are $ux+y=0$ and $(1/hu)\,x+y=0$.

If now we take (t, t') to be fixed, equation (4.2.2) is the equation which determines the axes $ux+y=0$ of the reflections which transform $tx+y=0$ into $t'x+y=0$. As a quadratic equation in u, equation (4.2.2) becomes
$$h(t+t')\,u^2 - 2(htt'+1)\,u + t + t' = 0,$$

of which, formally, the roots are

$$u = \frac{1}{h(t+t')}\{htt'+1 \pm \sqrt{\{(ht^2-1)(ht'^2-1)\}}\}. \qquad (4.2.3)$$

First, in the surdic net or real net, where $h = -1$, the roots always belong to the net, so that, as we shall see in more detail in the next Chapter, in these nets we can devise an iterative construction for lines in which one given line is the reflection of another.

On the other hand, in a finite net, the equation has solutions only if $(ht^2-1)(ht'^2-1)$ is a square, that is, only if ht^2-1 and ht'^2-1 are both squares or both non-squares. It follows that the set of lines $tx+y = 0$ through the origin ($x = 0$ is omitted for the moment) splits into two subsets:

S: lines for which ht^2-1 is a square, and
S′: lines for which ht^2-1 is a non-square,

which are such that a line $ux+y = 0$ can be found in which any two members of S or of S′ are each other's reflections, but no line can be found in which a member of S is the reflection of a member of S′.

To take account of $x = 0$ we may replace $t'x+y = 0$ by $x+\tau y = 0$, so that the equation (4.2.3) becomes

$$u = \{h(t\tau+1)\}^{-1}\{ht+\tau \pm \sqrt{[(ht^2-1)(h-\tau^2)]}\} \qquad (4.2.4)$$

and then take $\tau = 0$. We see that, in $\mathfrak{F}_q$, if -1 is a square (i.e. if $q = 4r+1$), then

$$y = 0 \in S \quad \text{and} \quad x = 0 \in S',$$

and if -1 is a non-square (i.e. if $q = 4r-1$), then both

$$y = 0 \in S' \quad \text{and} \quad x = 0 \in S'.$$

So long as we keep to genuine (geometrically constructable) reflections, we can reflect members of S into each other, and members of S′ into each other, but not members of S into members of S′. To complete the system of transformers we could adjoin a set of *pseudo-reflections* (for which we have at present no geometrical construction), and so create a set of transformers which reflect any line into any other. More economically we could adjoin one pseudo-reflection, $\mathscr{R}^*$ say. Then if $l \in S$ and $l' \in S'$ and $\mathscr{R}^*l' = m$, $m \in S$, we can find a line k such that $\mathscr{R}_k m = l$, and therefore $\mathscr{R}_l\mathscr{R}_k\mathscr{R}^*l' = l$. $\mathscr{R}_l\mathscr{R}_k\mathscr{R}^*$ is the pseudo-reflection which transforms l, $\in S$, into l', $\in S'$.

We have to separate the cases in which -1 is a square from those in which it is a non-square and we treat the simpler case first, namely, -1 is a square, so that we cannot specify h beyond the statement that it is a non-square. We take the pseudo-reflection $\mathscr{R}^*$ to be the transformer represented by a matrix $\mathbf{R}^*$, whose elements belong to a quadratic extension of $\mathfrak{F}_q$, which is of precisely the same pattern as a reflection matrix, and which transforms $y = 0\,(\in S)$ into $x = 0\,(\in S')$. We obtain $\mathscr{R}^*$ by substituting $t = 0, \tau = 0$ in equation (4.2.4), and find

$$u^* = 1/\sqrt{(-h)}.$$

With this value of u, equation (4.2.2), reduces to

$$htt'+1 = 0.$$

That is, $t'x+y = 0$ is the pseudo-reflection of $tx+y = 0$ if the ideal points $(\rho', 1, 0)$, $(\rho, 1, 0)$ on the lines are a pair in the involution

$$\rho\rho' = -h.$$

Thus, although the derivation of this involution required recourse to a quadratic extension of the field $\mathfrak{F}_q$, the involution finally obtained can be constructed within the net $\mathfrak{N}_q$.

The case in which -1 is a non-square requires some modification of the procedure above because $x = 0$ and $y = 0$ both belong to the subset S'. We may take $h = -1$, and use for $\mathscr{R}^*$ the transformer which transforms $y = 0$ into a line $x+gy = 0$, where g^2+1 is a non-square. We have

$$u^* = -g+\sqrt{(1+g^2)},$$

and on substitution of this value into equation (4.2.2), we find that $t'x+y = 0$ is the pseudo-reflection of $tx+y = 0$ under $\mathscr{R}^*$ if

$$tt'-g(t+t')-1 = 0.$$

Thus again the pseudo-reflection of a line is constructable within the framework of $\mathfrak{N}_q$.

While the argument has been presented largely in terms of lines through 0, the properties proved have in fact been expressible in terms of the ideal points on the lines. Thus the description of the system can be extended to all lines in the plane by compounding the reflections and pseudo-reflections with displacements, and reducing these by the 'three reflections' theorem, which, because of its algebraic form, can be applied immediately, by referring to the quadratic extension of the field, to the pseudo-reflections.

CHAPTER 4.3

LENGTH, AREA, ANGLE IN THE SURDIC AND REAL NETS

4.3.1 Length

The definition of length, which forms the core of this Chapter, depends on the iterative construction for the double-points of a hyperbolic involution developed in 2.5*C*4. This construction, while in Book 2 it was applied strictly to the process of extending the rational net $\mathfrak{N}_R$ by the adjunction of quadratic surds into the surdic net $\mathfrak{N}_S$, can clearly be applied equally to extend any net in which the elements can be arranged in betweenness order. In particular, of course, in the real net, the method provides an iterative construction based on named points for points already latent in the net. Because, essentially, we require 'lengths' to satisfy relations of betweenness we cannot define length in a finite net, even though we can define congruence.

In order to compare two segments on non-parallel lines we have first to devise a means of transforming one of them into a segment on the line containing the other—naturally the transformation we shall investigate is a congruence transformation, and we are faced then with the problem of finding the axis of the reflection which transforms the line containing one segment into the line containing the other. This is the problem which, in its algebraic form, we solved in Chapter 4.2. In geometrical terms the problem is that of constructing the ⁑angle-bisectors for two given lines, that is, of constructing the pair of lines through a point which both belongs to the orthogonality involution and harmonically separates a given pair of lines through the point. This is the problem that was solved in 2.5*C*4 (p. 170) and the essential part of the solution for the particular case under discussion is contained in 2.6*T*9, 10. We quote these and then express the theorem in terms of reflections.

2.6*T*9, 10 and 4.3*T*1 The axes of the reflections which transform one given line into another.

G $\mathscr{L}\wp|$AOB & OA $\not\perp$ OB
*C*1 H : H $\in$ OB & OH $\perp$ AH
K : K $\in$ OB & OA $\perp$ AK
*t*1 $\mathscr{B}|$OHK (2.6*T*4(ii))

(if we specify B so that $\angle O(AB)$ is an acute angular region (1.7D6), then $\{H, K\} \subset \ulcorner(OB))$

C2 $D : D \in \ulcorner O(H)$ & $s(OD/OH) = s(OK/OD)$ (2.5C4)
$D' : D' = \mathscr{D}_{DO}\, O$
$A' : A' = \mathscr{D}_{AO}\, O$
T9 $AD \perp A'D$ & $AD \perp AD'$
C3 $\{L, M\} : L = \text{mid-}\ulcorner AD \urcorner,\ M = \text{mid-}\ulcorner AD' \urcorner$
T10 $OL \perp AD$ & $OM \perp AD'$.

4.3T1 $D = \mathscr{R}_{OL} A$ & $D' = \mathscr{R}_{OM} A$
$\ulcorner O(H) = \mathscr{R}_{OL} \ulcorner O(A)$ & $\ulcorner O(\overset{\frown}{H}) = \mathscr{R}_{OM} \ulcorner O(A)$.

The construction described in this theorem cannot be carried out when $OB \perp OA$ (since ideal$|$K), but we can replace it by a construction in two steps.

4.3T2 T1 in case $OA \perp OB$.
G $\wp|OAB$ & $OA \perp OB$
S $\wp|G : G \in \angle O(AB)$
C1 $l, P : P \in \ulcorner O(G)$ & $\mathscr{R}_l A = P$
C2 $m, D : D \in \ulcorner O(B)$ & $\mathscr{R}_m P = D$
C3 $k : k = \langle O, \text{mid-}\ulcorner AD \urcorner \rangle$
T $\mathscr{R}_k A = \mathscr{R}_m \mathscr{R}_l A$ (4.1T12).

These theorems provide the solution to the problem of establishing a 'universal scale', that is, of determining a unique number (a member of the surdic or real field), which, when a base (unit) segment $\ulcorner OE \urcorner$ has been assigned, may be assigned to a given segment $\ulcorner PQ \urcorner$. In making the construction we do not in fact in the first place derive a unique number, since the process we are to use is reflection and there are two different lines in which we may reflect points of PQ into points of OE. However, from 4.3T1, if $(P', Q', P'', Q'') \in OE$ & $\mathscr{R}_l\{{}^{\mathscr{O}}P, Q\} = \{{}^{\mathscr{O}}P', Q'\}$ & $\mathscr{R}_m\{{}^{\mathscr{O}}P, Q\} = \{{}^{\mathscr{O}}P'', Q''\}$ (and $\#|lm$ so that $l \perp m$), then $s(P'Q'/P''Q'') = -1$. Consequently, we define *length* as a *positive quantity*, and when lengths of segments on the same line (or in any system in which sense is determinate and relevant) have to be considered, we replace the positive 'length' by a positive or negative 'magnitude' (for segments on parallel lines, the span-ratio), the signs being made to conform to the relevant senses.

4.3D1 Length.
G1 $\mathscr{L}\wp|OEF$ & $\mathfrak{N}_S(O, E, F)$ or $\mathfrak{N}_{\mathfrak{R}}(O, E, F)$
G2 $\wp|P, Q : \{P, Q\} \in OF$
C1 $l, P' : P' \in OE$ & $P' = \mathscr{R}_l P$

(l is either member of a pair of perpendicular lines)

$Q' : Q' = \mathscr{R}_1 Q$

*C*2 $s(P'Q'/OE)$

D length ⌜PQ⌝ in relation to ⌜OE⌝:

$s^+(PQ/OE) = |s(P'Q'/OE)|$.

The fundamental property of lengths in physical Euclidean geometry is that the sum of the ⁑lengths of any two side-segments of a triangle is greater than the ⁑length of the third. The 'length' defined above has this property; we have first to prove

4.3*T*3 The length of the hypotenuse of a right-angled triangle is greater than the length of either of the other side-segments.

G $\mathscr{L}\wp|OAH \;\&\; OH \perp AH$

C $D : D \in {}^{\ulcorner}O(H) \;\&\; s^+(OD/OA) = 1$ (2.6*T*9)

T $\mathscr{B}|OHD$

P *c* $U : U \in OD \;\&\; \text{ideal}|U$

*t*1 $AK \perp AO \;\&\; AH \perp AU \Rightarrow \mathscr{S}|OHKU$

$\Rightarrow \mathscr{B}|OHK$

*t*2 $s(OH/OD) = s(OD/OK) \;\&\; \mathscr{B}|OHK$

$\Rightarrow \mathscr{B}|OHDK$

T $\mathscr{B}|OHD$.

4.3*T*4 The sum of the lengths of two side-segments of a triangle is greater than the third.

G $\mathscr{L}\wp|PQR$

T $s^+(PQ/QR) + s^+(RP/QR) > 1$

P *c* $N : N \in QR \;\&\; PN \perp QR$

g (i) and *g* (ii) are alternatives

g (i) $\mathscr{B}|QNR$

t (i) $s^+(PQ/QR) > s(QN/QR)$ (*T*3)

$\& \; s^+(PR/QR) > s(NR/QR)$

$\& \; s(QN/QR) + s(NR/QR) = 1$

$s^+(PQ/QR) + s^+(PR/QR) > 1$

g (ii) $\mathscr{B}|QRN$

t (ii) $s^+(PQ/QR) > s(QN/QR) > 1$

T $s^+(PQ/QR) + s^+(PR/QR) > 1$.

Finally we can prove

4.3*T*5 The Pythagoras theorem.

G $\mathscr{L}\wp|PQR \;\&\; PQ \perp PR$

C $H : H \in QR \;\&\; PH \perp QR$

$D : D \in \ulcorner QR \urcorner \;\&\; s^{+}(QD/QP) = 1$

$D' : D' \in \ulcorner QR \urcorner \;\&\; s^{+}(RD'/RP) = 1$

*t*1 $\mathscr{B}|QD'HDR$

  p $\mathscr{B}|QHR \;\&\; s(QH/QD) = s(QD/QR)$
  $\&\; s(RH/RD') = s(RD'/RQ)$

*t*2 $s(QH/QR) = \{s^{+}(QP/QR\}^{2}$

  p $s(QD/QR) = s^{+}(QP/QR)$
  $\&\; s(QD/QR)\, s(QD/QH) = 1$
  $\Rightarrow s^{+}(QP/QR)\, s^{+}(QP/QH) = 1$
  $\Rightarrow \{s^{+}(QP/QR)\}^{2} s(QR/QH) = 1$

*t*3 $s(HR/QR) = \{s^{+}(RP/RQ)\}^{2}$

  p $s(RD'/RQ)\, s(RD'/RH) = 1$
  $\&\; s(RD'/RQ) = s^{+}(RP/RQ)$
  $\Rightarrow s^{+}(RP/RQ)\, s^{+}(RP/RQ)\, s(RQ/RH) = 1$

*t*4 *t*2 & *t*3

  $\Rightarrow \{s^{+}(QP/QR)\}^{2} + \{s^{+}(RP/RQ)\}^{2}$
  $= s(QH/QR) + s(HR/QR)$
  $= 1$

T $\{s^{+}(PQ/QR)\}^{2} + \{s^{+}(PR/QR)\}^{2} = 1.$

The Euclidean picture is now complete: we are at the stage at which we could introduce, if we wished to do so, rectangular Cartesian coordinates with the required formula for distance!

4.3.2 Area

There are two phases in the derivation of the definition of a number which is to correspond to the physical notion of the ⁑area of a region. The first is the devising of a means of determining a property of a region, momentarily to be called its 'content', which has the recognisable additive properties of ⁑area. The second is the establishing of a numerical comparison between 'contents', which, by relating the given region to a designated region of 'unit content', enables us to assign to the region a number, its geometrical area, which corresponds directly to the physical ⁑area of the physical ⁑region.

The relation between 'content' and 'area' is much the same as that between 'displacements' and 'span-ratio'. Displacements form a commutative group, and span-ratios are the elements of a field. In the same way 'contents' form a commutative group and 'areas' are, in the present context, members of the field generated by the square-root process on the rationals ($\mathfrak{F}_{S}$) or of the field of reals ($\mathfrak{F}_{\mathfrak{R}}$). There are good, and indeed obvious, reasons for writing the group operation for 'displacements' multiplicatively; there are equally good reasons for writing the group operations for 'contents' additively. Again, while we used $\mathscr{D}_{AB}$ and $\mathscr{D}_{BA}$ for inverse members of the displacement group, and we could equally well write $\mathscr{A}_{P_1 \dots P_n}$ and $\mathscr{A}_{P_n \dots P_1}$ for inverse members of the content group, we shall find it more convenient not to have to consider senses of description of the boundary of a region, and to write $\mathscr{A}^+_{P_1 \dots P_n}$ for the member of the group associated with the region $\mathfrak{R}_{P_1 \dots P_n}$, and $-\mathscr{A}^+_{P_1 \dots P_n}$ for its inverse in the group.

Throughout this section we shall be concerned with 'simple polygonal regions', and the word 'region' is nearly always to be interpreted in that sense. At the centre of the argument are the ideas of 'cross-segments' (1.7 *D* 10), and 'triangular dissection' (1.7 *D* 11).

4.3 *D* 2 Content

D Associated with a region $\mathfrak{R}_{P_1 \dots P_n}$ there is a unique element $\mathscr{A}^+_{P_1 \dots P_n}$ of an additive group,† to be called its *content*, with the properties:

(i) Congruent regions are associated with the same element of the group.

(ii) The sum of the contents of the two regions formed from a given region by adjoining a cross-segment to the boundary is the same as the content of the given region, i.e.

G $\mathfrak{R}_{P_1 \dots P_n}$ & $\ulcorner P_i P_j \urcorner$ a cross-segment & $i < j$

D (ii) $\mathscr{A}^+_{P_1 \dots P_i P_j P_{j+1} \dots P_n} + \mathscr{A}^+_{P_i P_{i+1} \dots P_{j-1} P_j} = \mathscr{A}^+_{P_1 P_2 \dots P_{n-1} P_n}$.

We have to prove that statement *D* (ii) is consistent, namely, that whatever methods of subdivision and recombination are adopted, the final sum $\mathscr{A}^+_{P_1 \dots P_n}$ is reached as the content of the region $\mathfrak{R}_{P_1 \dots P_n}$. We prove consistency in the following sequence of lemmas.

4.3 *T* 6 Simple dissection of a triangle.

G $\mathscr{L}\wp|ABC$ & $K \in \ulcorner BC \urcorner$

(i.e. a quadrilateral, boundary $\triangle ABKC$, with $\mathscr{L}|BKC$)

T $\ulcorner AK \urcorner$ is the only cross-segment of $\mathfrak{R}ABKC$ so that the relation

$$\mathscr{A}^+_{ABK} + \mathscr{A}^+_{ACK} = \mathscr{A}^+_{ABC}$$

cannot lead to an inconsistency.

4.3 *T* 7 Dissection of a convex quadrilateral region.

G $\mathscr{L}_3\wp|ABCD$ & $D \in \ulcorner AC(\tilde{B})$ & $C \in \ulcorner BD(\tilde{A})$

† An additive group is simply a commutative group with the symbol '+' used to represent the operation of combination.

(so that both ⌜AC⌝ and ⌜BD⌝ are cross-segments)

T $\mathscr{A}^{+}_{ACB}+\mathscr{A}^{+}_{ACD}=\mathscr{A}^{+}_{BDA}+\mathscr{A}^{+}_{BDC}$

P *c* K : K = BD ∩ AC (= ⌜BD⌝ ∩ ⌜AC⌝)

T $\mathscr{A}^{+}_{ACB}+\mathscr{A}^{+}_{ACD}=\mathscr{A}^{+}_{AKB}+\mathscr{A}^{+}_{CKB}+\mathscr{A}^{+}_{CKD}+\mathscr{A}^{+}_{AKD}$
$=\mathscr{A}^{+}_{BDA}+\mathscr{A}^{+}_{BDC}.$

⌜AC⌝ and ⌜BD⌝ are the only cross-segments of the quadrilateral.

For a non-convex quadrilateral region, say ℜABCD with D ∈ ⌜AC($\tilde{\mathrm{B}}$) but C ∈ ⌜BD(A), there is only one cross-segment, ⌜AC⌝, and the statement is consistent trivially.

4.3*T*8 Dissections of a convex hexagonal region.

G ℜ = ℜABCDEF with the property that the segment whose end-points are any pair of non-adjacent vertices is a cross-segment

T The sum (in the additive group) of the contents of the four triangular regions into which ℜ is split by a triangular dissection is independent of the choice of cross-segments determining the dissection

P There are three possible types of dissections of ℜ, exemplified by:

{⌜AC⌝, ⌜AD⌝, ⌜AE⌝}: triangular regions ABC, ACD, ADE, AEF,
{⌜AC⌝, ⌜AD⌝, ⌜DF⌝}: triangular regions ABC, ACD, ADF, DEF,
{⌜AC⌝, ⌜CE⌝, ⌜EA⌝}: triangular regions ABC, CDE, EFA, ACE.

If we take two distinct dissections, and one triangular region from each, then either:

(i) the regions have at least one common vertex, and their intersection is either empty or triangular or quadrilateral, or

(ii) they have no common vertex, and their intersection is exemplified by:

ℜABC ∩ ℜDEF = ∅,
ℜABD ∩ ℜCEF = convex quadrilateral region,
ℜACE ∩ ℜBDF = convex hexagonal region.

In all cases except the last (which is unique) the intersections of pairs of triangular regions, one from each of the two dissections, are either empty, or triangular regions, or quadrilateral regions; each quadrilateral region may be split into two triangular regions (in either of two ways, leading to the same sum, *T*7). Under the combination of the two triangular dissections, and the subsequent splitting of the quadrilateral regions, the region ℜ is dissected into a set S of triangular regions, which may be recombined in different associations to produce the two triangular dissections. The sum of the contents of the regions of the set S is the content $\mathscr{A} = \mathscr{A}^{+}_{\mathrm{ABCDEF}}$ of $\mathfrak{R}$, and (except for the one type of dissection that has still to be considered) $\mathscr{A}$ is independent of the choice of triangular dissection. The outstanding case is that of dissections by the sets of cross-segments {⌜AC⌝, ⌜CE⌝, ⌜EA⌝} and {⌜BD⌝, ⌜DF⌝, ⌜FB⌝}. Each of these may be combined with any other triangular dissection to produce only triangular and quadrilateral regions, and hence to lead to the same content $\mathscr{A}$ as that dissection determines. The value of $\mathscr{A}$ determined by either of the exceptional dissections is therefore the same as that determined by any other triangular dissection.

4.3*T*9 Under the assumption 4.3*D*2(ii) the content of a region $\mathfrak{R}_{\mathrm{P}_1 \ldots \mathrm{P}_n}$ is the sum of the contents of the triangular regions in a triangular dissection.

P Select any one of the triangular regions of the dissection, and adjoin successively triangular regions of the section in such a way that each has at least one side-segment common with some preceding member of the sequence. At each stage we obtain a new simple polygonal region, to which we may apply 4.3*D*2(ii) in the case where the adjoined cross-segment cuts off a triangular region.

This proves the theorem, but still does not prove the validity of the assumption in 4.3*D*2(ii). To prove this we have to prove for a general simple polygonal region the result that we proved in *T*8 for a convex hexagonal region.

4.3*T*10*X* The sums of the contents of the triangular regions formed in two distinct triangular dissections of a region $\mathfrak{R} = \mathfrak{R}_{\mathrm{P}_1 \ldots \mathrm{P}_n}$ are equal.

P Combine the two dissections and use the properties:

(i) Every vertex of a triangle in a triangular dissection of $\mathfrak{R}$ is a vertex of $\mathfrak{R}$.

(ii) No vertex of a triangle in one dissection lies inside any triangular region of the other dissection.

(iii) Every side-segment of a triangle in one dissection meets either two or none of the side-segments of any triangle in the other dissection.

(iv) Two triangular regions, one from each dissection, which have no common vertices have common a convex hexagonal region.

(v) The intersection of two triangular regions, one from each dissection, which have common one or two or three vertices, is either empty, or a triangular region or a quadrilateral region.

(vi) Any convex quadrilateral or convex hexagonal region can be dissected by cross-segments into triangular regions the sum of whose contents is independent of the selection of cross-segments.

We turn now to the second phase of the investigation, the prescription of a standard 'unit area', which will enable us to replace the numerically indefinite 'content' by a number representing the area. So much of the development here is similar to the path of development from displacements, through rational span-ratios to real span-ratios, that it is only necessary to indicate the general trend of the argument.

4.3 *D*3 The unit of area (the unit ⁑square).

*G*1 $\# \wp|$OE (⌜OE⌝ is the basis of measurement for length)

*G*2 $\wp|$ABCD & s(AB/OE) = s(AD/OE) = 1
& AB ⊥ AD & C = $\mathscr{D}_{AB}$D (= $\mathscr{D}_{AD}$B)

D $\mathscr{A}^+_{ABCD}$: the unit for measuring area.

Following the convention adopted with span-ratios we should first introduce a new symbol, say $\alpha^+(\mathfrak{R}_{P_1\ldots P_n}/\mathfrak{R}_{ABCD})$, derived from a relation between $\mathscr{A}^+_{P_1\ldots P_n}$ and $\mathscr{A}^+_{ABCD}$, but there is no loss in clarity in using instead of the α^+ symbol the content symbol $\mathscr{A}^+_{P_1\ldots P_n}$ again, on the understanding that $\mathscr{A}^+$ is no longer only a member of a group, but is a number derived ultimately from the basis ⌜OE⌝.

The steps now are: for a rectangular region,

$$\mathfrak{R}\text{PQRS} \ \&\ \text{PQ} \perp \text{PS} \ \&\ \text{R} = \mathscr{D}_{PQ}\text{S}.$$

(i) If s⁺(PQ/OE) and s⁺(PS/OE) are both integers, say m and n, then $\mathscr{A}^+_{PQRS}$ = mn (proof by dissecting $\mathfrak{R}$PQRS into unit squares).

(ii) The same for m, n rational.

(iii) The same for m, n real.

(iv) For a right-angled triangular region $\mathfrak{R}$PQR with PQ ⊥ QR. $\mathscr{A}^+_{PQR} = \frac{1}{2}\mathscr{A}^+_{PQRS}$ (S = $\mathscr{D}_{QP}$R; congruent regions have equal contents and therefore equal areas).

(v) For a triangular region $\mathfrak{R}$PQR.

G $\mathscr{L}\wp|$PQR
K : K ∈ QR & KP ⊥ QR

T $\mathscr{A}^+_{PQR} = \frac{1}{2}|$s(QR/OE) s(PK/OE)$|$

P *g*(i) and (ii) are alternatives
g(i) K ∈ ⌜QR⌝
t(i) $\mathscr{A}^+_{PQR} = \mathscr{A}^+_{PQK} + \mathscr{A}^+_{PRK}$
g(ii) K ∈ ⌜Q(R)
t(ii) $\mathscr{A}^+_{PQR} = \mathscr{A}^+_{PQK} - \mathscr{A}^+_{PRK}$

T $\mathscr{A}^+_{PQR} = \frac{1}{2}|$s(QR/OE) s(PK/OE)$|$.

(vi) Theorem (v) provides the means of finding the area of any triangular region: the area of any simple polygonal region may therefore be found as the sum of the areas of the constituent triangular regions in any triangular dissection.

4.3.3 The measure of an angular region

The pattern is by now so well established that no more than the briefest description is needed of the development of a method of assigning a measure to angular regions. As we did for areas, we begin by defining the 'content' $\measuredangle^{+}\mathrm{A(BC)}$ of the angular region $\angle\mathrm{A(BC)}$, which is defined in 1.7*D*6.

4.3*D*4 The content of an angular region.

Associated with an angular region $\angle\mathrm{A(BC)}$ there is a unique element $\measuredangle^{+}\mathrm{A(BC)}$ of an additive group, to be called the *content* of the region, with the properties:

(i) Congruent angular regions are associated with the same element of the group.

(ii) The sum of the contents of two adjacent angular regions within the same half-plane.

G $\mathscr{L}_3\wp|\mathrm{ABCD}\ \&\ \mathrm{D}\in{}^{\ulcorner}\mathrm{AB(C)}\ \&\ \mathrm{D}\notin\angle\mathrm{A(BC)}$
D $\measuredangle^{+}\mathrm{A(BC)}+\measuredangle^{+}\mathrm{A(CD)}=\measuredangle^{+}\mathrm{A(BD)}$.

(iii) The content Ω of a half-plane.

G $\mathscr{L}\wp|\mathrm{ABD}\ \&\ \mathscr{B}|\mathrm{DAB}$
$\mathrm{C}:\mathrm{C}\notin\mathrm{AB}$
D $\Omega:\Omega=\measuredangle^{+}\mathrm{A(BC)}+\measuredangle^{+}\mathrm{A(DC)}$
$(=\measuredangle^{+}\mathrm{A(BC)}+\measuredangle^{+}\mathrm{A(\tilde{B}C)})$.

(iv) The sum of the contents of two adjacent angular regions not within the same half-plane.

G $\mathscr{L}_3\wp|\mathrm{ABCD}\ \&\ \mathrm{D}\in{}^{\ulcorner}\mathrm{AB(\tilde{C})}\cap{}^{\ulcorner}\mathrm{AC(\tilde{B})}$
(i.e. $\mathrm{D}\in\angle\mathrm{A(\widetilde{BC})}$)
D $\measuredangle^{+}\mathrm{A(BC)}+\measuredangle^{+}\mathrm{A(CD)}=\Omega+\measuredangle\mathrm{A(D\tilde{B})}$.

4.3*T*11 *G* $\mathscr{L}_3\wp|\mathrm{ABCD}$ as in 4.3*D*4(iv)
T $\measuredangle^{+}\mathrm{A(BC)}+\measuredangle^{+}\mathrm{A(CD)}+\measuredangle^{+}\mathrm{A(BD)}=2\Omega$.

We may close the definition with the statement:

(v) $2\Omega=$ the identity element in the group of contents, that is, 2Ω is the same element as the 'zero' in the usual notation for additive groups.

We have now to assign numbers for the measures of angular regions, that is, we have to designate a unit angular region with which other regions may be compared. An obvious unit would be Ω. It must be borne in mind that when we were dealing with lengths and areas we could subdivide any given length or area in any way we chose; in particular, if we had, in relation to a given unit length, two segments of lengths r and s, $(r, s) \in \mathfrak{F}_{\mathfrak{R}}$, we could construct geometrically a segment of length r/s. With the measures of angular regions we have no such flexibility; effectively the only construction that is available besides direct addition is that of determining an angular region of measure $\frac{1}{2}\alpha$ when we are given one of measure α. Thus, if we are given $\mathscr{L}\wp|ABC$, we can construct L such that $\ulcorner A(C) = \mathscr{R}_{AL}\ulcorner A(B)$ (4.3T1), and then $\measuredangle^+A(BL) = \measuredangle^+A(CL) = \frac{1}{2}\measuredangle^+A(BC)$. We can therefore in general do no better than approximate to the measure of an assigned angle by performing the sequence of constructions symbolised in the following binary expressions:

$$\left(\frac{a_1}{2}+\frac{a_2}{4}+\ldots+\frac{a_n}{2^n}\right)\Omega > \measuredangle^+A(BC) > \left(\frac{a_1}{2}+\frac{a_2}{4}+\ldots+\frac{a_r}{2^r}\right)\Omega,$$

where $r < n$, $a_i \in \{0, 1\}$, $a_r = 1$, $a_{r+1} = \ldots = a_{n-1} = 0$, $a_n = 1$.

In order to obtain the usual degree measure of an angle we have to take $\Omega = 180$. To obtain the natural measure (radians) we may use as a basis the equivalent of the physical trigonometrical relation

$$\tan\theta > \theta > \sin\theta$$

which is true, over the range $0 < \theta < 1$ say, only when θ is measured in radians.

4.3T12 Definition of π as limiting form of some inequalities in measures of angular regions.

G $\quad \mathscr{L}\wp|OAB$ & $OA \perp OB$

$C1$ $\quad C : \mathscr{R}_{OC}\ulcorner O(A) = \ulcorner O(B)$ (4.3T2)

$C2$ $\quad H_1 : H_1 \in OC$ & $AH_1 \perp OC$

$K_1 : K_1 \in OC$ & $AK_1 \perp OA$

$D_1 : D_1 \in \ulcorner O(C)$ & $s(OH_1/OD_1) = s(OD_1/OK_1)$

$(\Rightarrow s^+(OD_1/OA) = 1)$

$t1$ $\quad s^+(AK_1/OA) = 1, \{s^+(OK_1/OA)\}^2 = 2$

$C3$ $\quad$ in succession, H_i, K_i, D_i $(i = 2, 3, \ldots, n, \ldots)$

$H_i : H_i = \text{mid-}\ulcorner AD_{i-1}\urcorner$ $(\Rightarrow AH_i \perp OH_i)$

$K_i : K_i = OH_i \cap AK_1$

$D_i : D_i \in \ulcorner H_iK_i\urcorner$ & $s(OH_i/OD_i) = s(OD_i/OK_i)$

$(\Rightarrow D_{i-1} = \mathscr{R}_{OK_i}A$ & $K_iD_{i-1} \perp OK_{i-1})$

$t2$ $\quad s^+(AH_i/OA) < s^+(AD_i/OA)$ (4.3T1)

$\Rightarrow s^+(AH_i/OA) < 2s^+(AH_{i+1}/OA)$

$$\Rightarrow \frac{s^+(AH_i/OA)}{\measuredangle^+O(AK_i)} < \frac{s^+(AH_{i+1}/OA)}{\measuredangle^+O(AK_{i+1})}$$

$$t3 \quad s^+(AK_i/OA) > s^+(AK_{i+1}/OA) + s^+(K_{i+1}D_i/OA)$$
$$\Rightarrow s^+(AK_i/OA) > 2s^+(AK_{i+1}/OA)$$
$$\Rightarrow \frac{s^+(AK_i/OA)}{\measuredangle^+O(AK_i)} > \frac{s^+(AK_{i+1}/OA)}{\measuredangle^+O(AK_{i+1})}$$

$$t4 \quad 1 < s^+(AK_n/AH_n) < 1 + \frac{1}{2^{n-2}}$$

$$p \quad 0 < \{s^+(AK_n/AH_n)\}^2 - 1$$
$$= \{s^+(H_nK_n/AH)\}^2$$
$$= \{s^+(AK_n/OA)\}^2$$
$$< \frac{1}{2^{2n-2}}\{s(AK_1/OA)\}^2 = \frac{1}{2^{2n-2}}$$

$$T \quad \frac{s^+(AK_1/OA)}{\frac{1}{2}\Omega} > \frac{s^+(AK_2/OA)}{\frac{1}{4}\Omega} > \dots > \frac{s^+(AK_n/OA)}{1/2^n\Omega} >$$
$$\dots > k > \dots > \frac{s^+(AH_n/OA)}{1/2^n\Omega} > \dots > \frac{s^+(AH_2/OA)}{\frac{1}{4}\Omega}$$
$$> \frac{s^+(AH_1/OA)}{\frac{1}{2}\Omega}$$
$$\& \; s(AK_n/AH_n) < 1 + \frac{1}{2^{n-2}}.$$

Now take $k = 1$, then

$$2^n s^+(AK_n/OA) > \Omega > 2^n s^+(AH_n/OA)$$

and Ω is the common limiting value of the two expressions. This is the number usually denoted by π. In physical trigonometrical terms π has been defined here as the common limiting value of $2^n \tan\phi_n$ and $2^n \sin\phi_n$, where ϕ_n is the angle obtained, by repeated halving, as $1/2^{n-1} \times$ right-angle.

EXCURSUS ON BOOK 4

EXCURSUS 4.1

CONGRUENCE WITHOUT DISPLACEMENTS

In the Excursus we are to develop a geometry in which perpendiculars depend on a polarity system (Excursus 2.4) instead of on an orthogonality involution as in Chapter 2.6. The system that is to be synthesised is in fact classical 'Non-Euclidean Geometry', but we do not propose to investigate it far enough to exhibit to any significant extent the properties of such a geometry. For these the reader might refer, for example, to H. F. Baker: 'Principles of Geometry', Volume 2, Chapter 5.

First let us recapitulate the properties that are specified in Chapter 2.6 for perpendicular lines:

(i) The pairs of perpendicular lines through a proper point belong to an involution.

(ii) There is through a proper point a single line perpendicular to a given line.

(iii) No proper line is perpendicular to itself.

The fourth property specified in Chapter 2.6, namely,

$$\mathrm{a} \perp \mathrm{c} \;\&\; \mathrm{b} \perp \mathrm{c} \Rightarrow \mathrm{a} \| \mathrm{b}$$

is of course no longer relevant.

We were able to satisfy these conditions (including the fourth) by prescribing an elliptic orthogonality involution on the ideal line, and defining perpendicular lines as lines whose ideal points are conjugate in this involution. We shall refer to this geometry as Affine Orthogonal Geometry and the system that is to be described in this Excursus as Projective Orthogonal Geometry.

Let us assume that, instead of an ideal line and an orthogonality involution $\mathscr{I}$, a polarity system $\mathscr{Q}$ (Excursus 2.4.2N1) has been prescribed, and that in relation to $\mathscr{Q}$ the following definition of perpendiculars has been proposed:

$$\mathrm{a} \perp \mathrm{b} \Leftrightarrow \mathrm{b} \supset \mathscr{Q}\mathrm{a},$$

i.e. a and b are conjugate in $\mathscr{Q}$ (Excursus 2.4.2N3). Perpendiculars so defined have the properties:

(i) The pairs of perpendicular lines through a point belong to an involution. If $\mathcal{Q}$ is of type EE (Excursus 2.4.4 (i)) there are no singular involutions in this system. If $\mathcal{Q}$ is not of type EE, and in any case if the net is finite, there are some points for which the involutions are singular, and these would need special consideration.

(ii) The line through P and perpendicular to l is PM where $\mathrm{M} = \mathcal{Q}\mathrm{l}$, and is unique except when $\mathrm{P} = \mathrm{M}$. This exceptional point, which is different for each line l, replaces the ideal points in Affine Orthogonal Geometry, which are exceptional for all lines.

(iii) A line l is self-perpendicular only if $\mathrm{l} = \mathcal{Q}\mathrm{l}$; so that, if the geometric net is $\mathfrak{N}_S$ or $\mathfrak{N}_R$, we can ensure that no line is perpendicular to itself by specifying that $\mathcal{Q}$ is of type EE. In a finite net $\mathfrak{N}_q$ there will be $q+1$ lines which are self-perpendicular in any given polarity (Excursus 2.4.4 (ii)).

Affine Orthogonal Geometry is the special case of Projective Orthogonal Geometry in which, if the geometric net is extended so that all involutions have double-points and there is on any line in the plane a pair of double-points of $\mathcal{Q}$ the locus of which is a conic, then the system of tangent-lines to $\mathcal{Q}$ is specialised to the system formed from lines which contain either of two given points. In algebraic terms, the condition

$$\alpha u_i u_j + \beta v_i v_j - 1 = 0$$

(with $\alpha < 0$, $\beta < 0$ for type EE) for perpendicular lines in relation to $\mathcal{Q}$ is replaced by

$$\alpha u_i u_j + \beta v_i v_j = 0$$

(with α/β a non-square) for $\mathcal{J}$.

The proposed definition of perpendiculars therefore provides a reasonable starting point for an investigation. We observe at the outset that there is a significant difference between the affine and projective systems: in Affine Orthogonal Geometry points and lines behave quite differently, while in Projective Orthogonal Geometry there is an exact duality between them. We can define therefore in relation to $\mathcal{Q}$ not only pairs of perpendicular lines, but also pairs of 'perpendicular points'.

Exc. 4.1 *D*1 Perpendicular lines and points.

G Polarity system $\mathcal{Q}$

D(i) $\mathrm{l} \perp \mathrm{m} \Leftrightarrow \mathrm{m} \supset \mathcal{Q}\mathrm{l}$

($\Leftrightarrow \mathrm{l} \supset \mathcal{Q}\mathrm{m}$; l and m are conjugate)

D(ii) $\mathrm{P} \perp \mathrm{Q} \Leftrightarrow \mathrm{Q} \in \mathcal{Q}\mathrm{P}$.

Under this definition we can satisfy condition (ii) with only a slight modification, namely,

Exc. 4.1 *T*1 The perpendicular through a point to a line

G l, P

T (i) $\# | (l, \mathscr{Q}P) \Rightarrow$ there is a single perpendicular through P to l

T (ii) $l = \mathscr{Q}P \Rightarrow$ every line through P is perpendicular to l.

We could use the fourth property of perpendiculars in Affine Orthogonal Geometry to *define* parallels in Projective Orthogonal Geometry, but we shall obtain a very different-looking system, in that parallelism is defined differently for every point in the plane. In physical terms there is a 'direction' at each point P which is parallel to a given line l, and every line, at one point of itself, satisfies the condition of being parallel to l.

Exc. 4.1 *D*2 Parallel lines.

*G*1 Polarity system, $\mathscr{Q}$

*G*2 (i) P, l & $\# | (l, \mathscr{Q}P)$

*G*3 (i) $p : p \supset P$

D (i) $p \| l \Leftrightarrow \mathscr{Q}p \in \langle P, \mathscr{Q}l \rangle$

*G*2 (ii) P, l & $l = \mathscr{Q}P$

D (ii) Every line through P is parallel to l (and perpendicular to l!).

Since in Projective Orthogonal Geometry we now have relations of perpendicularity and of harmonic conjugacy, we may define an operation of 'reflection'. In Affine Orthogonal Geometry we defined a transformer $\mathscr{R}_l$ thus:

G P, l

C $p : p \supset P \,\&\, p \perp l$

$N : N = p \cap l$

$M : M \in p$ & ideal$|M$

D $\mathscr{R}_l P : \mathscr{R}_l P = \mathscr{H}_{MN} P$.

This operation can be adapted to Projective Orthogonal Geometry by replacing the pair {M, N} by the pair $\{M, \mathscr{Q}M \cap p\}$, which are conjugate in the induced involution on p. Moreover we can equally well define a dual operation of 'reflection in a point'.

Exc. 4.1 *D*3 Reflections in a line and in a point.

	$\mathscr{R}_l P$	$\mathscr{R}_L p$
G	l	L
S	$P : \# \| P, \mathscr{Q}l$	$p : \# \| p, \mathscr{Q}L$
C	$M : M = \mathscr{Q}l$	$m : m = \mathscr{Q}L$
	$N : N = PM \cap l$	$n : n = \langle p \cap m, L \rangle$
D	$\mathscr{R}_l P = \mathscr{H}_{MN} P$	$\mathscr{R}_L p = \mathscr{H}_{mn} p$.

There are no exceptional lines or points. The fixed elements in a reflection transformation are the axis of reflection and the point which is its pole; explicitly:

Exc. 4.1*T*2 Fixed elements in a reflection.

T $R_l P = P \Leftrightarrow$ either $P \in l$ or $P = \mathscr{Q}l$.

It should be noticed that, while we have referred to 'reflection in a line', there are in fact two lines which play equivalent parts in the definition. In Affine Orthogonal Geometry the position is different: the two corresponding lines are the selected line l and the fixed ideal line u and in no sense can any point be defined as '$\mathscr{R}_u P$'. In Projective Orthogonal Geometry we have the theorem

Exc. 4.1*T*3 The two lines which, in relation to a given point, determine a reflection.

G $l, P \;\&\; P \notin l \;\&\; \# |(P, \mathscr{Q}l)$
C $M : M = \mathscr{Q}l$
$N : N = PM \cap l$
$m : m = \mathscr{Q}N$
T $\mathscr{R}_l P = \mathscr{R}_m P$.

The second axis of reflection, m, depends not only on l but also on P, or, more precisely on $N = l \cap \langle P, \mathscr{Q}l\rangle$; m is the line $\mathscr{Q}(l \cap \langle P, \mathscr{Q}l\rangle)$.

By renaming the points and lines we can set up the two dual systems of reflections together.

Exc. 4.1*T*4 Involutions of reflected elements.

G $\mathscr{Q}, A, a \;\&\; \mathscr{Q}A = a \;\&\; A \notin a$
$B : B \in a$
*C*1 $b : b = \mathscr{Q}B$
$c, C : c = AB, C = a \cap b$
(ABC is self-polar in $\mathscr{Q}$, Excursus 2.4.2*N*4)
S $P : P \in AB$
*C*2 $p : p = CP$
*C*3 $Q : Q = \mathscr{H}_{AB} P$
$q : q = \mathscr{H}_{ab} p$
T(i) $Q = \mathscr{R}_a P = \mathscr{R}_b P$
T(ii) $q = \mathscr{R}_A p = \mathscr{R}_B p$.

There is no need therefore to consider the two types of reflections separately. For many purposes we shall need to consider only sets of reflections of points of a collinear set. We take as basis for reflections on a line l the involution induced by $\mathscr{Q}$ on l, and from this, by joining points of l to the point $\mathscr{Q}$l, we obtain both the reflections of lines of the plane in relation to point-pairs on l and the reflections of points of the plane in relation to line-pairs through $\mathscr{Q}$l. In particular it should be remarked that when reflections involving only the lines of a concurrent set are under consideration, we obtain precisely the same figure in the Projective as in the Affine Geometry. Thus

	in A.O.G.	in P.O.G.
*G*1	C, a : a ∈ C	
*G*2	$\mathscr{J}$: $\mathscr{J}$ ∈ c & ideal|c	$\mathscr{Q}$: c = $\mathscr{Q}$C
*C*1	B : B = c ∩ a	B : B = c ∩ a
	A : A = $\mathscr{J}$B (∈ c)	A : A = c ∩ $\mathscr{Q}$B
S	l : l ⊃ C	
*C*2	L : L = c ∩ l	
	M : M = $\mathscr{H}_{AB}$L	
D	m = CM = $\mathscr{R}_a$l.	

Since the 'three-reflections' theorem (4.1*T*16) can be stated in terms of a collinear set and pairs in involution in the set, it follows that it is valid also in Projective Orthogonal Geometry. That is,

Exc. 4.1*T*5 The three reflections theorem.

G	l ; $\mathscr{J}$ the involution induced on l by $\mathscr{Q}$
*S*1	# |LMN ∈ l
*C*1	L′, M′, N′ : {$^{\mathscr{O}}$L′, M′, N′} = $\mathscr{J}${$^{\mathscr{O}}$L, M, N}
*S*2	P : P ∈ l
*C*2	S : S = $\mathscr{R}_N\mathscr{R}_M\mathscr{R}_L$P
*C*3	{K, K′} : K′ = $\mathscr{J}$K & PS $\mathscr{H}$ KK′
T(i)	{K, K′} is unique and linearly constructable (4.1*T*15)
T(ii)	$\mathscr{R}_N\mathscr{R}_M\mathscr{R}_L = \mathscr{R}_K = \mathscr{R}_L\mathscr{R}_M\mathscr{R}_N$ (4.1*T*18).

This theorem therefore states: any set, ordered by betweenness, of three pairs of perpendicular {lines through a point / points on a line} determine a fourth, and the resultant of the sequence of reflections determined by the three pairs is a reflection in the fourth pair.

Looking back over the evolution of the idea of congruence in Affine Orthogonal Geometry we can see the vital part played in the early stages by the operation of 'displacement', which we later recognised as the resultant of two specially related reflections. 'Displacements' are distinguished from 'rotations' only by the property that the point of intersection of the pair of axes of reflection concerned is ideal. But, *ipso facto*, displacements, unlike rotations, are independent of the selection of the orthogonality involution.

In Projective Orthogonal Geometry (at least in the system in which no point lies on its polar) no pair of reflections has any property which distinguishes it from other pairs. However, there are sets of reflections which transform a given point into points of a collinear set, namely, the reflections in the lines concurrent in the point of which the line is the polar. This set of reflections will enable us at least to begin to set up a 'scale' on the line, but we shall meet the same difficulty as we met in establishing a measure for angles in Affine Orthogonal Geometry, namely, that the only operation of subdivision we can perform is halving. Putting the matter a little differently, we may say that in Affine Orthogonal Geometry we are faced with two separate problems:

(*a*) *the linear problem*: given $\# \wp|PQ$ to construct l such that

$$Q = \mathscr{R}_l P \quad (4.1C1),$$

(*b*) *the non-linear problem*: given $\# \wp|pq$ to construct l such that

$$q = \mathscr{R}_l p \quad (2.6T9,\ 10 \text{ and } 4.3T1).$$

In Projective Orthogonal Geometry these problems are equivalent and non-linear. In terms of points all in the same collinear set, the problem is:

G $\quad$ l; $\mathscr{J}$ the involution on l induced by $\mathscr{Q}$

S $\quad$ $\{P, Q\} \subset l$

C $\quad$ $\{A, B\} \subset l : \mathscr{R}_A P = Q\ (= \mathscr{R}_B P)$.

That is, we have to find the pair {A, B} such that

$$PQ\ \mathscr{H}\ AB,$$
$$\&\ \mathscr{J}A = B.$$

This is exactly the problem to which an iterative solution is described in 2.5*C*4.

Exc. 4.1*C*1 Construction for '‡equal segments' on a line.

*G*1 $\quad$ l; $\mathscr{J}$ the involution induced on l by $\mathscr{Q}$

*G*2 $\quad$ $\# |ABA' \in l$

*C*1 $\quad$ $L, M : M = \mathscr{J}L \ \&\ AA'\ \mathscr{H}\ LM \quad (\Rightarrow A' = \mathscr{R}_L A = \mathscr{R}_M A)$

*C*2 $\quad$ $B'' : B'' = \mathscr{R}_L B$

*C*3 $\quad$ $B' : B' = \mathscr{R}_{A'} B''$

T $\quad$ $\mathscr{R}_{A'} \mathscr{R}_L \{{}^{\circ}A, B\} = \{{}^{\circ}A', B'\}$

D $\quad$ $s(AB/A'B') = 1$.

We have to show that this definition is consistent; the relation between ⌜AB⌝ and ⌜A′B′⌝ is clearly reflexive (i.e. valid when $A = A'$, $B = B'$) and symmetric ($s(AB/A'B') = s(A'B'/AB)$); it has to be proved transitive.

Exc. 4.1 *T*6 $s(AB/A'B') = 1 \;\&\; s(A'B'/A''B'') = 1$
$\Rightarrow s(AB/A''B'') = 1$

C $L' : \mathscr{R}_{L'}A' = A''$

T
$$\begin{aligned}\{A'', B''\} &= \mathscr{R}_{A''}\mathscr{R}_{L'}\{A', B'\}\\ &= \mathscr{R}_{A''}\mathscr{R}_{L'}\mathscr{R}_{A'}\mathscr{R}_{L}\{A, B\}\\ &= \mathscr{R}_{A''}\mathscr{R}_{L''}\{A, B\} \quad \text{(Excursus 4.1} \,T\,5)\\ &\Rightarrow s(AB/A''B'') = 1.\end{aligned}$$

Given E_0, E_1 it is a simple matter to construct a set of 'integer' points by successive reflections, thus

$$E_{r+1} = \mathscr{R}_{E_r} E_{r-1},$$

but to obtain a true scaling on the line we have to find transformations $\mathscr{T}_r$ such that

$$\{E_r, E_{r+1}\} = \mathscr{T}_r\{E_{r-1}, E_r\}.$$

We can do this by adapting the construction of Excursus 4.1 *C* 1 to the case $A' = B$.

Exc. 4.1 *C*2 The integer net.

G $\{E_0, E_1\} \in l \;\; : \mathscr{J}$ induced on l by $\mathscr{Q}$

C 1 $L_1, M_1 : M_1 = \mathscr{J}L_1 \;\&\; E_0E_1 \;\mathscr{H}\; L_1M_1$
$E_2 : E_2 = \mathscr{R}_{E_1}\mathscr{R}_{L_1}E_1$

*C*r $L_r, M_r : M_r = \mathscr{J}L_r \;\&\; E_{r-1}E_r \;\mathscr{H}\; L_rM_r$
$E_{r+1} : E_{r+1} = \mathscr{R}_{E_r}\mathscr{R}_{L_r}E_r$

T $\{^{\mathscr{C}}E_r, E_{r+1}\} = \mathscr{R}_{E_r}\mathscr{R}_{L_r}\{^{\mathscr{C}}E_{r-1}, E_r\}.$

Exercise 1 In $\mathfrak{N}_{11}$ take $\mathscr{J}$ to be the involution determined by $xx' = 2$. Prove that the sequence of points determined as in Excursus 4.1 *C*2 from $E_0(x = 0)$ and $E_1(x = 1)$ is

E_0	$x = 0$	conjugate	U	E_6
E_1	1		$x = 2$	E_7
E_2	5		7	E_8
E_3	8		3	E_9
E_4	4		6	E_{10}
E_5	9		10	E_U

As a consequence of Excursus 4.1*C*1 and the three reflections theorem (*T*5), the sequence of relations

$$\{{}^{\mathcal{O}}E_r, E_{r+1}\} = \mathscr{R}_{E_r}\mathscr{R}_{L_r}\{{}^{\mathcal{O}}E_{r-1}, E_r\}$$

leads to the theorem

Exc. 4.1*T*7 $s(E_rE_s/E_{r+t}E_{s+t}) = 1$ for any three integers r, s, t.

If the net is finite the construction Excursus 4.1*C*2 is exhaustive and all points in a primary net are accounted for. (The reader could experiment with $\mathfrak{N}_9$.) If the net is not finite we cannot construct, as we did in 2.2*C*2 the points corresponding to the reciprocals of the integers. That is to say, given E_0 and P (and $\mathscr{J}$) we have no means of constructing, for $n > 2$, the sequence of points $P_1, P_2, \ldots$, where $P_{r+1} = \mathscr{R}_{P_r}P_{r-1}$ and $E_0 = P_0$ and $P = P_n$. When $n = 2$ we have to find the pair of points $\{M, \mathscr{J}M\}$ such that $P = \mathscr{R}_M E_0$, and this is the construction 2.5*C*4 that we have already used in Excursus 4.1*C*1. That is

Exc. 4.1*T*8 Points constructable in a non-finite net.

G $\# |E_0E_1$ & $\mathscr{J}$

T The only points that can be constructed are those corresponding to elements

$$n + \tfrac{1}{2}a_1 + \tfrac{1}{4}a_2 + \tfrac{1}{8}a_3 + \ldots$$

of $\mathfrak{F}_R$, where n is an integer and $a_i \in \{0, 1\}$.

This is the same situation as that which we met in devising means of measuring the content of an angular region (Section 4.3.3), a result that we should have expected, since, within $\mathscr{Q}$, there is complete duality between relations among collinear points and relations among concurrent lines.

We have established two groups of congruence transformations for a given polarity $\mathscr{Q}$, one based on reflections of points into points (in relation to perpendicular line-pairs) and the other on reflections of lines into lines (in relation to perpendicular point-pairs). By applications of the 'three reflections' theorem we may reduce any sequence of reflections to either the identity transformation, or a single reflection, or a rotation, or the resultant of the reflections in three non-concurrent lines (or non-collinear points). There is clearly much in this system which would warrant further investigation, but we shall close this Excursus by establishing the vital algebraic property of two ordered pairs of points on a line which satisfy the geometric condition $s(AB/A'B') = 1$.

Exc. 4.1*T*9 The cross-ratio property of 'equal segments' defined in relation to a polarity.

*G*1 l & $\mathscr{J}$ on l

*C*1 the double-points, D, D′ of $\mathscr{J}$, by extending the net if necessary

*G*2 # |AA′B ∈ l

*C*2 L : $\mathscr{R}_L A' = A$

B′ : $B' = \mathscr{R}_{A'} \mathscr{R}_L B$

T $\text{R}(AB/DD') = \text{R}(A'B'/DD')$

P *c* Select a reference system on l such that $L = E_0$, $\mathscr{J}L = U$ & ideal|U. Assume $\mathscr{J}$ consists of the pairs {x, x′} for which $xx' = h$. {$^{\mathscr{O}}$D, D′} are given by {$^{\mathscr{O}}\delta, -\delta$}, where $\delta^2 = h$, and {$^{\mathscr{O}}$A, A′, B, B′} are given by x = {$^{\mathscr{O}}$a, a′, b, b′}

*t*1*X* $$b' = \frac{2ah - b(a^2+h)}{2ab - (a^2+h)}$$

*t*2*X* $$\text{R}(a, b/\delta, -\delta) = \frac{[(ab-h)+\delta(a-b)]^2}{(a^2-h)(b^2-h)} = \text{R}(a', b'/\delta, -\delta)$$

T $\text{R}(AB/DD') = \text{R}(A'B'/DD')$

(While straightforward, the arithmetic verifying *t*1 and *t*2 is fairly formidable!)

Since $$\text{R}(AB/DD')\,\text{R}(BC/DD') = \text{R}(AC/DD'),$$
in order to obtain an additive system of lengths from these cross-ratios, we should have to take
$$\text{length } AB = k \log \text{R}(AB/DD')$$
for some fixed scale factor k. In particular, in Euclidean geometry we obtain the measure of an angle in this way. Thus, consider lines
$$OA: y = mx, \quad m = \tan\theta,$$
$$OA': y = m'x, \quad m' = \tan\theta',$$
in a rectangular Cartesian system. The lines are perpendicular if $mm' = -1$, and the involution of perpendicular lines has double-lines if we extend the field of the reals by adjoining the unit i with $i^2 = -1$, and extend the real geometric net by adjoining the line OD. Then
$$\text{R}(OA, OA'/OD, OD') = \text{R}(m, m'/i, -i) = \frac{(m^2-1-2im)(m'^2-1+2im')}{(m^2+1)(m'^2+1)} = e^{2i(\theta'-\theta)}.$$
Thus
$$\measuredangle^{+}O(AA') = |\theta'-\theta| = |\tfrac{1}{2}i \log \text{R}(OA, OA'/OD, OD')|.$$
The basic unit (the measure Ω defined in 4.3*D*4 (iii)) is again π.

RECAPITULATION

Three interwoven strands of development may be discerned running through this book and the Excursuses, all three originating in the same three main axioms concerning the set of objects called 'points' and the set of subsets ('collinear sets') called 'lines'. Briefly the first two of the axioms may be expressed as:

'Any two distinct points determine a line and the line is determined by any two distinct points of itself.'

'Any two distinct lines have a single common point.'

The third has two forms, which, although verbally similar, are quite distinct; written together as $\left\{\begin{matrix} A\mathscr{I}\,3, \text{ Desargues} \\ A\mathscr{I}\,4, \text{ Pappus} \end{matrix}\right\}$, they may be stated thus:

'When, among a given set of $\left\{\begin{matrix} \text{nine} \\ \text{ten} \end{matrix}\right\}$ points, the points of each of $\left\{\begin{matrix} \text{nine} \\ \text{eight} \end{matrix}\right\}$ sets of three points are prescribed to be collinear, then there is a $\left\{\begin{matrix} \text{tenth} \\ \text{ninth} \end{matrix}\right\}$ set of three points collinear.'

We found that, while

$$A\mathscr{I}\,4 \text{ (Pappus)} \Rightarrow A\mathscr{I}\,3 \text{ (Desargues)},$$

but not conversely, the form of axiom that was almost always used was the Desarguesian, and the Pappus axiom emerged only at a quite late stage.

Besides these axioms we introduced the separation axioms, of which $A\mathscr{S}\,1, 2$ were merely the prescription of the properties of a separation relation (with no necessary reference to points and collinear sets), and $A\mathscr{S}\,3$ embodied separation in the geometry by relating the separation properties of sets of points belonging to different collinear sets. We found that no finite system satisfies $A\mathscr{S}3$ when it is supplemented by the curious little axiom $A\mathscr{S}4$ which is necessary to eliminate the net $\mathfrak{N}_2$ (and $\mathfrak{N}_{2^r}$).

The first of the three strands of development might be called 'Making congruence respectable'. This strand forms the main subject of Books 1, 2 and 4, almost without reference either to Book 3 or to any Excursus. It leads out from $A\mathscr{I}\,1, 2, 3^*$ (the simply-special Desargues) and $A\mathscr{S}\,1, 2, 3, 4$, and requires the extension of the primary net $\mathfrak{N}_R$ only to the surdic net $\mathfrak{N}_S$ by the adjunction of successive square roots. It can be generated by these axioms and constructions from any set of four points no three collinear. It provides a complete model for physical Euclidean geometry.

The second strand can be traced through the first five Chapters of Book 1 and Books 2 and 3. Effectively the system evolved from the set of geometrical axioms shows that:

$A\mathscr{I}$ 1, 2, 4 (Pappus) are equivalent to the algebraic axioms for a commutative field.

$A\mathscr{I}$ 1, 2, 3 (Desargues) are equivalent to the algebraic axioms for a field not specified to be either skew or commutative.

The third strand, which in general runs through the Excursuses, is the development of the finite geometries, very largely of geometries over Galois fields. While mathematicians have explored pretty thoroughly the general principles of these geometries, the detailed structure of particular systems still includes sizable tracts of virgin territory. They provide the basis for many quite attractive investigations among incidence relations, and the interpretation of manipulative results for three-by-three matrices over Galois fields. A small hint is given too, in Excursuses 1.3 and 3.3, of one of the more baffling of geometrical problems, namely, what manageable systems are there which satisfy $A\mathscr{I}$ 1, 2 but not $A\mathscr{I}$ 3 and yet have significant geometrical structure?

The whole book and its triple network of development may be summarised in the chart on page 316.

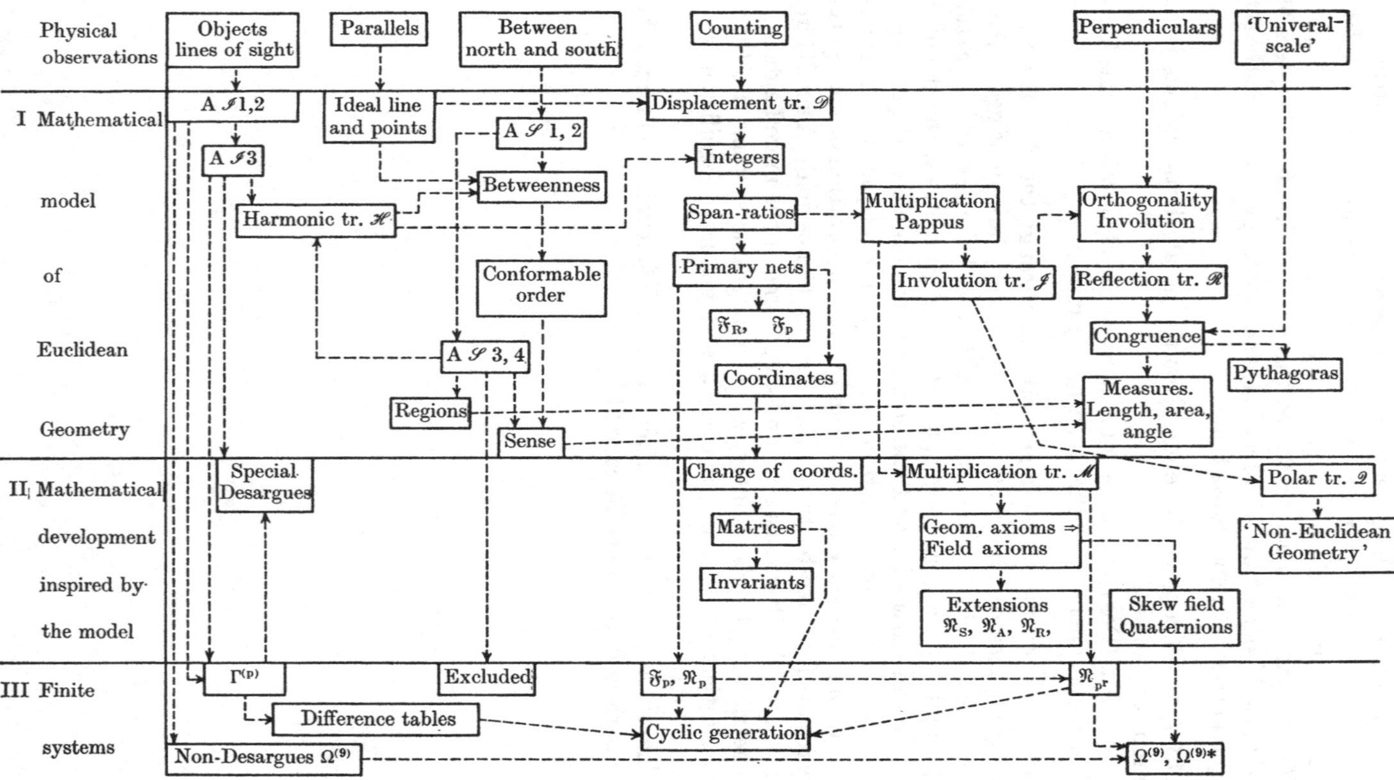
Physical observations
I Mathematical model of Euclidean Geometry
II Mathematical development inspired by the model
III Finite systems
Objects lines of sight
Parallels
Between north and south
Counting
Perpendiculars
'Univeral-scale'
A ℐ1,2
Ideal line and points
Displacement tr. 𝒟
A 𝒮 1, 2
A ℐ3
Integers
Betweenness
Harmonic tr. ℋ
Span-ratios
Multiplication Pappus
Orthogonality Involution
Primary nets
Involution tr. 𝒥
Reflection tr. ℛ
Conformable order
$\mathfrak{F}_R$, $\mathfrak{F}_p$
Congruence
A 𝒮 3, 4
Coordinates
Pythagoras
Regions
Measures. Length, area, angle
Sense
Special Desargues
Change of coords.
Multiplication tr. ℳ
Polar tr. 𝒬
Matrices
Geom. axioms ⇒ Field axioms
'Non-Euclidean Geometry'
Invariants
Extensions $\mathfrak{R}_S$, $\mathfrak{R}_A$, $\mathfrak{R}_R$,
Skew field Quaternions
$\Gamma^{(p)}$
Excluded
$\mathfrak{F}_p$, $\mathfrak{R}_p$
$\mathfrak{R}_{p^r}$
Difference tables
Cyclic generation
Non-Desargues $\Omega^{(9)}$
$\Omega^{(9)}$, $\Omega^{(9)*}$

DIAGRAMS

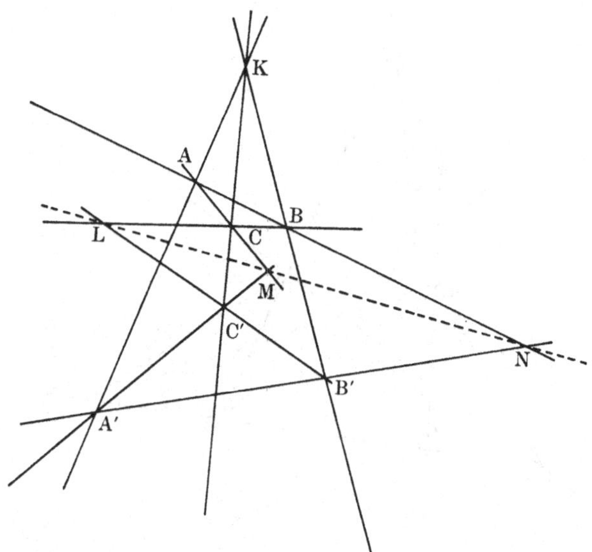

$A\,\mathscr{I}\,3$ (Desargues)

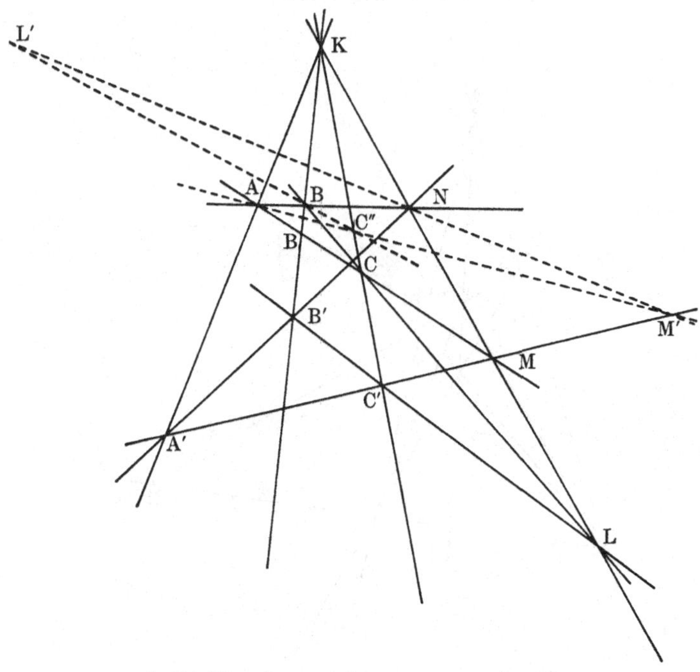

$A\,\mathscr{I}3^*$ (Simply-special Desargues) and $1.3\,T\,2$

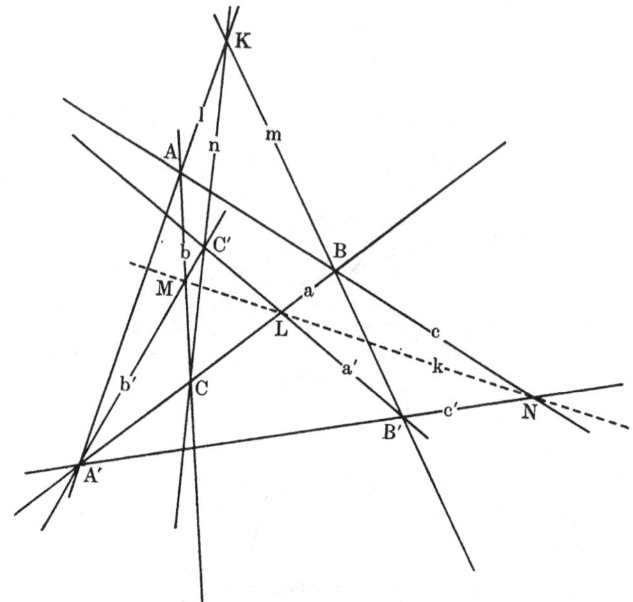

Simply-special Desargues and dual 1.3 T 3

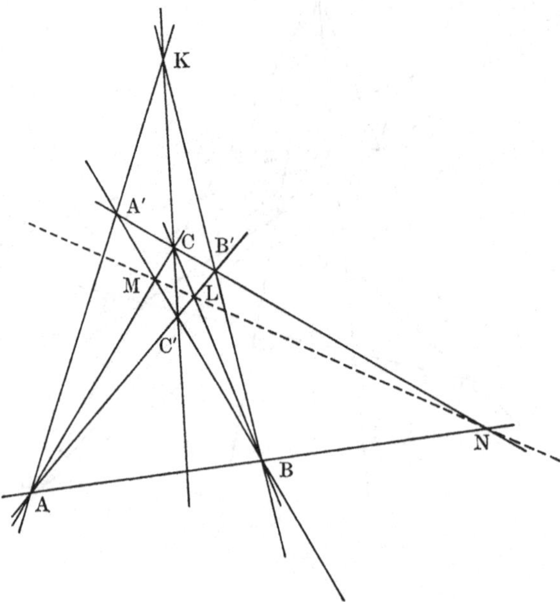

Triply-special Desargues (for quadruply-special, $K \in LM$)

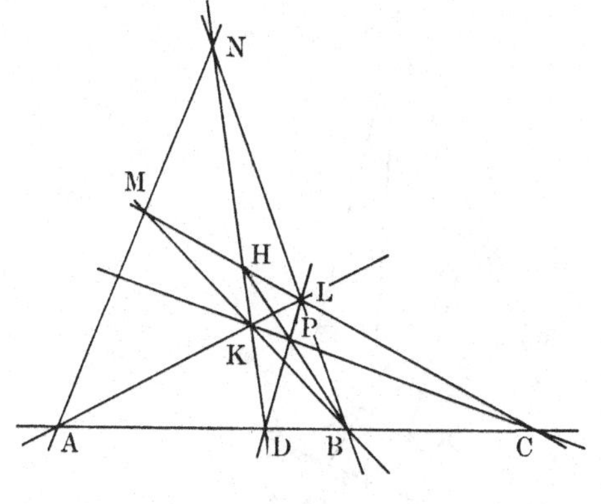

1.4*T*3

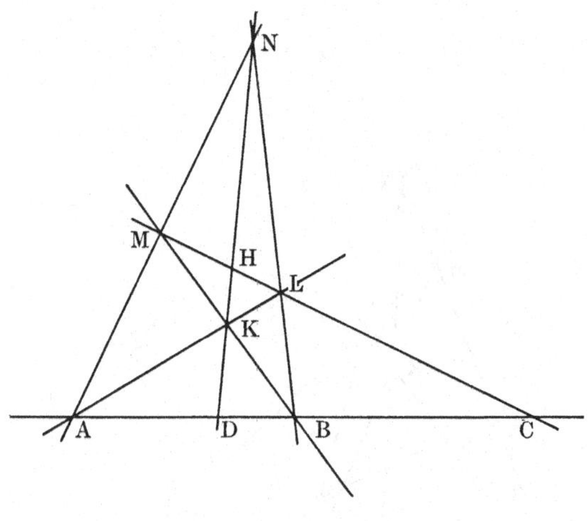

1.4*C*1

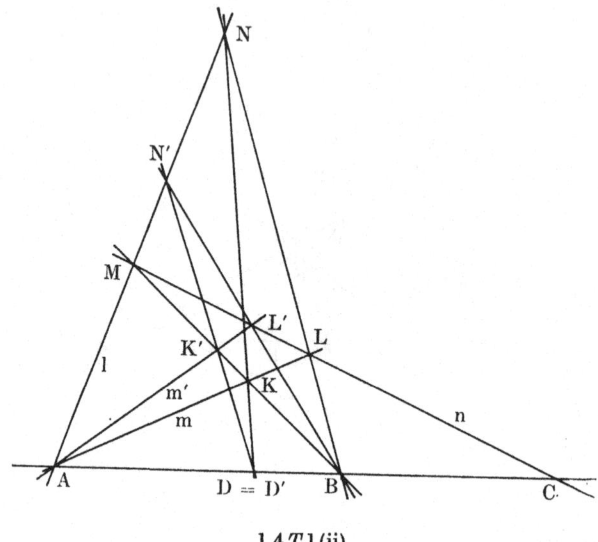

1.4 T 1 (ii)

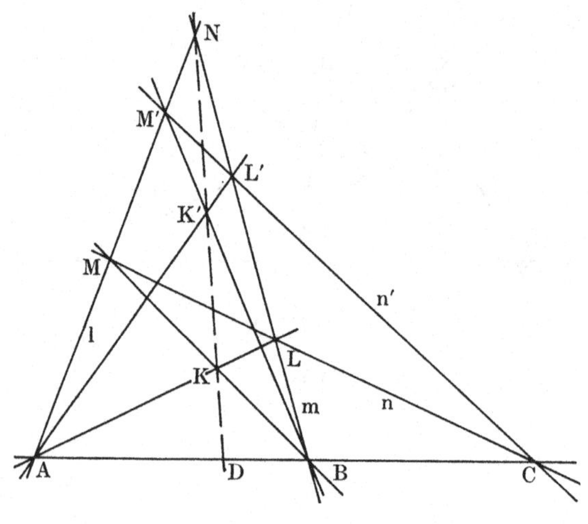

1.4 T 1 (i)

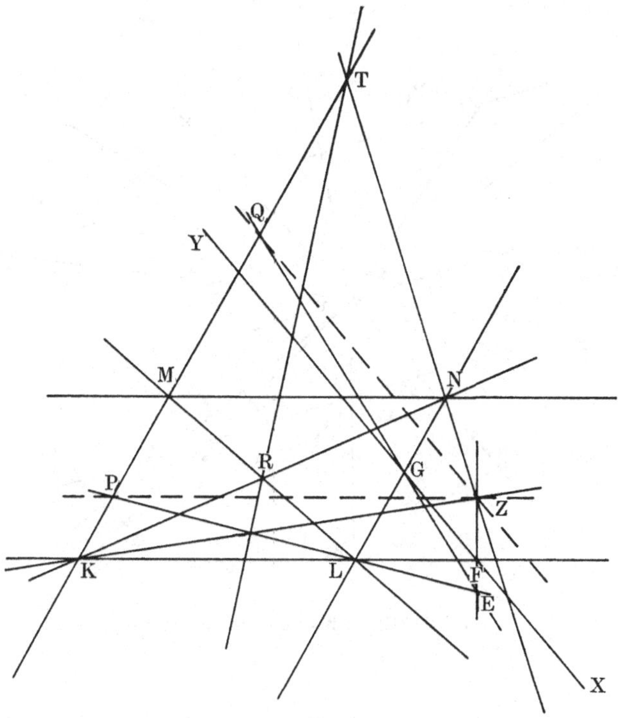

1.5 T1 and 2

$$1.5T1 \quad \begin{Bmatrix} \mathrm{T} \\ \mathrm{R} \\ \mathrm{S} \end{Bmatrix} \begin{Bmatrix} \mathrm{K} & \mathrm{L} \\ \mathrm{Z} & \mathrm{P} \\ \mathrm{N} & \mathrm{M} \end{Bmatrix} \Rightarrow \mathscr{P}|\mathrm{KL} \cap \mathrm{MN} \cap \mathrm{PZ}$$

$$1.5T2 \quad \mathrm{E}\begin{Bmatrix} \mathrm{F} & \mathrm{Z} \\ \mathrm{L} & \mathrm{P} \\ \mathrm{G} & \mathrm{Q} \end{Bmatrix} = \mathscr{L}|\langle \mathrm{LG} \cap \mathrm{PQ}, \mathrm{FL} \cap \mathrm{ZP}, \mathrm{FG} \cap \mathrm{ZQ} \rangle \text{ i.e. ideal } |\mathrm{XY} \cap \mathrm{ZQ}$$

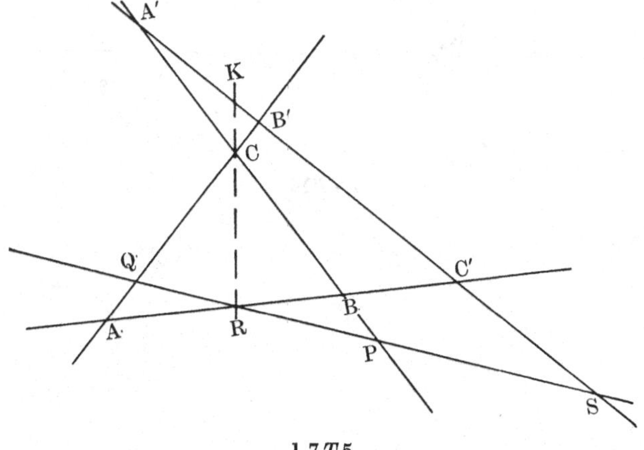

1.7 T5

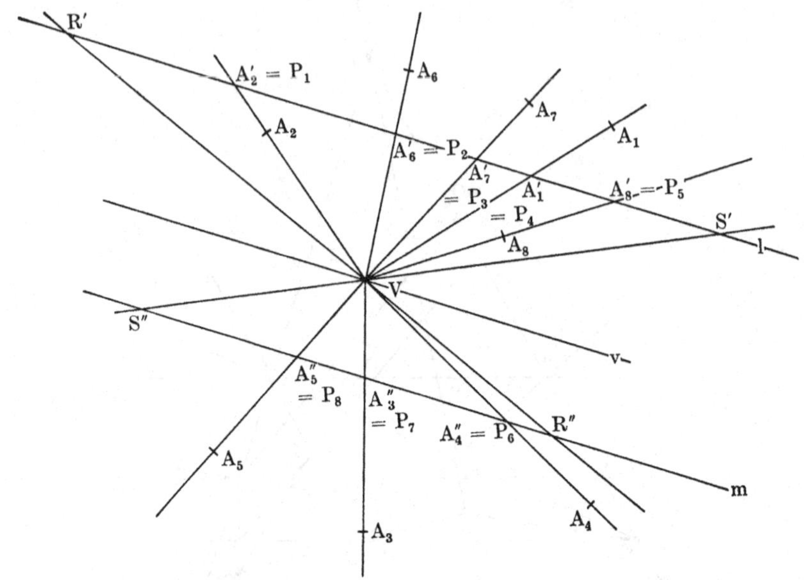

1.7 T 13 (i) for n = 8, r = 5, s = 3

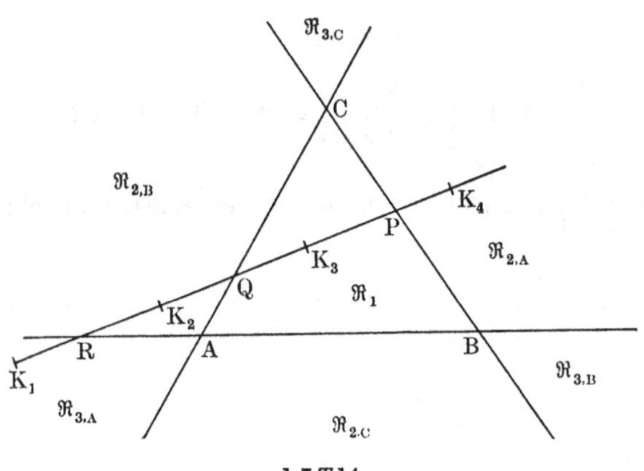

1.7 T 14

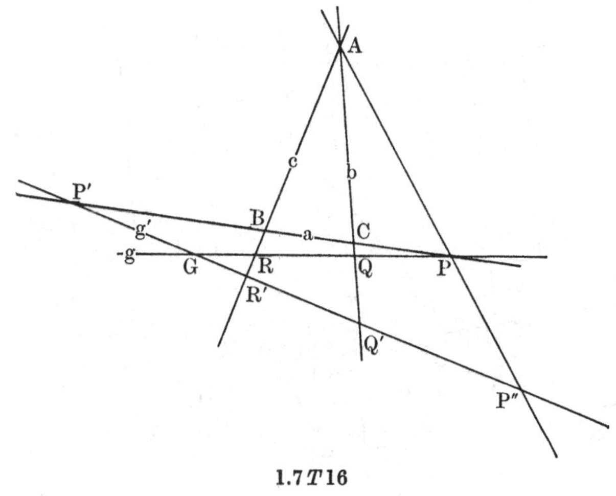

1.7 *T* 16

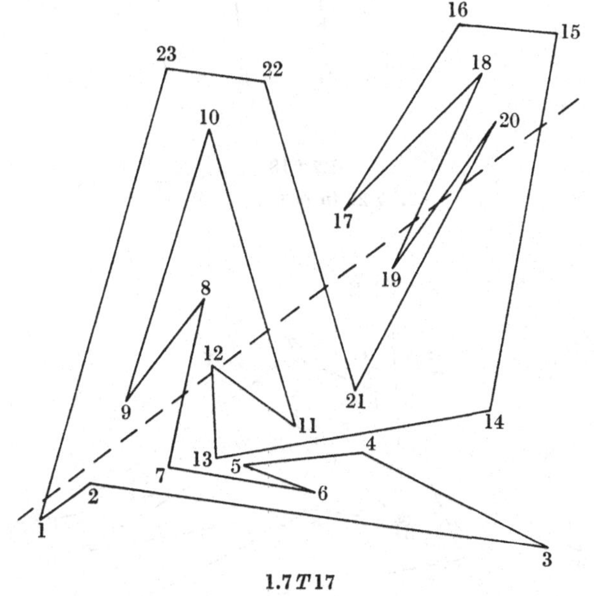

1.7 *T* 17

1 2 3 4 5 6 7
8 9 10
11
12
13 14
15 16 17 18
19
20
21
22 23

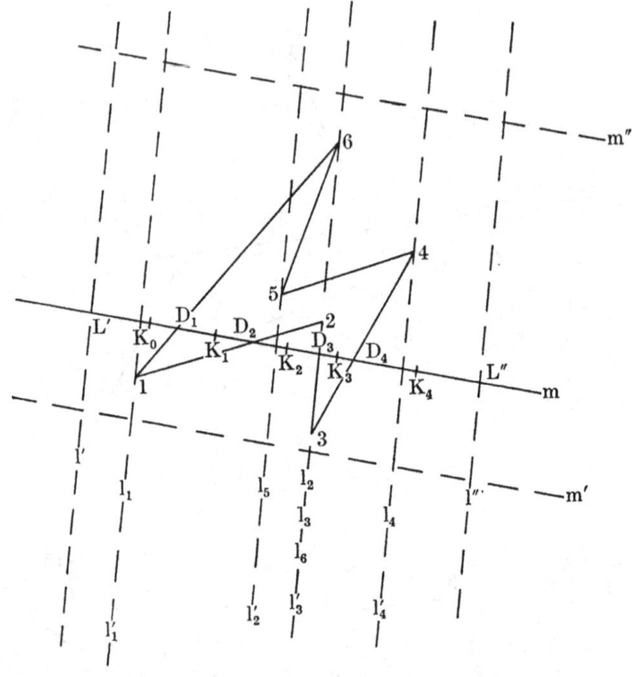

1.7 *T* 18

1.7 *T* 20 (n = 6, $\rho_1 = 5$)

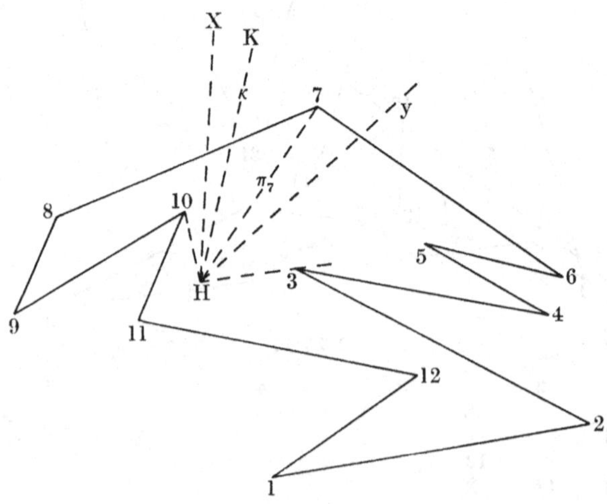

1.7 *T* 19 r = 10, s = 7, t = 3

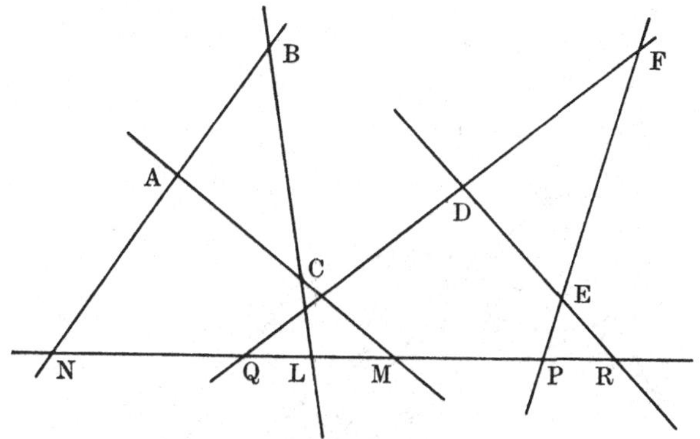

1.8D3 ∂(LM/PQ) ∂(LN/PR) ∂(MN/QR) = JJJ = J
$\Rightarrow$ ∂(ABC/DEF) = J

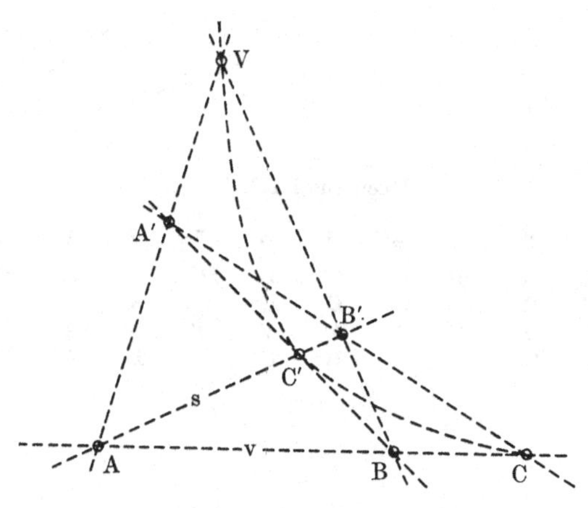

Excursus 1.1.1. $\Gamma^{(2)}$

	0	1	2	3	4	5	6
0	V	A	B	A′	C	C′	B′
1	A	B	A′	C	C′	B′	V
3	A′	C	C′	B′	V	A	B

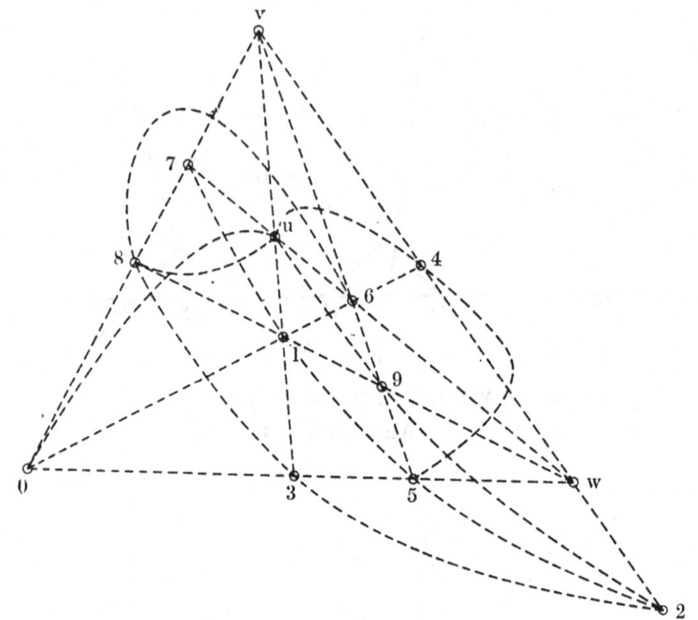

Excursus 1.1.2. $\Gamma^{(3)}$

	0	1	2	3	4	5	6	7	8	9	u	v	w
0	A	B	A″	B″	C	C″	D	D′	C′	A′	V	B′	D′
1	B	A″	B″	C	C″	D	D′	C′	A′	V	B′	D′	A
4	C	C″	D	D′	C′	A′	V	B′	D′	A	B	A″	B″
6	D	D′	C′	A′	V	B′	D′	A	B	A″	B″	C	C″

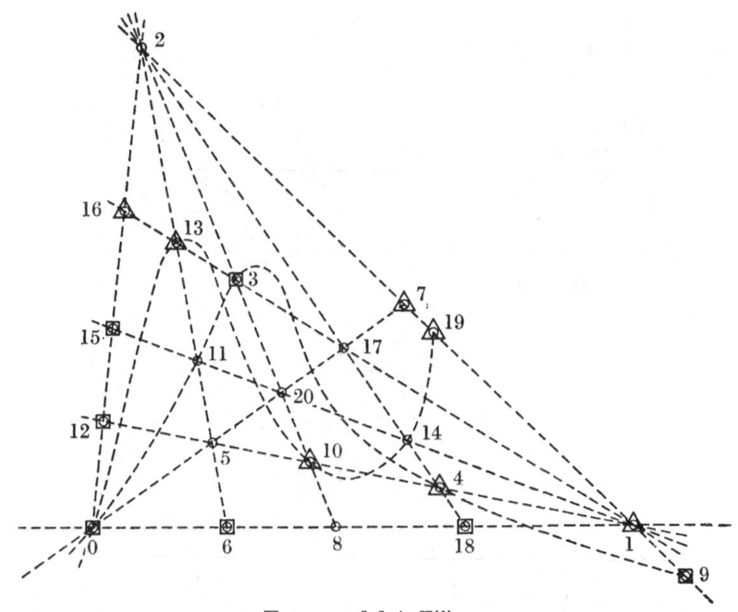

Excursus 1.1.4. $\Gamma^{(4)}$

	0	1	2	3	4	5	6	7	8	9	10	11	12	13	14	15	16	17	18	19	20
0	A_1	B_2	D_1	D_3	C_2	A_2	C_3	A_3	D_4	J′	D_2	B_4	W	J	B_3	B_1	C_1	A_4	U	C_4	V
1	B_2	D_1	D_3	C_2	A_2	C_3	A_3	D_4	J′	D_2	B_4	W	J	B_3	B_1	C_1	A_4	U	C_4	V	A_1
6	C_3	A_3	D_4	J′	D_2	B_4	W	J	B_3	B_1	C_1	A_4	U	C_4	V	A_1	B_2	D_1	D_3	C_2	A_2
8	D_4	J′	D_2	B_4	W	J	B_3	B_1	C_1	A_4	U	C_4	V	A_1	B_2	D_1	D_3	C_2	A_2	C_3	A_3
18	U	C_4	V	A_1	B_2	D_1	D_3	C_2	A_2	C_3	A_3	D_4	J′	D_2	B_4	W	J	B_3	B_1	C_1	A_4

(Points of geometries $\Gamma^{(2)}$ marked □, △, ○.)

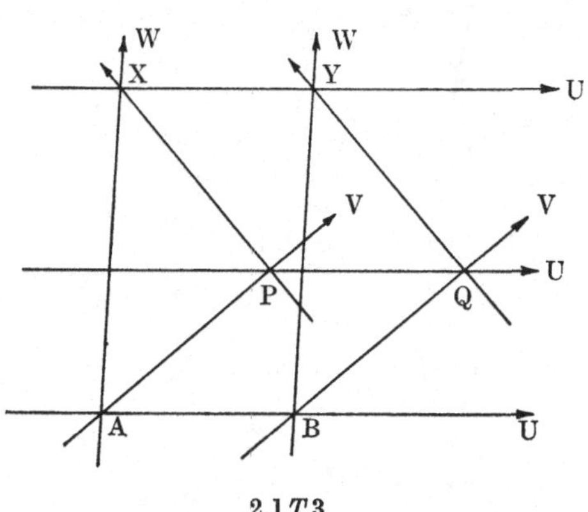

2.1T3

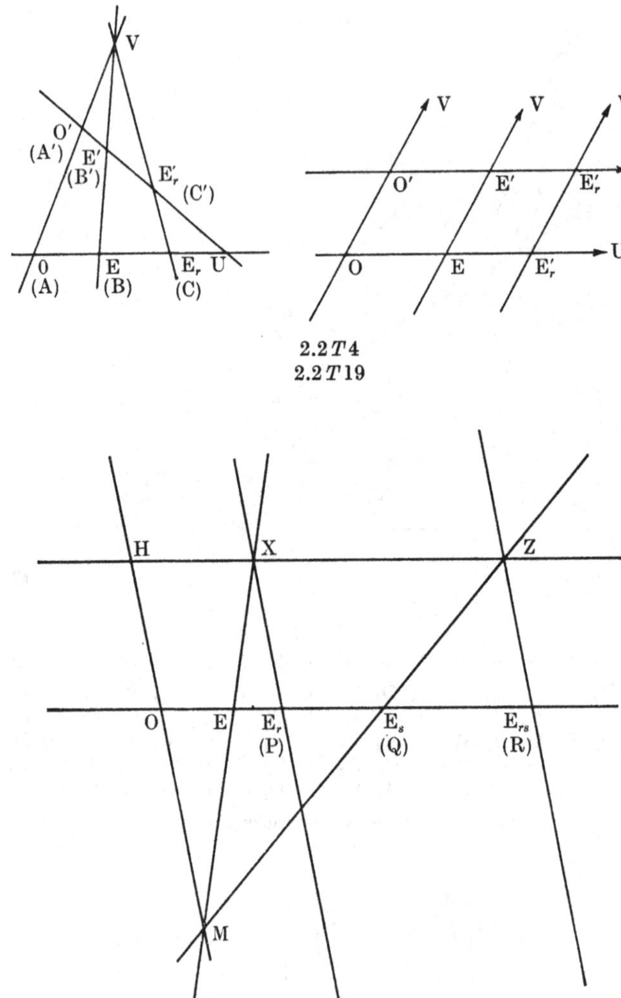

2.2*T*4
2.2*T*19

2.2*T*5
2.2*C*4

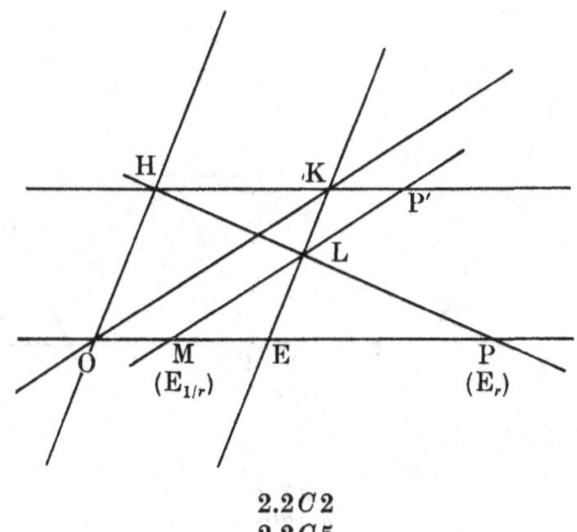

2.2*C*2
2.2*C*5

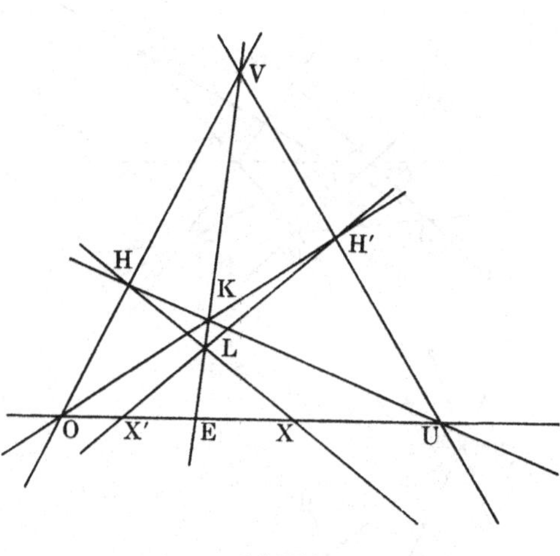

2.2*T*20

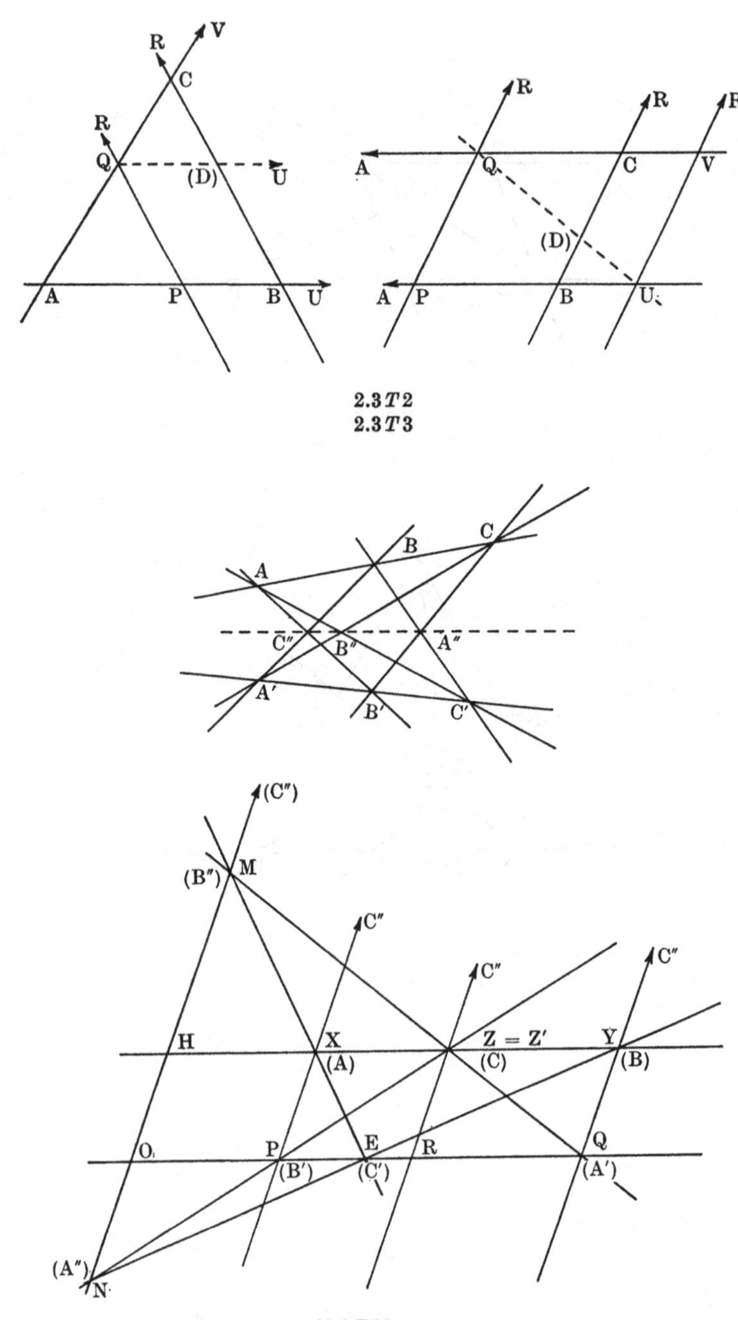

2.3 T 2
2.3 T 3

{2.3 T 10
{A $\mathscr{I}$ 4 (2.3 T 10)

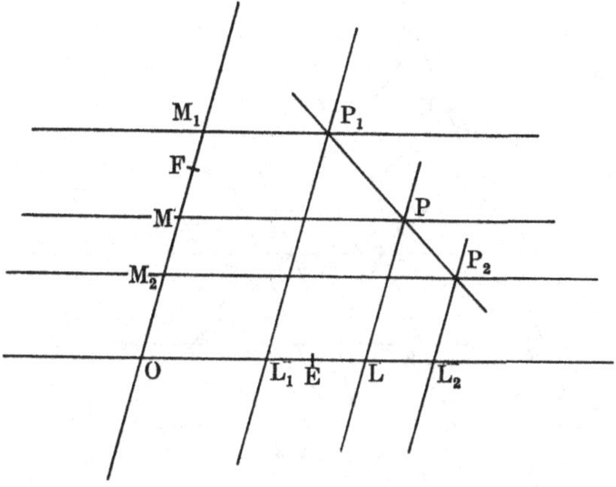

$2.4T2, 3$

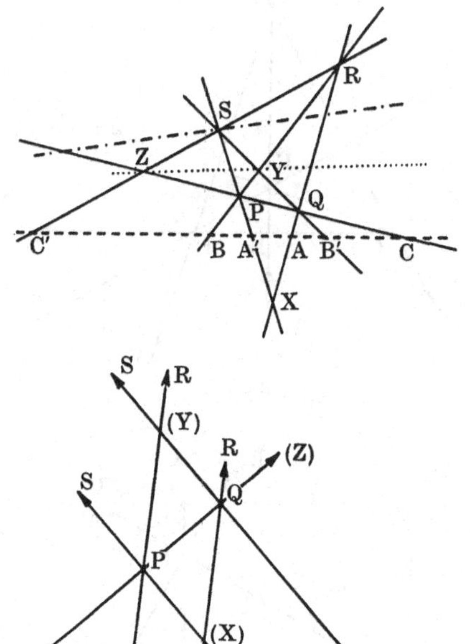

$2.5C1$, $2.5C2$, $2.5T2$, $2.5N7$

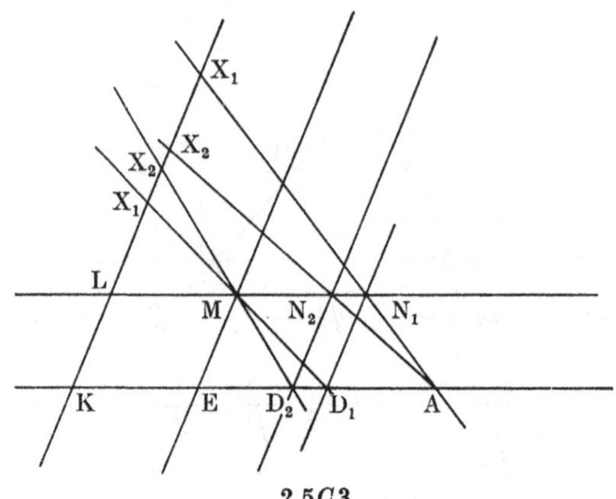

2.5C3

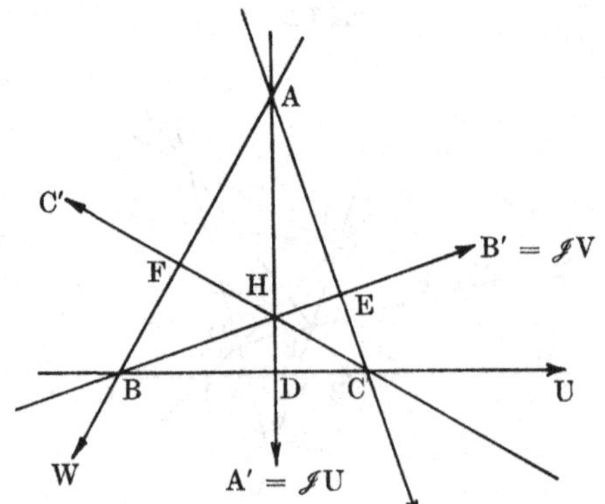

2.6T5

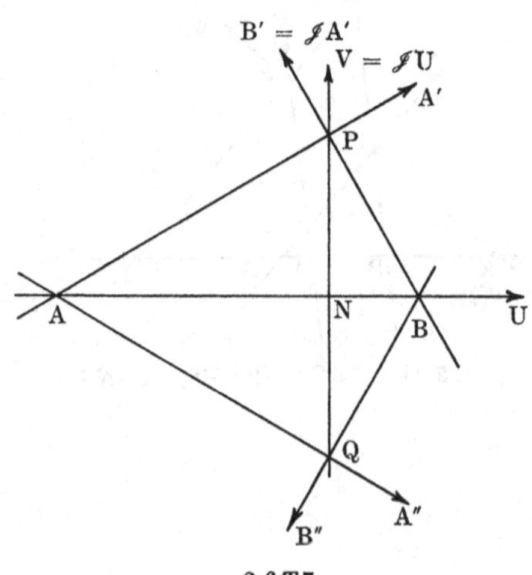

2.6T7

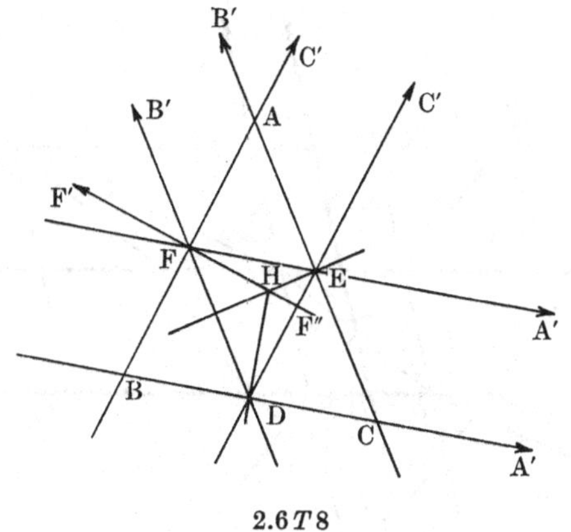

2.6 *T* 8

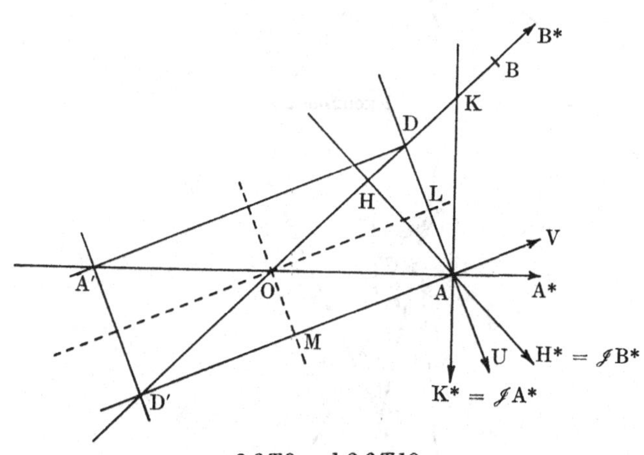

2.6 *T* 9 and 2.6 *T* 10

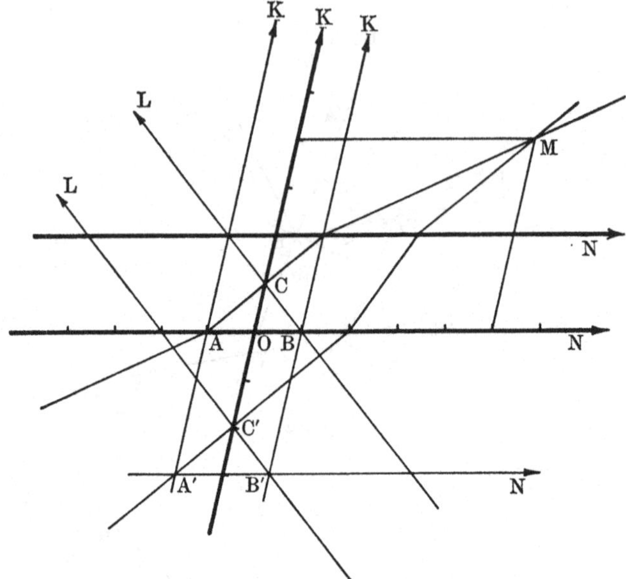

Excursus 2.2

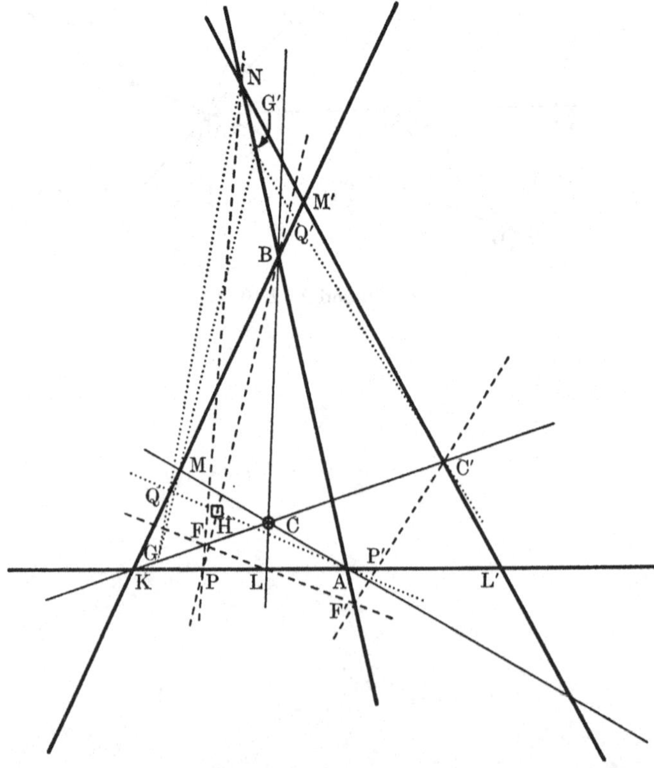

Excursus 2.4. Exercise 1

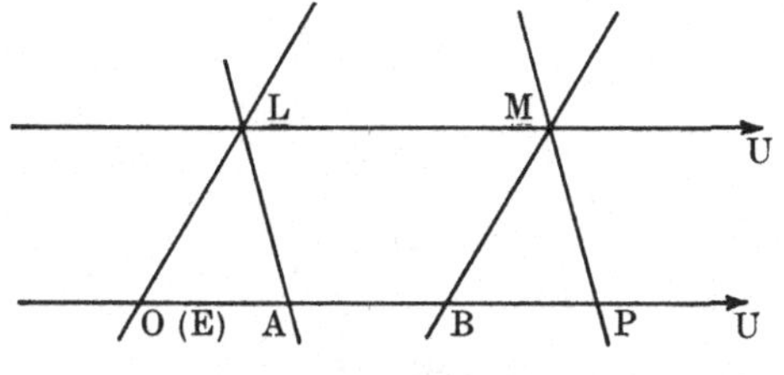

3.1D1 $\mathrm{P} = \mathscr{D}_{\mathrm{A}}\mathrm{B}$

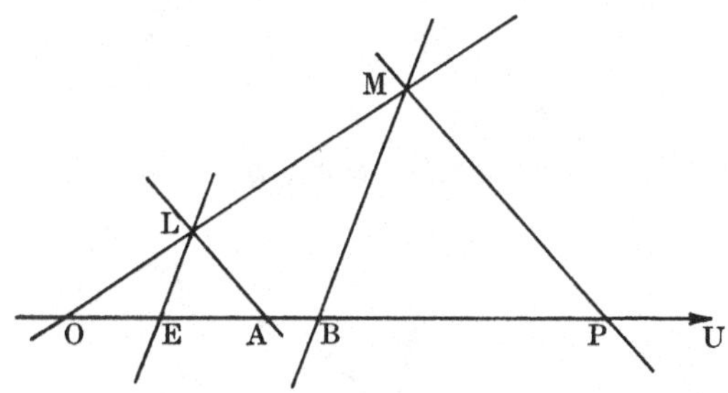

3.2D1 $\mathrm{P} = \mathscr{M}_{\mathrm{A}}\mathrm{B}$

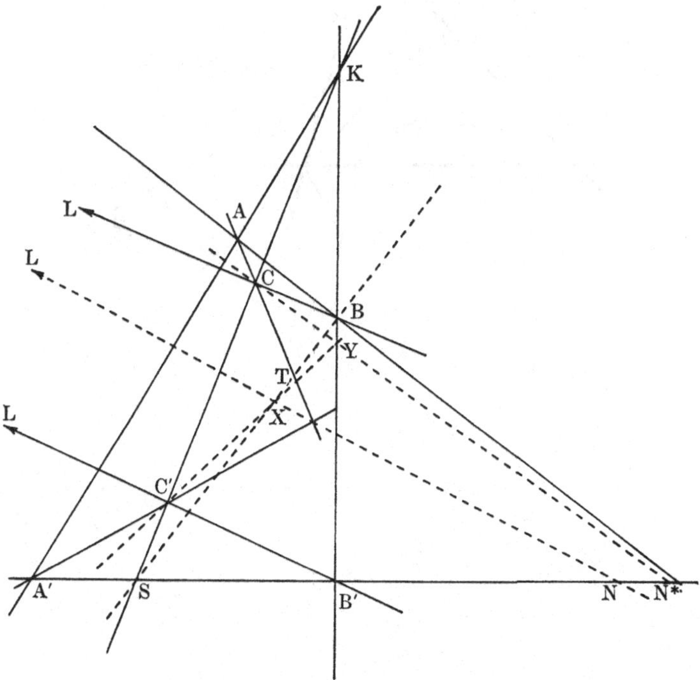

3.5T2 (Pappus ⇒ Desargues)

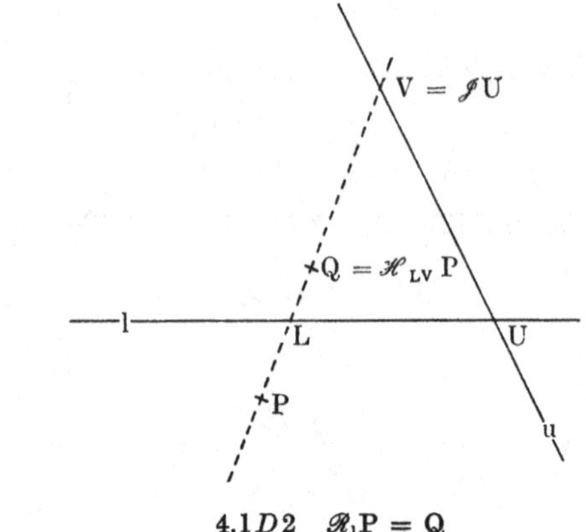

4.1D2 $\mathscr{R}_l$P = Q

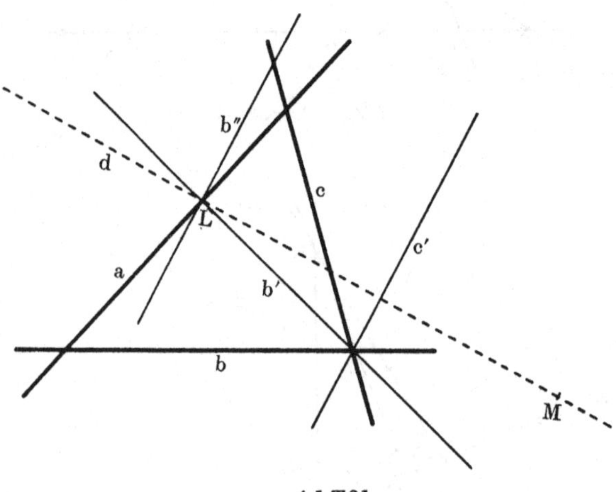

4.1T21

NOTATIONS

(See also **1.2**, page 17)

A	Axiom: $A\mathscr{I}$, of incidence, $A\mathscr{S}$, of separation
$\mathscr{A}^+$	Area, 301 (Content, 298)
$\mathscr{B}, \not\mathscr{B}$	Between, Not between, 52, 53
C	Construction, 20
$\mathfrak{C}$(OE, OF)	Co-ordinate system, 146
D	Definition, 20
$\partial(\,.\,.\,/\,.\,.\,) = \mathrm{I}, \mathrm{J}$	Relative sense of two collinear point-pairs, 80
$\partial(\,.\,.\,.\,/\,.\,.\,.\,) = \mathrm{I}, \mathrm{J}$	Relative sense of two coplanar point-triplets, 83
EE, EH, HH	Types of polarities, 211
$\mathfrak{F}$	Field, 128, 130
$\mathfrak{f}$	Unit in extension field
G, g	Given (data in theorems)
$\mathscr{H}$	Harmonic relation, 36
$\mathscr{H}_{\mathrm{AB}}$	Harmonic transformer, 35
$\mathscr{I}$	Identity transformer, 119
$\mathscr{I}_{\mathrm{K;A,A'}}$	Involutory transformer centre K, 161
$\mathscr{I}_{\mathrm{A,A';B,B'}}$	Involutory transformer det. by {A, A'}, {B, B'}, 162
$\mathscr{L}\vert$	Collinear set of points, 19
$\not\mathscr{L}\vert$	Set of points not all collinear, 19
$\not\mathscr{L}_3\vert$	Set of points no three collinear, 19
$\mathscr{M}_{\mathrm{A}}$	Multiplication transformer, 224
$\mathfrak{N}$	Net, 130
$\mathscr{O}\vert$	Ordered set, 81
$\{^{\mathscr{O}}...\}$	Set ordered as in printed order, 17
P, p	Proof of theorem of lemma
$\mathscr{P}\vert$	Concurrent set of lines, 19
$\wp\vert$	Set of proper points or lines, 40
$Q(\,.\,.\,.\,.\,/\,.\,.\,.\,)$	Complete quadrangle, 163
$\mathscr{Q}$	Polarity transformer, 203
$\mathfrak{R}$	Region, 70, 76
$\mathscr{R}_1$	Reflection transformer, 275
R(../..)	Cross-ratio, 136
R, as subscript	$\mathfrak{F}_{\mathrm{R}}$, field of rationals, 130
$\mathfrak{R}$, as subscript	$\mathfrak{F}_{\mathfrak{R}}$, field of reals, 238
s(../..)	Span-ratio, 132; s^+(../..), 296
S, s	Select (element chosen arbitrarily), 21
$\mathscr{S}, \not\mathscr{S}$	Points are (are not) in the designated separation relation, 44, 47

T, t	Theorem, lemma (number of or statement of), 21
X	Exercise, 21
$\Gamma^{(m)}$	Finite (cyclic) plane of order m, 85, 93
Γ_{ijk}	Invariant of non-collinear points, 156
Δ	Triangle, 70
ϖ	Coplanar set, 105
Ω	Unit of measure of angular region, 302
$\Omega^{(9)}$, $\Omega^{(9)*}$	Veblen–Wedderburn geometries, 99, 265
$\#\|\ldots$	Set of distinct elements, 17
$\|$	Parallel, 39
$\overline{\overline{\wedge}}$	In perspective with, 57
$\ulcorner A(\mathrm{P})$, $\ulcorner \mathrm{A}(\tilde{\mathrm{P}})$, $\ulcorner \mathrm{AB} \urcorner$	Half-line, complementary half-line, segment, 63
$\ulcorner \mathrm{a}(\mathrm{P})$, $\ulcorner \mathrm{a}(\tilde{\mathrm{P}})$, $\ulcorner \mathrm{ab} \urcorner$	Half-plane, complementary half-plane, strip, 66
$\angle$	Angular region, 68
$\sphericalangle$	Measure of angular region, 302

INDEX